AF598136

NEUROMETHODS

Series Editor
Wolfgang Walz
University of Saskatchewan
Saskatoon, SK, Canada

For further volumes:
http://www.springer.com/series/7657

The Making and Un-Making of Neuronal Circuits in *Drosophila*

Edited by

Bassem A. Hassan

VIB Center for the Biology of Disease, Center for Human Genetics
VIB and University of Leuven School of Medicine
Leuven, Belgium

Editor
Bassem A. Hassan
VIB Center for the Biology of Discase
Center for Human Genetics VIB
University of Leuven School of Medicine
Leuven, Belgium

ISSN 0893-2336 ISSN 1940-6045 (electronic)
ISBN 978-1-61779-829-0 ISBN 978-1-61779-830-6 (eBook)
DOI 10.1007/978-1-61779-830-6
Springer New York Dordrecht Heidelberg London

Library of Congress Control Number: 2012939966

© Springer Science+Business Media, LLC 2012
All rights reserved. This work may not be translated or copied in whole or in part without the written permission of the publisher (Humana Press, c/o Springer Science+Business Media, LLC, 233 Spring Street, New York, NY 10013, USA), except for brief excerpts in connection with reviews or scholarly analysis. Use in connection with any form of information storage and retrieval, electronic adaptation, computer software, or by similar or dissimilar methodology now known or hereafter developed is forbidden.
The use in this publication of trade names, trademarks, service marks, and similar terms, even if they are not identified as such, is not to be taken as an expression of opinion as to whether or not they are subject to proprietary rights.

Cover Illustration Caption: Following traumatic incision, regenerating *Drosophila* circadian LNv neurons in the CNS grow back to their original target sites and are seen approach the remnants of their former distal stumps (right top). In this image the LNv neurons have been topographically colour coded and superimposed onto a *Drosophila* half brain showing the optic lobe arborisation from the contralateral GFP expressing LNv neurons. This image was acquired with a Zeiss Lumar V12 and a Zeiss LSM510 META NLO and processed with Adobe photoshop CS2. Image by Marta Koch, Derya Ayaz and Bassem A. Hassan.

Printed on acid-free paper

Humana Press is part of Springer Science+Business Media (www.springer.com)

Preface to the Series

Under the guidance of its founders Alan Boulton and Glen Baker, the Neuromethods series by Humana Press has been very successful since the first volume that appeared in 1985. In about 17 years, 37 volumes have been published. In 2006, Springer Science + Business Media made a renewed commitment to this series. The new program will focus on methods that are either unique to the nervous system and excitable cells or need special consideration to be applied to the neurosciences. The program will strike a balance between recent and exciting developments like those concerning new animal models of disease, imaging, in vivo methods, and more established techniques. These include immunocytochemistry and electrophysiological technologies. New trainees in neurosciences still need a sound footing in these older methods in order to apply a critical approach to their results. The careful application of methods is probably the most important step in the process of scientific inquiry. In the past, new methodologies led the way in developing new disciplines in the biological and medical sciences. For example, Physiology emerged out of Anatomy in the nineteenth century by harnessing new methods based on the newly discovered phenomenon of electricity. Nowadays, the relationships between disciplines and methods are more complex. Methods are now widely shared between disciplines and research areas. New developments in electronic publishing also make it possible for scientists to download chapters or protocols selectively within a very short time of encountering them. This new approach has been taken into account in the design of individual volumes and chapters in this series.

Wolfgang Walz

Preface

The small fruit fly, *Drosophila melanogaster*, has for over a century now had a large impact on biological and biomedical research. From the description of the rules of heredity to the elucidation of signals necessary for development, *Drosophila* research has greatly contributed to—and often lead to—the emergence of important paradigms in our understanding of development, physiology, behavior, and disease. The use of *Drosophila* as a model system for the genetic control of almost every biological process one can think of has meant that we now have a detailed understanding of the genetic logic underlying this animal's development, as well as of the cellular organization of most of its tissues and organs. It is therefore rather surprising that our knowledge of the fly brain has lagged significantly behind our understanding of other aspects of its development, physiology, and function. In part, this is due to the sheer complexity of the brain. In that sense, despite a significantly reduced size in comparison to the mouse brain, for example, the fly brain is exquisitely complicated in its connectivity pattern and compartmentalization. Related to that is the fact that studying the brain requires sophisticated tools, both genetic and technical, to carefully dissect its development, physiology, function, and eventually degeneration and to eventually link these different aspects to the genetic logic(s) underlying them.

The past decade has witnessed a massive expansion in the extent and sophistication of the genetic tool kit of *Drosophila*. Because of the generic nature of genetic approaches, brain research has benefited significantly from essentially all of these approaches. In addition, increasingly clever and technologically sophisticated behavioral assays have led to an increase in the number of laboratories studying behavioral encoding using *Drosophila* as a model system. Much more recent developments and expansions of these tools mean that we are on the verge of a veritable revolution in our understanding of the genetic, cellular, and network properties of the formation and function of the fly brain.

This book is intended as a general introduction to the ideas and concepts behind the methods, tools, and tricks that *Drosophila* neuroscientists utilize to decode the secrets of their favorite enigma. It is by no means intended to be an exhaustive survey of the approaches used in studying the fly brain. Rather, it provides cutting-edge examples of the tools and techniques used today in the research aimed at understanding the fly brain as a model for understanding how brains are organized and how they encode innate behaviors and learn new ones. The major purpose is to equip current and future *Drosophila* neurobiologists with a set of cutting-edge tools and the ideas underlying them so that fly neurobiologists are able not only to exploit these techniques but also to develop them further to suit their particular needs.

In this sense, there are perhaps two aims of this volume that are essential to emphasize in introducing it. The first idea is that the major unifying focus of the different chapters is the concept of a neuronal circuit, defined as a series of synaptically connected neurons subservient to a particular behavioral modality. This is why, after an introduction to the general anatomy of the fly brain and its compartmental organization, all chapters dealing with anatomical analysis focus on cellular and subcellular morphologies. These chapters range from presenting the reader with currently available genetically encoded neuronal markers to methods for generating labeled single neurons, imaging them at very high resolution and reconstructing those images, eventually into circuit diagrams. Although each of these chapters uses examples in different neuronal lineages and at different developmental stages, the basic tools and methodologies apply equally well to any neuron or circuit the reader wishes to study. Once assembled, a circuit functions via synaptic communication between its members in order to allow the animal to encode, produce, and learn behaviors. The classical approach to circuit function is electrophysiological recording of neuronal activity. An emerging, and increasingly powerful, alternative is visualization of synaptic activity through genetically encoded indicators. Part II of the book provides two powerful examples of these two approaches. Importantly, one chapter deals with studying neuronal physiology at larval stages and the other in the adult brain. This is in order to cover the two major forms of the *Drosophila* life cycle that are heavily used by researchers. In a very similar vein, Part III of the book provides assays for larval as well as adult behavioral paradigms.

The second major aim of the book is to have a current and futuristic "feel" to it. To this end, two elements were incorporated into the book's concept. First, the choice of contributors is intended to reflect a relatively "young" group of fly neurobiologists who are nonetheless at the forefront of developing tools for the study of brain development, anatomy, and function. In fact, the lead authors of almost all chapters describe approaches that they themselves have developed and that are setting new standards for neuronal circuit analysis. Here, it is important to emphasize that there are many, many more such talented scientists in the field developing at least equally exciting approaches. However, it is simply impossible to include the entire spectrum of tools currently available. Thus, this book by no means claims comprehensiveness, but rather provides a relatively limited selection of many equally good approaches, but which nonetheless covers a broad spectrum of modern concepts and techniques. The choice of the contributors is also intended to provide methods that are constantly being updated by the authors themselves. We hope that this allows the user direct access to the source of further information about an insight into most of the methods. Second, we included a final Part IV which includes chapters on relatively recent developments that are still finding their way into broader use and are still being further optimized by many workers in the field. Specifically, we hazard to predict that more hardcore molecular, ex vivo and even in vitro approaches that have proven very powerful in gaining insights into the working of mammalian neurons will become increasingly used in *Drosophila* neurobiology research to complement the classical in vivo genetic approaches. On the subject of prediction, we also assume the risk of predicting that developmental, anatomical, physiological, and behavioral vision research in *Drosophila* will likely rise to significant prominence in the medium-term future of fly neuroscience. This is why one of the two chapters in Part II and both chapters of the behavioral section apply their tools strongly or exclusively to the visual system. The fly visual system has lagged behind other models mainly due to the complexity and density of its connections. However, the rapid progress being made at mapping the entire fly brain connectome means that the complexity of the visual system will change from being a foreboding into being an enticing problem to solve.

In closing, and on behalf of all the authors, we hope that readers will find in this book the information and tools necessary to carry out their current experiments and—more importantly—further advance the progress of the *Drosophila* neurobiology field and neurobiology in general.

Leuven, Belgium **Bassem A. Hassan**

Acknowledgments

We acknowledge Dr. P. Robin Hiesinger for Figure 10 and Dr. Carmen Francis for help in establishing the in situ hybridization protocol. The work in the laboratory was supported by grants from the IWT (Agency for Innovation by Science & Technology), FWO (grant numbers G.0654.08 and G071412N) and VIB.

Contents

Part IV Futures

Contributors

ANDREA H. BRAND • *The Gurdon Institute and Department of Physiology, Development and Neuroscience, University of Cambridge, Cambridge, UK*

PATRICK CALLAERTS • *Laboratory of Behavioral and Developmental Genetics, Department of Human Genetics, KU Leuven, Leuven, Belgium; VIB Center for the Biology of Disease, KU Leuven, Leuven, Belgium*

ELIZABETH E. CAYGILL • *The Gurdon Institute and Department of Physiology Development and Neuroscience, University of Cambridge, Cambridge, UK*

CHIH-CHIANG CHAN • *Department of Physiology and Green Center for Systems Biology, UT Southwestern Medical Center, Dallas, TX, USA*

THOMAS R. CLANDININ • *Department of Neurobiology, Stanford University, Stanford, CA, USA*

JASON CLEMENTS • *Laboratory of Behavioral and Developmental Genetics, Department of Human Genetics, KU Leuven, Leuven, Belgium; VIB Center for the Biology of Disease, KU Leuven, Leuven, Belgium*

JAN FELIX EVERS • *Department of Physiology, Development and Neuroscience, University of Cambridge, Cambridge, UK*

ANA CLARA FERNANDES • *VIB, Center for the Biology of Disease, Leuven, Belgium; Laboratory of Neuronal Communication, Programs in Molecular and Developmental Genetics and Cognitive and Molecular Neuroscience, Center for Human Genetics, KU Leuven, Leuven, Belgium*

DARYL M. GOHL • *Department of Neurobiology, Stanford University, Stanford, CA, USA*

KATRINA S. GOLD • *The Gurdon Institute and Department of Physiology, Development and Neuroscience, University of Cambridge, Cambridge, UK*

BASSEM A. HASSAN • *VIB Center for the Biology of Discase, Center for Human Genetics VIB and University of Leuven School of Medicine, Leuven, Belgium; Center for Human Genetics, KU Leuven School of Medicine, Leuven, Belgium; Janelia Farm Research Campus, Howard Hughes Medical Institute, Ashburn, VA, USA*

P. ROBIN HIESINGER • *Department of Physiology and Green Center for Systems Biology, UT Southwestern Medical Center, Dallas, TX, USA*

CHIH-FEI KAO • *Department of Neurobiology, University of Massachusetts Medical School, Worcester, MA, USA*

MARTA KOCH • *Department of Molecular and Developmental Genetics, VIB, Leuven, Belgium; Center for Human Genetics, Katholieke Universiteit Leuven, Leuven, Belgium*

BARBARA KÜPPERS-MUNTHER • *Cellartis, Göteborg, Sweden; University Skövde, Systems Biology Research Center, Skövde, Sweden*

MATTHIAS LANDGRAF • *Department of Zoology, University of Cambridge, Cambridge, UK*
TZUMIN LEE • *Janelia Farm Research Campus, Howard Hughes Medical Institute, Ashburn, VA, USA*
MARIANA LOPEZ-MATAS • *EMBL-CRG Systems Biology Unit, Center for Genomic Regulation, UPF, Barcelona, Spain*
MATTHIEU LOUIS • *EMBL-CRG Systems Biology Unit, Center for Genomic Regulation, UPF, Barcelona, Spain*
MORAEA PHILLIPS • *EMBL-CRG Systems Biology Unit, Center for Genomic Regulation, UPF, Barcelona, Spain*
ANDREAS PROKOP • *Wellcome Trust Centre for Cell-Matrix Research, University of Manchester, Manchester, UK*
XIAO-JIANG QUAN • *Laboratory of Neurogenetics, Department of Molecular and Developmental Genetics, VIB, Leuven, Belgium; Center for Human Genetics, KU Leuven School of Medicine, Leuven, Belgium*
ARIANE RAMAEKERS • *Laboratory of Neurogenetics, Department of Molecular and Developmental Genetics, VIB, Leuven, Belgium; Center for Human Genetics, KU Leuven School of Medicine, Leuven, Belgium*
DIERK F. REIFF • *Department for Systems and Computational Neuroscience, Max-Planck-Institute for Neurobiology, Martinsried, Germany; Department for Animal Physiology/Neurobiology, Albert-Ludwigs University Freiburg, Institute for Zoology, Freiburg, Germany*
NATALIA SÁNCHEZ-SORIANO • *Wellcome Trust Centre for Cell-Matrix Research, University of Manchester, Manchester, UK*
MARION A. SILIES • *Department of Neurobiology, Stanford University, Stanford, CA, USA*
SIMON SPRECHER • *Department of Biology, Institute of Developmental and Cell Biology, University of Fribourg, Fribourg, Switzerland*
VALERIE UYTTERHOEVEN • *VIB, Center for the Biology of Disease, Leuven, Belgium; Laboratory of Neuronal Communication, Programs in Molecular and Developmental Genetics and Cognitive and Molecular Neuroscience, Center for Human Genetics, KU Leuven, Leuven, Belgium*
PATRIK VERSTREKEN • *VIB, Center for the Biology of Disease, Leuven, Belgium; Laboratory of Neuronal Communication, Programs in Molecular and Developmental Genetics and Cognitive and Molecular Neuroscience, Center for Human Genetics, KU Leuven, Leuven, Belgium*
W. RYAN WILLIAMSON • *Department of Physiology and Green Center for Systems Biology, UT Southwestern Medical Center, Dallas, TX, USA*
LIESBETH ZWARTS • *Laboratory of Behavioral and Developmental Genetics, Department of Human Genetics, KU Leuven, Leuven, Belgium; VIB Center for the Biology of Disease, KU Leuven, Leuven, Belgium*

Part I

Anatomy

Chapter 1

Deciphering the Adult Brain: From Neuroanatomy to Behavior

Liesbeth Zwarts, Jason Clements, and Patrick Callaerts

Abstract

The *Drosophila* brain with an estimated 100,000 neurons provides at once an excellent opportunity to describe a complex brain in great detail and to identify the genetic and neurobiological basis of a wide array of behaviors. Furthermore, the sequencing of the genome with the concurrent identification of the elaborate homology of human and *Drosophila* genes and numerous functional studies have established the high relevance of *Drosophila* to understand diseases of the human brain. An overview is provided of available techniques to visualize neurons and neural circuitry and to study their function in behavior. Various resources are listed, and future and emerging technologies are discussed.

Key words: Adult brain, Neuroanatomy, Behavior

1. Introduction

The brain of *Drosophila melanogaster* consists of an estimated 100,000 neurons. This number places the fly brain in terms of complexity in between the *Caenorhabditis elegans* nervous system consisting of 302 neurons and the mammalian and human brain with up to 10^{10} neurons. The wealth of techniques available to study any aspect of *Drosophila* neurogenetics makes it a model of choice for many questions related to development, function, and disease of the nervous system. It must therefore not come as a surprise that major efforts are under way to map and identify every neuron in the larval and adult *Drosophila* brain. It can be expected that the insights stemming from these efforts will further enhance our ability to dissect the neural circuitry and the genetics controlling various behaviors. Furthermore, it will allow studies in great

Bassem A. Hassan (ed.), *The Making and Un-Making of Neuronal Circuits in Drosophila*, Neuromethods, vol. 69,
DOI 10.1007/978-1-61779-830-6_1, © Springer Science+Business Media, LLC 2012

detail of the functions of genes that have been implicated in human disorders related to the nervous system such as neurodegenerative and psychiatric disorders.

In this chapter, we provide an overview of techniques that are currently available to study the adult brain. In the first part of the chapter, we address the range of techniques and markers that are available to visualize neurons and neural circuits. In a second part, we give an overview of techniques that can be used to study the function of neurons and neural circuits, as they control behavior. In the final part, we discuss future perspectives with respect to future and emerging technologies that will further help increase our understanding of how the fly brain works. The emphasis has been placed on providing brief descriptions of methods that are available to study the fly brain and how it functions in behavior. For a subset of these, elaborate protocols are provided, whereas for others the primary references in which detailed protocols can be found are listed. In addition, we provide overview tables of markers and lists of resources where reagents of all kinds and information can be found.

2. Visualizing Neurons and Neural Circuitry

Since the pioneering work of Ramon y Cajal, fly brains have been studied for more than 100 years using a variety of increasingly sophisticated techniques. Some of these techniques are not used frequently anymore, but they are highly valuable and often represent first and easy access to information concerning fly brain structures upon experimental or genetic perturbation. Therefore, following the description of adult brain structure, we systematically describe techniques ranging from the older histochemical to the more recent transgenic-based.

2.1. The Adult Fly Brain

The adult *Drosophila* brain consists of three neuromeres termed protocerebrum, deutocerebrum, and tritocerebrum that are further subdivided into various regions. The nomenclature that is used here is based on the canonical naming system (1) intended to unambiguously define precise locations in the brain (Fig. 1).

Brain regions as defined by Ito et al. and described with additional details in the FLYBRAIN Neuron Database (2).

(a) Optic lobes
- Lamina (LA): lamina proper (LAP)
  - Lamina dorsal rim area (LADRA)
  - Plexiform lamina (PLLA)
  - Accessory lamina (ALA)

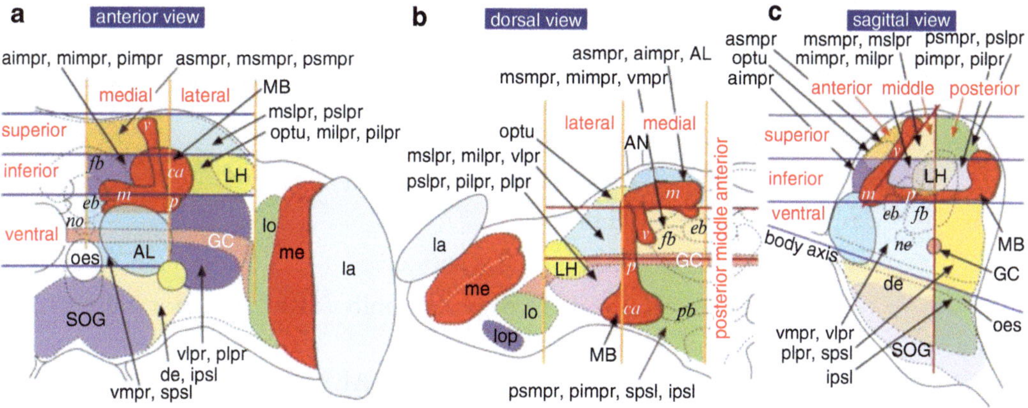

Fig. 1. *Drosophila* adult brain subdivisions. *Blue*, *orange*, and *red lines* define the horizontal, sagittal and coronal planes that were used by Otsuna and Ikeo (1) to divide the brain. *la* lamina; *me* medulla; *lo* lobula; *lop* lobula plate; *a* anterior; *m* middle; *p* posterior; *s* superior, *i*, inferior; *v* ventral; *lpr* lateral protocerebrum; *mpr* medial protocerebrum; *optu* optic tubercle; *de* deutocerebrum; *psl* posterior slope; *oes* esophagus; *GC* great commissure; *AL* antennal lobe; *SOG* subesophageal ganglion; *LH* lateral horn; *MB* mushroom body (*ca* calyx; *p* pedunculus; *m* medial lobe; *v* vertical lobe); *CC* central complex (*fb* fan-shaped body; *eb* ellipsoid body; *no* noduli; *pb* protocerebral bridge). From Otsuna and Ito [1], with permission.

- Medulla (ME): medulla proper (MEP)
  - Medulla dorsal rim area (MEDRA)
  - Plexiform medulla

    Outer medulla (MEO): medulla layer 1–6 (M1–6)

    Serpentine layer (SPL, M7)

    Inner medulla (MEI): medulla layer 8–10 (M8–10)
  - Accessory medulla (AME)
- Lobula complex (LOX): lobula (LO): lobula layer 1–6 (LO1–6)
  - Lobula plate (LOP): lobula plate layer 1–4 (LOP1–4)

(b) Mushroom bodies (MB)

- Calyx complex (CAX)
  - Calyx (VA): medial calyx (MCA); lateral calyx (LCA)
  - Accessory calyx (ACA)
- Pedunculus (PED)
  - Pedunculus neck (PEDN); pedunculus divide (PEDD)
- Spur (SPU)
- Vertical lobes (VL): vertical γ lobe (VγL); α′ lobe (α′L); α lobe (αL); ap lobe (apL)
- Medial lobes (ML): γ lobe (γL); β′ lobe (β′L); βp lobe (βpL); trauben (TRA)

(c) Central complex (CX)
- Central body (CB)
  - Fan-shaped body (FB)
  - Ellipsoid body (EB)
- Protocerebral bridge (PB)
- Noduli (NO)

(d) Lateral complex (LX)
- Bulb (BU): superior bulb (SBU); inferior bulb (IBU); anterior bulb (ABU)
- Lateral accessory lobe (LAL): upper lateral accessory lobe (ULAL); lower (LLAL); gall

(e) Ventrolateral neuropils (VLNP)
- Optic tubercle (OTU): upper unit (UU); lower unit (LU)
- Ventrolateral protocerebrum (VLP): anterior (AVLP); inferior (IVLP); posterior (PVLP)
- Posteriorlateral protocerebrum (PLP)

(f) Lateral horn (LH)

(g) Superior neuropils (SNP)
- Superior lateral protocerebrum (SLP); intermediate (SIP); medial (SMP)

(h) Inferior neuropils (INP)
- Clamp (CL): superior clamp (SCL); inferior clamp (ICL)
- Crepine (CRE): rubus (RUB)
- Antler (ATL)
- Inferior bridge (IB)

(i) Antennal lobe (AL)
- Antennal lobe hub (ALH)
- Antennal lobe glomeruli (GL or ALGL)

(j) Ventromedial neuropils (VMNP)
- Ventral complex (VX): vest (VES); epaulette (EPA); gorget (GOR)
- Posterior slope (PS): superior posterior slope (SPS); inferior posterior slope (IPS)

(k) Periesophageal neuropils (PONP)
- Saddle (SAD): antennal mechanosensory and motor center (AMMC)
- Flange (FLA)
- Cantle (CAN)
- Prow (PRW): dorsal pharyngeal sensory center (DPS)

(l) Subesophageal ganglion (SOG)

- Ventral pharyngeal sensory center (VPS)
- Anterior maxillary sensory center (AMS)
- Posterior maxillary sensory center (PMS)
- Labellar sensory center (LBS)

2.2. Methods

2.2.1. Mass Histology

Heisenberg and Böhl (3) devised this method to simultaneously process many adult fly heads for histological analysis. This method makes use of fly collars (see Fig. 2) for easy manipulation and orientation of fly heads. Briefly, after mounting adult flies in fly collars, they are fixed in Carnoy's fixative, followed by dehydration and embedding in paraffin. Subsequently, 7-μm sections are made that can be used for different applications. Note that depending on the downstream application, different fixatives may be used.

2.2.2. Autofluorescence

Due to fixation and processing, the sections of fly heads become autofluorescent. After deparaffination, the sections can be dehydrated and covered with coverslips. Brain structures can be analyzed directly using epifluorescence microscopy (Fig. 3).

2.2.3. Silver Staining of Sections

A standard histological technique to study insect brains is silver staining. Two commonly used methods to study *Drosophila* brains are the Holmes–Blest method (4) and Bodian's method (5) (Fig. 4).

2.2.4. Golgi Staining

The Golgi stain is a classical method in which entire neurons in fixed tissue can be stained for the visualization of the neuron and its neurites. Upon impregnation with potassium dichromate and silver nitrate, microcrystals of silver chromate form in the neuron via a stochastic and unknown mechanism producing a purplish stain against a yellow background (Fig. 5). A frequently used

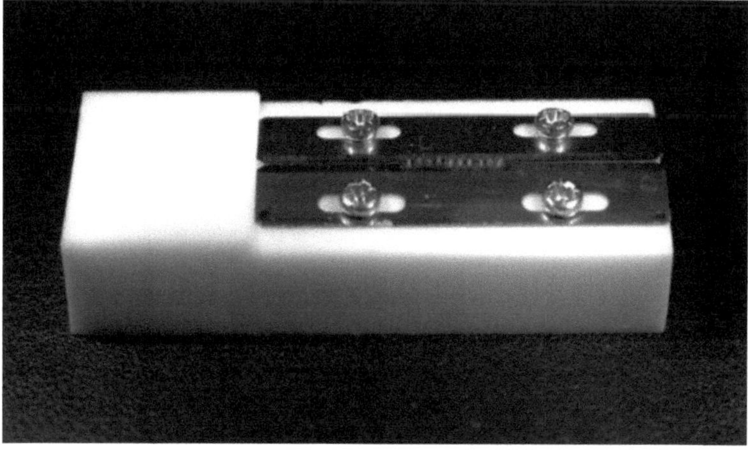

Fig. 2. Fly collar for mass histology.

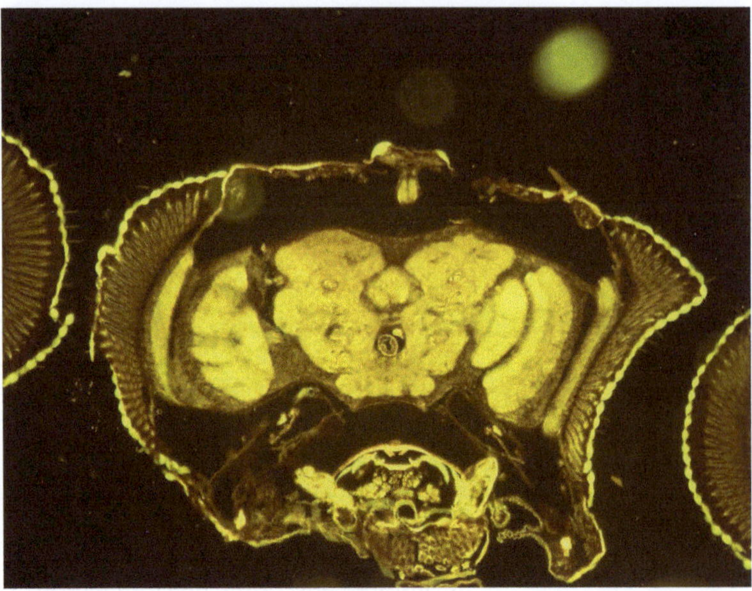

Fig. 3. Autofluorescence of sectioned *Drosophila* brain.

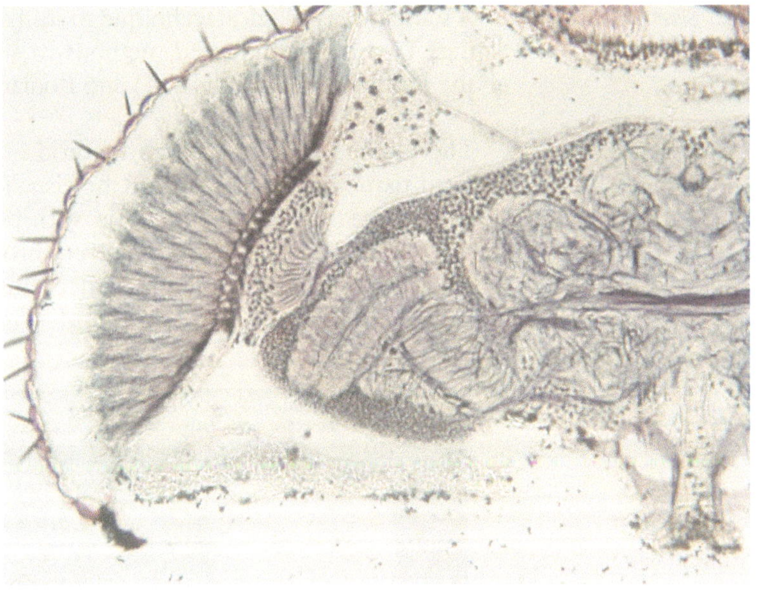

Fig. 4. Bodian stain of *Drosophila* sections.

variant of the method (6) was originally described by Fischbach and Götz (7) as modified by them from the original protocol by Colonnier (8).

2.2.5. Immunohistochemistry

Immunohistochemistry allows the direct visualization of proteins in the adult brain and can as such be used to study (1) brain

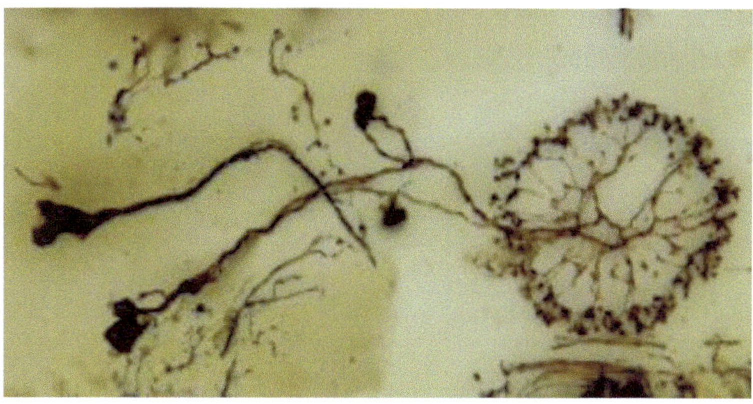

Fig. 5. Golgi stain revealing ellipsoid body neurons. From FLYBRAIN. An online atlas and database of the *Drosophila* nervous system (http://www.flybrain.org).

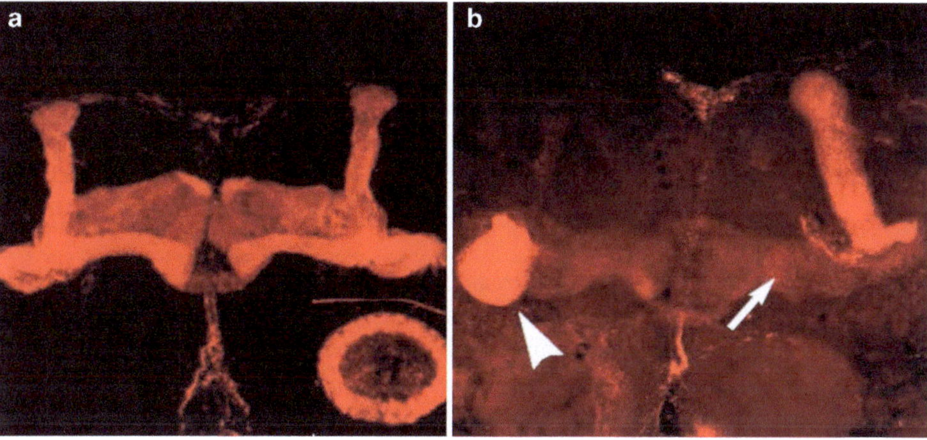

Fig. 6. Immunohistochemistry with the 1D4 anti-fasciclin2 monoclonal antibody. (**a**) Wild-type brain with mushroom body alpha, beta, and gamma lobes. *Inset* is ellipsoid body. (**b**) Mutant brain with axonal stalling (*arrowhead*) and missing beta lobe (*arrow*) phenotypes.

morphology and (2) neuropil-specific and (3) (sub)cellular localization of proteins. In the 1980s, several monoclonal antibody libraries have been generated, of which a number of antibodies have been characterized (Fig. 6). In addition, over the past decades, many research groups have generated polyclonal and monoclonal antibodies that are being used as markers. More recently, a number of companies are making systematic efforts to test whether antibodies against vertebrate antigens cross-react with (homologous) *Drosophila* proteins. These efforts make that an increasing number of antibody reagents are now available for studies in *Drosophila*. Below, we describe a protocol for whole-mount brain immunohistochemistry. Furthermore, Table 1 presents a number of very frequently used primary antibodies and their use. Of note is also that antibodies against neurotransmitters and neuropeptides have been used

Table 1
Frequently used antibodies

nc82 (α-Bruchpilot)	Presynaptic marker Neuropil marker	1:20	DSHB
4F3 (α-discs large)	Postsynaptic marker Neuropil marker	1:20	DSHB
Elav-9F8A9 (α-elav) Rat-Elav-7E8A10 (α-elav)	Neuronal marker (nuclear)	1:20 1:20	DSHB
8D12 (α-repo)	Glial cell marker (nuclear)	1:20	DSHB
24B10 (α-chaoptin)	Photoreceptor marker	1:30	DSHB
1D4 (α-fasciclin 2)	Mushroom body marker (αβ and γ neurons) Central complex marker (ellipsoid body)	1:50	DSHB
Rabbit α-GFP		1:500	Invitrogen Cat. No. A6455
Mouse α-GFP		1:500	Roche Cat. No. 11 814 460 001

successfully in *Drosophila*. The latter often have discrete expression patterns, making them excellent landmarks when studying the *Drosophila* brain (9–12).

Protocol: Immunostaining of Adult Brains (13, 14)

1. Dissect brains in ice-cold 1× PBS for up to 1 h. Collect in microcentrifuge tube with PBS on ice.
2. Fix brains for 15 min in 500 μL 4% formaldehyde at room temperature.
3. Wash brains three times in 500 μL PBS for 10 min (be careful that brains do not get stuck to pipette tip!)
4. Preincubate brains for at least 15 min in 500 μL PAXD (PBS containing 5% BSA, 0.3% Triton X-100, 0.3% sodium deoxycholate).
5. Incubate overnight in PAXD containing primary antibody (alternative: 1 h at room temperature).
6. Wash thoroughly in PAXD, making 4–5 changes throughout the day.
7. Incubate overnight in PAXD containing secondary antibody (alternative: 1 h at room temperature).
8. Wash thoroughly in PAXD, making 4–5 changes throughout the day.
9. Mount brains in a drop (10–15 μL) of Vectashield (Vector Laboratories) (Fig. 7).

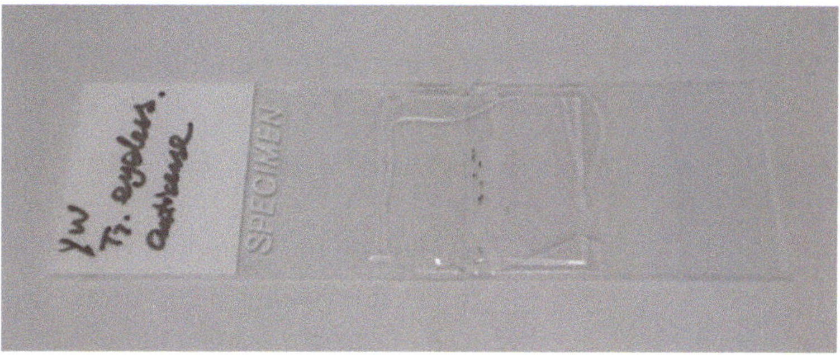

Fig. 7. Mounting of fly brains.

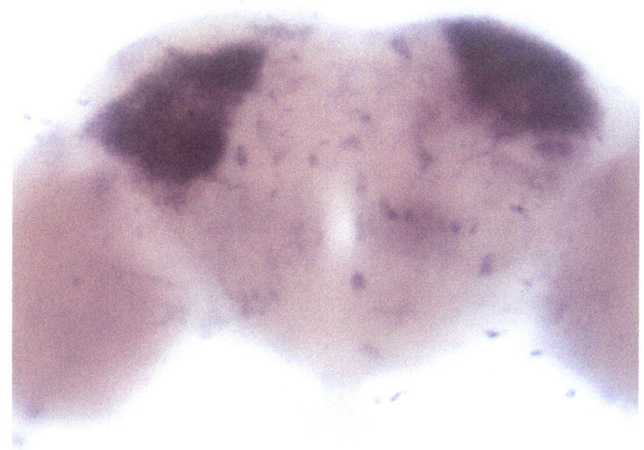

Fig. 8. In situ hybridization with eyeless antisense probe.

2.2.6. In Situ Hybridization (see Fig. 8)

A complementary method to immunohistochemistry is in situ hybridization, which allows the analysis of transcript distribution of a gene of interest. Although in situ hybridization is not limited by the availability of reagents (such as antibodies), it often has variable success owing to e.g., low copy number of a transcript. In addition, the method is more labor-intensive than immunohistochemistry. We have extensively and successfully used the following protocol for the localization of mRNA in adult *Drosophila* brains.

Protocol: In Situ Hybridization (15)

DIG RNA Labeling

1. Linearize the template DNA at a suitable site. Avoid using 3′-overhang producing enzymes
2. Purify the linearized template DNA using QiaQuick PCR purification kit (Qiagen)
3. Label the linearized and purified template DNA (1 μg) using the DIG RNA labeling kit (Roche) (= >24 μL, ±42 ng/μL)

4. Add: 26 μL MilliQ
   (a) 30 μL Na_2CO_3 (200 mM)
   (b) 20 μL Na_2HCO_3 (200 mM)
5. Incubate at 60°C for *X*min where $X = (\text{Lo} - \text{Ld})/(0.11 \times \text{Lo} \times \text{Ld})$
   (a) Lo = original length in kb, Ld = desired length in kb
   (b) Desired length is 200 bases
6. Add: 5 μL 10% acetic acid
   (a) 11 μL NaAc (3 M, pH 6)
   (b) 1 μL tRNA (10 mg/mL)
   (c) 1.2 μL $MgCl_2$ (1 M)
   (d) 300 μL EtOH
7. Incubate for 4–16 h, −20°C
8. Centrifuge at 13,000 rpm, 15 min, 4°C
9. Remove supernatant and dry the pellet
10. Resuspend probe in 50 μL (= >200 ng/μL)
11. The RNA transcripts can be analyzed for size by agarose gel electrophoresis and ethidium bromide staining
12. Determine labeling efficiency (Roche kit)

Tissue Dissection, Fixation, and Hybridization

1. Dissect tissues in PBS
2. Wash tissues in 200 μL PBT
3. Fix for 60 min in 1 mL PP + 0.1% sodium deoxycholate. Rinse once in 1 mL PBT. Wash 5× 5 min in 1 mL PBT
4. Proteinase K treatment (optional):
   (a) Incubate in 100 μL of proteinase K solution depending on the tissue:
       - Inverted *third instar*: 15 μg/mL; 2 min at 37°C
       - Adult *abdomen*: 15 μg/mL; 3 min at 37°C
       - Adult *thorax*: 10 μg/mL; 2 min at 37°C
       - Adult *brain*: 10 μg/mL; 2 min at 37°C
       - Whole mount *embryo*: 40 μg/mL; 3 min at RT
   (b) Add 900 mL of 2 mg/mL glycine in PBT to quench ProtK
   (c) Wash once with 1 mL of 2 mg/mL glycine in PBT for 1 min
   (d) Wash 2× 5 min in PBT
   (e) Refix 1× 20 min in PP
   (f) Wash 6× 5 min in PBT

5. Wash for 10 min in 1:1 PBT–Hybrix
6. Wash for 10 min in 200 μL Hybrix (do not shake)
7. Prehybridize in 200 μL Hybrix for 3 h at 60°C
8. Meanwhile, denature the RNA probe (use final concentration of 0.25 ng/μL (= >25 ng in 100 μL Hybrix)) in a separate tube by incubating at 100°C for 5–10 min. Next, cool on ice for 5 min and spin
9. Remove 200 μL prehyb buffer from tissues and add 100 μL of the denatured RNA probe
10. Hybridize overnight at 60°C (in oven, no water bath, no rotation necessary)
11. Prewarm wash solutions (60°C): Hybrix–PBT 5:0/4:1/3:2/2:3/1:4/0:5

    Wash tissues in 200 μL of the above solutions for 20 min each at 60°C
12. Wash 4× 20 min in 300 μL of PBT at RT

Antibody Preabsorption

1. Dissect and fix 20 inverted larvae
2. Centrifuge antibody for 5′ at 10,000 rpm
3. Add anti-DIG-AP (1:200) in PBT (no BSA), shake at 4°C, overnight

Antibody Treatment

1. Block tissues for 2 h in blocking buffer (1× PBT, 1% blocking reagent (Roche))
2. Incubate with 500 μL 1:2,000 anti-DIG-AP in blocking buffer for 1 h
3. Wash 3× 20 min in 1 mL PBT
4. Wash ON at 4°C in PBT
5. Wash 1× 20 min in 1 mL PBT
6. Transfer to 24-well plate before adding staining solution.
7. Wash in staining buffer 3× 3 min. Transfer to 24-well plate before adding staining solution. Remove staining buffer and add staining solution.
8. Incubate in AP substrate (5 μL/mL NBT + 3.75 μL/mL BCIP in staining buffer) in the dark. The reaction takes minutes to several hours to develop depending on target transcript abundance. Monitor color development (brief exposure to light is ok)
9. Wash 3× 5 min in PBT to stop reaction

10. Clear in the following:
    (a) 30% glycerol in PBS (30′)
    (b) 50% glycerol in PBS (30′)
    (c) 70% glycerol in PBS (30′)

Reagents

- RNase-free water
- LiCl 4 M
  - Use RNase-free water!!
- Hydrolysis buffer
  - Use RNase-free water
  - 60 mM Na_2CO_3
  - 80 mM $NaHCO_3$
  - Adjust to pH 10.2
  - Aliquot and store at –20
- Tris 1 M pH 7.5
  - Use RNase-free water!!
- PBS stock 10×
  - Use RNase-free water!!
  - 1.3 M NaCl
  - 30 mM NaH_2PO_4
  - 70 mM Na_2HPO_4□$2H_2O$
  - Adjust to pH 7.0
- Tween-20 10% stock
  - In RNase-free water!!
  - Use molecular biology grade Tween!!
- Paraformaldehyde (PFA) stock 20%
  - Dissolve 10 g PFA in 40 mL RNase-free water
  - Add 300 μL 5 M NaOH (made in RNase-free water)
  - Incubate at 65°C until dissolved
  - Add RNase-free water up to 50 mL
  - Store in aliquots at –20°C, thaw each aliquot only once
- SSC 20×
  - In RNase-free water!!
  - 3 M NaCl
  - 0.3 M Sodium citrate
  - pH 7

- tRNA stock 20 mg/mL
  - Dissolve tRNA in TE buffer pH 7.6 containing 0.1 M NaCl
  - Extract twice with an equal volume of phenol
  - Extract twice with an equal volume of chloroform
  - Recover RNA by precipitation with 2.5 volumes of 100% ethanol at RT
  - Dissolve RNA at a concentration of 20 mg/mL
  - Store at −20°C in small aliquots
- ssDNA stock 10 mg/mL
  - Dissolve 100 mg of salmon sperm DNA in RNase-free water by stirring for 2 h at RT
  - Adjust NaCl concentration to 0.1 M
  - Extract once with equal volume of phenol, recover aqueous phase
  - Extract once with equal volume of phenol–chloroform, recover aqueous phase
  - Shear the DNA by passing it 12 times rapidly through a 17-gauge hypodermic needle
  - Precipitate DNA by adding 2 volumes of ice-cold 100% ethanol and recover by centrifugation
  - Dissolve in 5 mL RNase-free water, measure DNA concentration, and adjust to 10 mg/mL
  - Boil the solution for 10 min and store at −20°C in aliquots. Just before use, heat the solution for 5 min in a boiling water bath and chill quickly on ice
- PBT
  - PBS
  - % Tween (from 10% stock)
  - Filter the solution
- PP (make fresh)
  - 4% paraformaldehyde (from 20% stock) in PBT
- Hybrix
  - 50 mL formamide (molecular biology grade)
  - 25 mL 20× SSC
  - 0.5 mL tRNA from 20 mg/mL stock
  - 1 mL ssDNA from 10 mg/mL stock
  - 45 µL Heparin from 0.11 g/mL stock
  - 1 mL Tween-20 from 10% stock
  - 22.5 mL RNase-free water

- Proteinase K
  - 15 μg/mL in PBT
- Staining buffer
  - 01. M Tris–HCl, pH 9.5
  - 0.1 M NaCl

2.2.7. Enhancer Trapping

The enhancer trap technique was first described by O'Kane and Gehring (16). The initial method made use of LacZ as a reporter gene. Since then, numerous variants have been generated in which other reporters are used (e.g., GAL4, GFP, and others). In this method, a transposable element, usually a P-element, is used to introduce the reporter gene in the genome. The specific expression pattern of the reporter is dependent on enhancer elements in the genome, as the minimal promoter present in the vector is insufficient to activate transcription on its own. Due to the semirandom nature of transposable element insertions, enhancer trap insertions throughout the genome have been generated. Following the initial description of the enhancer trap technique, many large-scale efforts have been made to generate and identify collections of enhancer trap lines (17–21). Several of these efforts have gone towards identifying enhancer trap lines with (selective) expression in the brain and continue to date (Fig. 9).

It should be noted, however, that the reporter's expression pattern may not fully reproduce the expression pattern of the gene in which it is inserted. The insertion site of the transposable element can be determined using plasmid rescue or inverse PCR. In addition to identifying a gene's (partial) expression pattern, the reporter can also be used to study factors influencing its expression, making enhancer traps ideal tools for, e.g., identifying transcription factors that activate or repress a particular gene's expression.

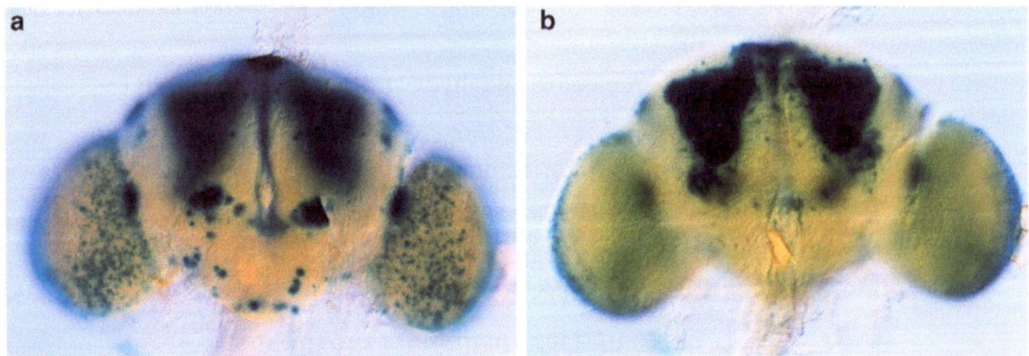

Fig. 9. OK107-GAL4 enhancer trap line driving expression of UAS-LacZ revealed by β-galactosidase histochemical staining. (**a**) Anterior view, (**b**) posterior view.

2.2.8. The GAL4-UAS System

The GAL4-UAS system is a technique enabling the expression of a gene of choice in a defined spatiotemporal pattern. GAL4 is a transcription factor from yeast that was adapted for use in *Drosophila* by Brand and Perrimon (22). In short, when the expression of GAL4 is under the control of a particular enhancer (e.g., an enhancer that is activated in all mushroom body neurons), the GAL4 line can be used to express any gene of interest in the same pattern by cloning the coding cDNA behind the UAS sequence. When both components are present in the same fly, the GAL4 protein that is expressed will bind to the UAS sequence, activating the expression of the gene of interest. More recently, it has become possible to more finely control activation of GAL4-mediated gene expression via use of the Gal80 protein, which binds GAL4 and inhibits it from binding the UAS sequence.

This technique has revolutionized all aspects of *Drosophila* research, including studies of the brain. Via the UAS–GAL4 system, it is possible to study morphology (e.g., via expression of GFP or *lacZ*) (Fig. 10), gene function (via rescue or overexpression experiments on the one hand, or the expression of RNAi or dominant-negative proteins on the other), neuronal function (via activating or inhibiting neurons, or by ablating them entirely) (23, 24). In recent years, several comprehensive studies have been published describing collections of GAL4 lines with specific expression in different parts of the brain. These have been included in Table 2 and the appropriate references are listed.

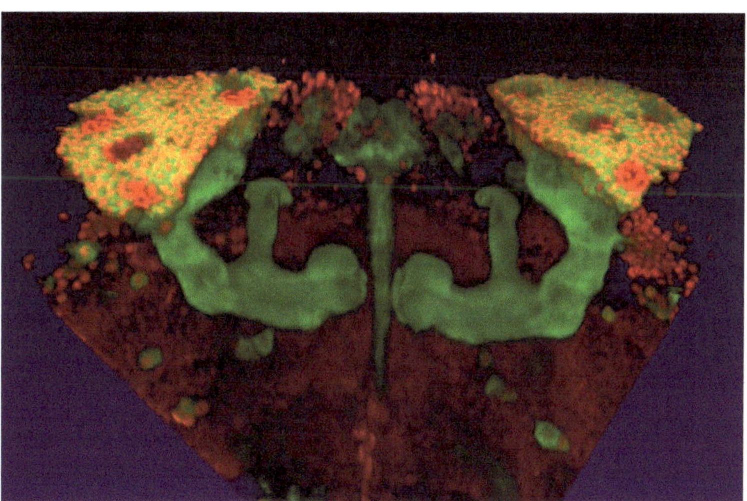

Fig. 10. The GAL4-UAS system. OK107-GAL4 driving UAS-GFP-CD8 reporter gene (*green*) counterstained with anti-dachshund nuclear protein (*red*).

Table 2
GAL4 lines

GAL4 line	Specificity	References
Cha-GAL4	Cholinergic neurons	(25)
Gad1-GAL4	GABAergic neurons	(26)
VGlut-GAL4	Glutamatergic neurons	(27)
Trh-GAL4	Serotonergic neurons	(28)
TH-GAL4	Dopaminergic neurons	(29)
Tdc2-GAL4	Octopaminergic neurons	(30)
Ddc-GAL4	Serotonergic and dopaminergic neurons	(31)
	Mushroom body neuron GAL4	(32, 33)
	Gustatory neuron GAL4	(34)
	Visual projection neuron GAL4	(1)
	Auditory neuron GAL4	(35)
	Olfactory neuron GAL4	(36)
	Central complex GAL4	(37)

3. Functioning Neurons and Neural Circuits and *Drosophila* Behavior

3.1. A Fly's Behavioral Repertoire

The genetic analysis of *Drosophila* behavior and of the brain structures controlling it was strongly influenced by two types of studies. The isolation of behavioral mutants by countercurrent distribution (38) and of three clock mutants with different phenotypes that are all caused by mutations in the same gene, *period* (39), showed that isolation of genetic mutants affecting specific behaviors was possible. On the other hand, the isolation of structural brain mutants by Heisenberg and colleagues has given insight into e.g. the role of the mushroom bodies in learning (40), the role of the central complex in walking behavior (41, 42), and the role of the optic lobes in vision (43), thereby linking specific parts of the brain to specific behaviors. Since these studies, the behavioral repertoire of *Drosophila* has proven to be amazingly complex. This is reflected in the range of techniques described below.

3.2. Methods

3.2.1. Isolating Mutants

Mutants can be obtained in numerous ways. Many of the mutants first studied were the result of spontaneous mutations due to insertions of transposable elements or were induced by X-ray irradiation, which produces larger chromosomal aberrations. The use of chemical mutagens (e.g., ethylmethanesulfonate) in conjunction with screening for viable but behaviorally abnormal offspring or for adult flies with structural brain defects heralded the onset of behavioral genetics. P-element mediated mutagenesis is another commonly used approach. One advantage of P-element mediated mutagenesis is that these alleles are often hypomorphic as the P-element inserts itself in the 5′ UTR of genes with high frequency, affecting the expression level of the gene product. Furthermore, the locus of many of the P-element insertions can be determined by use of plasmid rescue or inverse PCR, greatly facilitating the characterization of the allele. Finally, P-element insertions can be remobilized. This may result in imprecise excisions that can be strong or null alleles. Alternatively, this may result in precise excisions that then allow to demonstrate that the observed mutant phenotype is the result of the original P-element insertion. More recent techniques include homologous recombination and a range of recombineering techniques.

3.2.2. Altering Gene Expression Levels

In addition to using mutants isolated as described in the previous paragraph, the function of a gene is nowadays frequently studied by transgenic RNA interference (RNAi) to knock down the expression of a given gene in a spatiotemporally controlled manner. Several genome-wide RNAi collections have been established to facilitate its use (44).

3.2.3. Modulating Neural Activity

Neuronal Silencing

Just as a null or hypomorphic allele can be critical for understanding the requirement and function of a given gene, neuronal silencing has become an indispensible tool for understanding the function of a neuron in a particular process or behavior. One of the first tools to block neuronal activity was the use of $shibire^{ts1}$, a temperature-sensitive mutant of the dynamin gene (45). At permissive temperature, $shibire^{ts1}$ has no affect on neuronal function. However, when shifted to restrictive temperature, endocytosis of vesicles at the synapse is blocked, disrupting the normal release of neurotransmitters at the synapse.

Excitability of a neuron is largely determined by the activity of its ion channels. Thus, by manipulation of the ion channels present on the neuronal membrane, it is possible to hyperpolarize it to the degree that it can no longer generate action potentials. The most widely used approach for membrane hyperpolarization is the overexpression of K^+ channels, such as Kir2.1, Shaker, or DORK via the UAS-GAL4 system (46–48). Alternatively, knocking down or reducing the function of existing Na^+ or Ca^{2+} channels will also result in a net hyperpolarization of the membrane potential.

Neuronal Activation

Manipulation of ion channels can also be used to activate neurons. Not surprisingly, these are the reciprocal manipulations used to silence neuronal activity. In the first experiments to artificially activate neurons, K^+ channel activity was hindered by the expression of dominant-negative proteins (Shaker K^+ dominant-negative, Shaw K^+ dominant-negative), RNAi-mediated knockdown of K^+

channels, or overexpression of Na^+ channels such as the bacterial-derived voltage-gated Na^+ channel NaChBac, resulting in membrane depolarization and hyperexcitation of the neuron (46–48). More recently, techniques have been developed which allow neurons to be artificially activated with a degree of temporal control. This can be accomplished by overexpression of a temperature-responsive cation channel (TrpA1, which is activated at elevated temperatures, or TrpM8, which is activated at reduced temperatures (48–50)).

Optogenetics

Optogenetics is a developing series of techniques in which light-sensitive channel proteins are expressed neuronally, allowing either the activation or silencing of the neuron upon illumination. In *Drosophila*, the first use of a light-activated, but technically demanding, genetic system for neuronal activation was with P2X2, an ATP-gated ionotropic purinoceptor (51). More recently, the most widely used channel for controlled activation of neurons is channelrhodopsin-2 (52). This cation channel from the green alga *Chlamydomonas* is activated by blue light, when in the presence of all-*trans* retinal which can be administered to the fly via the food, and leads to membrane depolarization. Advantages of channelrhodopsin-2 are that it is a monomer, making it easy to express via the GAL4 system, and its rapid activation and deactivation, which occurs in the millisecond range. While optogenetic tools for neuronal inactivation have also recently been developed, such as the halorhodopsin chloride pump NpHR, these have yet to be used in *Drosophila* (53).

3.3. Analyzing Behavior

3.3.1. Aggression

Aggression is a well-known trait throughout the animal kingdom. It is essential for a variety of elementary functions like acquisition of food or mates and defense against predators. By contrast, excessive aggression requires high-energy depending acts that are evolutionary unfavorable.

Aggression in *Drosophila* has been well characterized (see Table 3) and has been shown to depend on a variety of environmental factors, sex, and previous history of the individual flies (54–61). Several neurotransmitters have been shown to elicit changes in aggressive behavior and also different brain structures, such as the mushroom bodies, the *fruitless* circuit, and different neurons located in the subesophageal ganglion, play a role in the control of this behavior (28, 62–73). Furthermore, genome-wide studies showed the involvement of a large number of genes involved in a variety of pleiotropic functions (72, 74–77).

Methods

The different setups to study aggression vary at multiple levels, i.e., the number of flies tested, their social experience, hunger state, presence of food or mates, and arena size. However, in all setups, the following subsets of aggressive encounters can be observed (59, 74, 78).

Table 3
Subsets of aggressive encounters

Males/females

Offensive action	Description
Charge/approach	Rapid approach/directed movement of one fly to another
Wing threat	One fly quickly raises both wings to a 45° angle towards opponent
Kicking/fencing	Leg extension from one fly to another resulting in physical contact
Chasing	One fly runs after the other
Lunging	One fly rears up on hind legs and snaps down on the other

Males		Females	
Offensive action	**Description**	**Offensive action**	**Description**
Boxing/tussling	Both flies rear up on hind legs and strike the opponent with forelegs/grip each other with their front legs	Head butt	Head-butting at another female

In general, three main setups have been described:

- 8-male assay: This assay makes use of 3–7-day-old, socially experienced males which are kept in an incubator with a fixed day night cycle and a constant temperature. Flies are analyzed up to a maximum of 5 h after the start of the day light cycle, which allows testing of 45 replicates per day.

  Males are placed in groups of eight, 24 h before testing and are not anesthetized during this period. 2 h after the start of the day light cycle, flies are transferred to an empty vial (H: 95 mm, diameter: 25 mm) by tapping them down. The empty vial is closed using a foam plug on which a drop of food can be applied (Fig. 11). After the desired starvation period, usually 90 min, the flies are tapped down after which the foam plug is turned, resulting in access to the drop of food. The total test time takes 4 min, including the acclimation period. At the start of the assay, the first test vial is tapped down, the foam plug is turned and flies are allowed to acclimate during 1 min 50 s, during the following 10 s, the seconds vial is prepared and the first vial is transferred to the test surface. This surface is preferably white and immobilizes the vial. The number of aggressive encounters of the flies in the first vial is scored during 1 min 50 s, which is also the acclimation period for the second vial. This procedure is repeated for the following 45 vials.
- 2-Male arena assay: This assay allows the analysis and/or recording of the aggressive behavior of a pair of males on a

Fig. 11. Aggression: 8-male assay for the instant analysis of aggressive encounters during a period of 2 min.

food cup that can contain a virgin female and that is placed centrally in a fighting arena (79) (Fig. 12). Flies are usually analyzed during a 30 min period. This setup is compatible with CADABRA software that allows the automatic scoring of lunging, tussling, wing threats, and chasing (80).

- 2-Male plate assay: This assay allows the simultaneous recording of the aggressive behavior of pairs of males in 35 round wells of a test plate (81) (Fig. 13).

Fig. 12. Aggression: 2-male arena assay with centrally placed food cup. From Mundiyanapurath et al. [79], with permission.

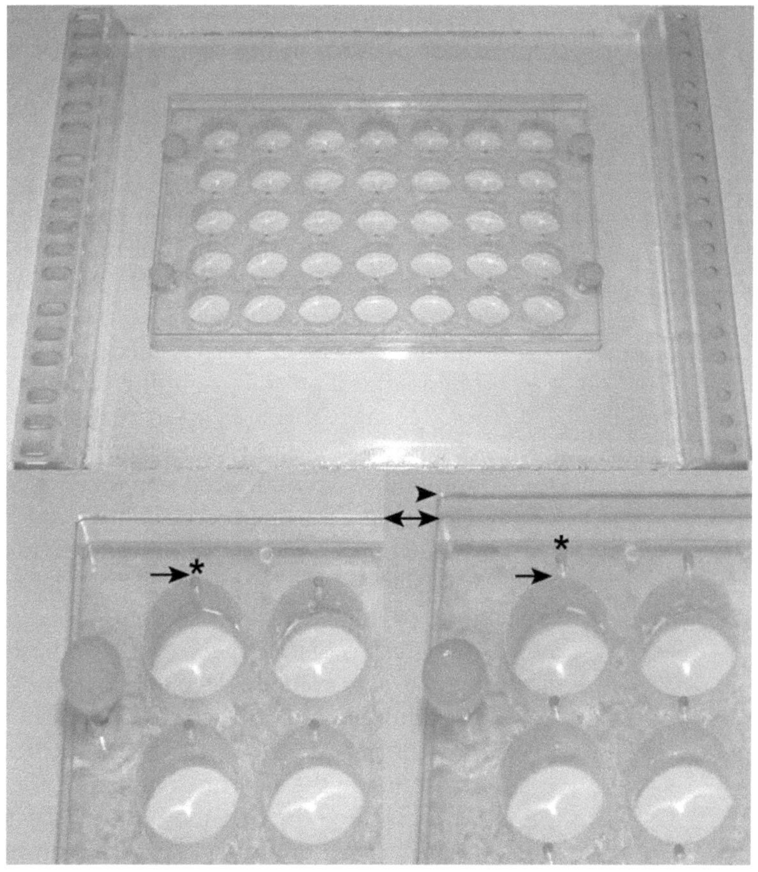

Fig. 13. Aggression: 2-male plate assay allowing the simultaneous videotaping of aggressive encounters of 35 pairs of male flies. From Dierick [81] with permission.

3.3.2. Circadian Behavior and Sleep

Circadian clocks are responsible for the regulation of rhythmic processes in animals, plants, fungi and even some prokaryotes. They allow the organism to adapt to and anticipate environmental changes in light, temperature, food, and mate availability and are necessary to optimally synchronize these different processes. In *Drosophila* they are responsible for multiple rhythmic outputs such as sleep, light response, learning and memory, feeding and metabolism, courtship and mating, immunity, eclosion, locomotor activity, and olfaction (82, 83). Circadian rhythms are characterized by four parameters: (1) presence of a self-sustaining clock, (2) approximately 24 h rhythms, (3) light and temperature entrainment, and (4) temperature compensation meaning stability over a wide temperature range (83).

The principal molecules involved in the circadian clock act in a negative feedback loop that creates a rhythmic wave of gene expression. Two basic helix-loop-helix transcription factors, *clock* and *cycle*, directly activate the transcription of *period* and *timeless*. These two proteins enter the nucleus where they inhibit the transcription of *clock* and *cycle*, thus resetting the circadian cycle. This clock can be entrained by alterations in light and temperature cycles. The clock neurons in the central brain are subdivided into two main types, the dorsal neurons, DN1–3, and the lateral neurons. These lateral neurons are further subdivided into the dorsolateral neurons and the large and small ventrolateral neurons, which, with the exception of one cell, express Pigment dispersing factor. Other known signaling molecules in the clock circuit include glutamate, Pdf, neuropeptide F, and Neuropeptide-like precursor 1 (84).

Sleep is closely related to the circadian clock and also the methods to analyze sleep patterns and circadian rhythmicity show a significant overlap. Sleep is defined by changes in different electrophysiological parameters in the brain, but also by different behavioral changes that are characterized by the following criteria. First of all, this behavior is characterized by periods of quiescence or immobility. Second, sleep implies an increase in arousal threshold. Third, there is a rapid reversibility to the waking state. Finally, sleep is under homeostatic regulation, meaning that periods of extended wakefulness will be followed by compensatory increases in sleep time or intensity (85). All these different sleep criteria have been investigated in *Drosophila* and have been shown to be remarkably parallel to vertebrate sleep parameters.

Multiple brain structures have been shown to influence sleep. First of all, due to the close relation between sleep and circadian rhythmicity, the neurons regulating the clock are involved (86). These include the PDF expressing lLNv's and the sLNv's. Further, also other neuropils, including the mushroom bodies and the pars intercerebralis region, more precisely Dilp2 and Spitz and EGFR ligand neurons, have been implicated in sleep regulation (86). Finally, virtually all of the main neurotransmitters, including GABA, acetylcholine and the monoamines serotonin, dopamine and octopamine, the invertebrate counterpart of noradrenaline, have been shown to influence sleep in species ranging from vertebrates to *Drosophila* (86, 87).

Methods

Quiescence-Homeostatic Regulation

Quiescence and activity are parameters that are commonly used to study both circadian rhythmicity and sleep patterns. Different methods have been applied to investigate these parameters in *Drosophila*. The same methods have also been used to show the homeostatically regulated increase in quiescence upon sleep deprivation (88–90).

- Ultrasound technology: This approach is used to detect very small movements of the fly's head, wings, and limbs (88).
- High magnification video monitoring: This technique is capable of detecting even small respiratory changes and allows the precise evaluation of sleep behavior (89).
- Trikinetics activity monitor: This apparatus monitors the crossing of the midline of a 64-mm tube by individual flies and allows a more high-throughput analysis (91). This method has been shown to correspond very well to the more precise but labor-intensive methods described above and is also the main technique used to analyze circadian rhythmicity in flies (85).

Arousal Threshold

Flies have been shown to have increased arousal thresholds after a quiescence period of more than 5 min which result in unresponsiveness to environmental changes including visual, tactile, and olfactory stimuli as well as to interactions with conspecifics (85). A method using vibratory stimuli was described in more detail (88).

Electrophysiology

Electrophysiological correlates of sleep can be analyzed by recording field potentials in unanesthetized flies. These field potentials reflect the general decrease in activity during periods of immobility and increased arousal thresholds (92).

3.3.3. Arousal

Arousal is defined by increased activity, sensitivity to sensory stimuli, and distinct patterns of brain activity (93). The two extremes of this behavior are attention and sleep, which can be regarded as the highest and lowest states of arousal, respectively. Arousal is a crucial factor for the performance of complex behaviors. Selective attention, for instance, is necessary to filter important sensory input from other cues, a process that can be critical for survival.

As sleep and arousal are closely related, the brain structures involved in both processes show overlap. The sLNvs, for example, which are regarded as the most important pacemaker neurons of the circadian clock, play an important role in light-induced arousal (47). Further, also dopamine signaling is involved in both processes (94, 95). However, despite this close relation between both behaviors, it seems that there are different subtypes of these behaviors that are independently regulated by distinct neural circuits. Mutations in the D1 dopamine receptor, for instance, lie at the basis of a decreasing nocturnal arousal, but also induce an opposite increase in startle-induced locomotion. These two types of arousal are also inversely influenced by cocaine, providing more evidence for a separate regulation of both arousal states. Furthermore, mutant screens for startle induced arousal indicate that the genetic mechanisms underlying arousal are very complex and consist of genes involved in numerous processes, including brain function or development of integrative brain structures such as the mushroom bodies and the central complex (96).

Methods

Startle-Induced Locomotion

- Tap down assay: This assay makes use of 3–7-day-old, socially experienced flies which are kept in an incubator with a fixed day night cycle and a constant temperature. Flies are analyzed within a 4 h time window that starts 2 h after the start of the day light cycle.
- Single flies are transferred to a standard food vial (H: 95 mm, diameter: 25 mm) at least 24 h before testing after which they are not anesthetized anymore. Flies receive a mechanical stimulus by tapping the vial lightly and subsequently placing it horizontally (Fig. 14). Locomotor behavior is quantified as the number of seconds each fly is active during an observation period immediately following the disturbance. The length of this observation period usually varies between 30 and 45 s.
- Air puff assay: This assay makes use of an air puff as a mechanical stimulus (94).

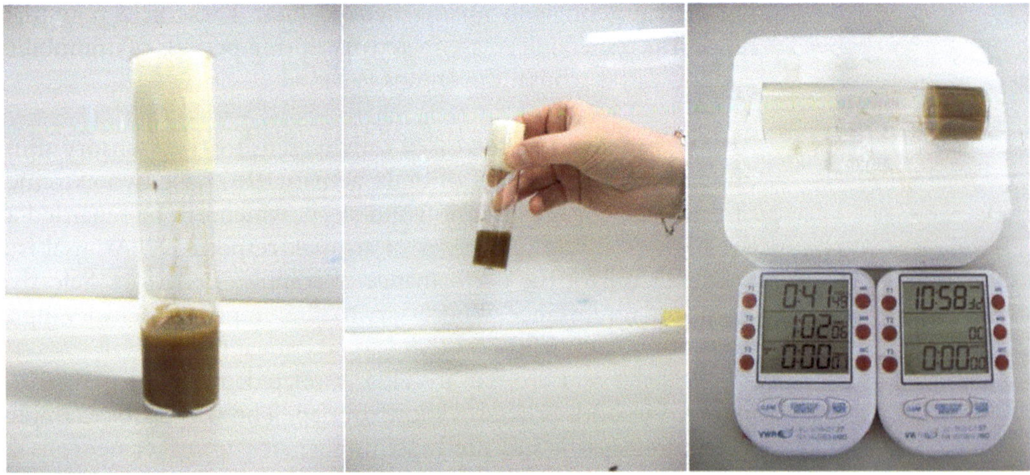

Fig. 14. Startle-induced locomotion assay. An individual fly is aroused by gentle tapping of the vial on the table followed by subsequent monitoring of locomotion during a period of 45 s.

Selective Visual Attention

- Flight simulator: The flight simulator that is used to study flight behavior (see "Methods" in Sect. 3.3.6.1) can be applied to provide flies the choice between different visual stimuli (97).
- Visual attention maze: an adaptation of the Hirsch classification maze, used for the study of gravitaxis behavior, can also be used to study selective visual attention. The maze, which is see-through, is fixed above a monitor on which various patterns can be played. The tendency of flies to follow a certain pattern can be analyzed (98).

3.3.4. Courtship

Courtship behavior is present among all animals and is essential for survival. In each species, other rituals are present which precede actual copulation. In *Drosophila*, this ritual consists of tapping on the abdomen or cuticle of the potential female mate with the foreleg, vibrating with one wing and thus generating a courtship song, followed by circling around the female. These behaviors elicit rejection, expressed by wing flicking and running away, or approval, expressed by immobility, by the female. Immobility will be the signal for the male to start licking the female genitalia and finally to attempt to copulate.

Copulation success is influenced by multiple factors. Courtship is strongly circadianly influenced leading to differences in success rate depending on the time of copulation. Further, also age, experience, and environment, including the size of the test chamber, have been shown to play a role. Finally, also correct sex and species recognition is important. This process requires the correct interpretation of various sensory inputs by both sexes. These inputs include auditory, visual and tactile inputs to correctly read all the courtship cues and olfactory and gustatory input to interpret pheromone signaling (99).

Many of the brain structures that are involved in these processes show distinct sexual dimorphisms. The genes involved in the sex determination pathway, such as *fruitless* and *doublesex*, are crucial for sex-specific behaviors and seem to play an important role in the sexual identity of the involved neurons (100). These sex-specific neurons are located in higher integration centers in the central brain as well as in the different layers of the olfactory and gustatory system and the abdominal ganglion (101).

Methods

Both mating frequency and changes in the different parameters of the courtship routine can be analyzed.

Mating Frequency

Mating frequency can be very easily analyzed by allowing groups of flies to mate after which the females are transferred to individual vials. After a few days, it can be examined whether the females produced offspring and thus mated (102).

Courtship Behavior

Courtship behavior can be analyzed in mating chambers. Different shapes and sizes of these chambers have been described. However, it has been observed that smaller chambers increase the chance of mating. Pairs of flies are transferred into these chambers after which different courtship parameters can be recorded and scored. The previously mentioned CADABRA software can be used to automatically score different subsets of courtship behaviors (80). Different courtship parameters have been described, the most commonly used ones are listed in Table 4 (107).

3.3.5. Feeding Behavior/ Gustatory Behavior

Feeding is a vital behavior for all organisms. It is necessary to facilitate growth, to insure survival and to meet reproductive requirements. Feeding can be regulated by multiple factors such as metabolic requirements, feeding status and different sensory inputs, including olfactory and gustatory signals. Furthermore, this behavior includes multiple decision-making processes, some obvious, e.g., hungry or not hungry, and others more complex, e.g., choosing between different food sources and determining whether to eat reduced-quality resources with possibly novel or aversive tastes in order to survive (108).

Feeding behavior has been shown to be controlled by genes involved in insulin signaling or glycine cleavage. Further, also different neuropeptides have been implicated. The propeptide hugin is an important mediator of feeding behavior in both larvae and adult flies where it is involved in the decision making processes that precede feeding initiation (109–115). These decisions are also influenced by another neuropeptide, NPF (116–119). The vertebrate homologues of both neuropeptides, neuromedin U and NPY, respectively, share a conserved role in feeding regulatory mechanisms.

In larvae, neurons of the subesophageal ganglion, the mushroom bodies and the median neurosecretory cells of the protocerebrum are involved in the processing of inputs involved in feeding

Table 4
Commonly used courtship parameters

Parameter	Description
Latency (103)	Time to initiate courtship
Courtship duration (103)	The time before copulation occurs
Copulation duration (104)	The time during which the flies copulate
Courtship index (CI) (105)	The percentage of the observation period that a males spends on courting a female
Sex appeal parameter (SAP) (106)	The time a male vibrates its wing towards a female

behavior from both external gustatory receptor neurons and internal pharyngeal chemosensory organs (113). In adult flies, gustatory receptors are located on the proboscis, legs, wings and vaginal plate sensilla (120). Information from these receptors is primarily conducted via gustatory receptor neurons towards the SOG. How this information is further processed in the brain is less well understood, but SOG neurons have been shown to project towards multiple regions in the brain, including the antennal lobe, the lateral horn, and the mushroom bodies (121).

Methods

Feeding behavior in *Drosophila* can be studied during both larval and adult stages. While larval feeding is an almost continuous process required to allow a huge mass increase, adult feeding is more complex (108). Furthermore, in adults, this behavior is sex-dependent, as egg production implies higher biosynthetic needs in females (122).

Larvae

- Analysis of food intake: this analysis is very easy thanks to the translucentness of larvae which allows an easy visualization of dyed food (115).
- Feeding rate: this rate can be examined as the number of cephalopharyngeal sclerite retractions in two consecutive 1-min intervals (123).

Adult Flies

- Two-choice preference test: consists of the choice, under dark conditions, between two tastes that are differentially colored. Scoring for the preferred food source can be performed through the semitransparent abdomen of the fly (124).
- Quantification of food intake: food intake can be assessed by color spectrophotometry of homogenized flies that were fed dyed food (125).
- Proboscis extension assay: relies on the reflex of flies to extend their proboscis upon recognition of food by the taste receptors on the forelegs (126). Different tastes can be presented to the fly. This approach has previously been combined with in vivo imaging of a calcium indicator to visualize neurons involved in taste recognition in the adult brain (127).
- Feeding frequency: a proboscis extension assay has been described in which the number of extensions and touching of the food surface by the proboscis was measured during a 90-min period (125).

3.3.6. Flight Behavior/Locomotor Behavior/Equilibrium

Locomotor behavior and flight behavior in flies both require a complex coordination. Both behaviors depend on the integration of various inputs, such as visual cues or proprioceptive information, to ensure accurate movement and equilibrium.

One of the key structures in the fly brain responsible for the interpretation of these sensory inputs and the generation of an appropriate locomotor response is the central complex, but also the mushroom bodies have been shown to be involved (128, 129). Furthermore, correct locomotion and flight behavior requires the correct connections with both locomotor and flight muscles to insure coordinated movement. Most of the main neurotransmitters have been shown to be involved in different parts of these signaling processes.

Methods

Flight Behavior

- Free flight behavior: Different three-dimensional tracking software packages have been described which allow the analysis of a freely flying fly, or even complex multicamera setups which allow the three-dimensional tracking of multiple flies simultaneously (130–132). These systems allow, for example, the analyses of the influence of various sensory inputs on flight control.
- Flight simulator: This "virtual reality flight simulator" allows the investigation of flight behavior and corresponding changes in visual input. In this system, flies are glued to a small steel wire that is attached to a torque meter. In this manner, they are allowed to virtually fly in a cylindrical arena. The patterns on the walls of this arena can be altered real-time according to the flies movements allowing both to adjust the pattern to the flies movements (open loop) as to impose a certain pattern on the fly (closed loop) (133, 134).

Locomotion

- Buridan's paradigm: This is one of the first reported locomotion assays. Flies are located in a circular arena with two inaccessible landmarks opposite to each other. These landmarks induce spontaneous walking of the fly from one target to the other, a behavior that can persist for hours (128, 135).
- Free locomotion: Different tracking software packages have been developed that allow the analysis of flies freely walking in an arena (136–138).
- Gait analysis: An automated assay to evaluate step-resolved walking data has been developed which allows gait analysis. This setup allows the recording of leg tip positions on a glass plate on which the fly can freely walk around (139).
- Gap crossing: this assay can be used to study climbing behavior. The gap crossing abilities and the pertinent decision making of individual flies are recorded (140).

3.3.7. Olfactory Avoidance

Olfactory avoidance responses to repellant odorants are essential for survival, and chemosensory behavior in general is critical for food localization, food intake, interactions with reproductive partners, and localization of oviposition sites.

The olfactory system of *Drosophila* is one of the best-characterized chemosensory systems, consisting of olfactory neurons on the third antennal segments and the maxillary palps. The individual neurons express a unique odorant receptor from a repertoire of 60 odorant receptor genes together with the common Or83b receptor, which is essential for transport and insertion of odorant receptors in the chemosensory dendritic membranes (141, 142). Olfactory information is transmitted via these neurons to 43 glomeruli in the antennal lobe where olfactory information is encoded in a spatial and temporal pattern of glomerular activation (143). From here, the chemosensory information is sent to central brain structures such as the lateral horn of the protocerebrum and the mushroom bodies where the higher order integration and processing of these inputs takes place (26).

Methods

Groups of five socially experienced flies of the same sex and age (usually 4–7 days) are transferred to empty vials (25×95 mm) 3–6 h prior to testing. The used vials are marked at 3 and 6 cm from the bottom (Fig. 15).

The vials are placed on their side at the start of the test to exclude positive geotactic influences. To start the test, a Q-tip that has been dipped in distilled water or odorant is inserted besides the cotton plug and pushed down until the 6 cm mark (~1 cm below the cotton plug). The flies are allowed to recover from this disturbance during 15 s after which after ten 5-s intervals the number of

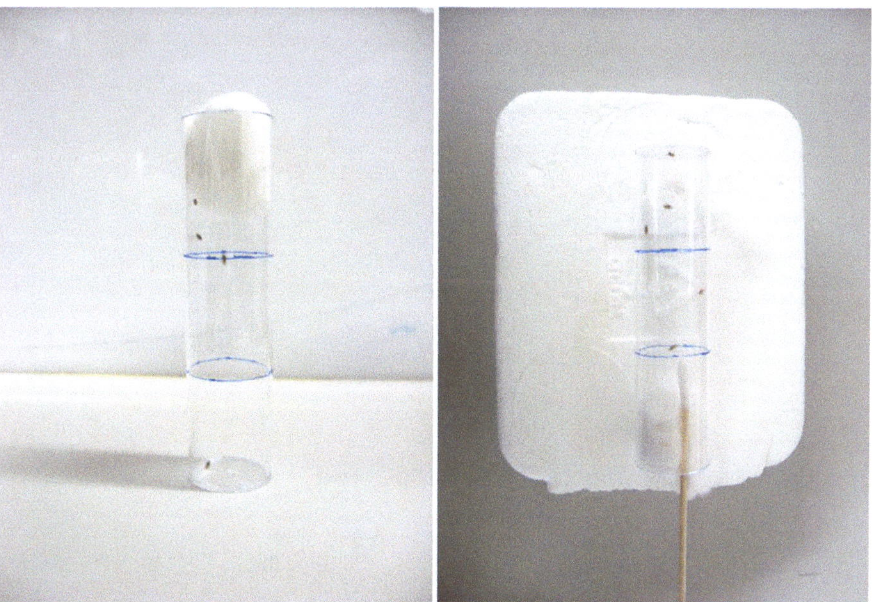

Fig. 15. Olfactory avoidance assay to determine olfactory behavior. A Q-tip is inserted in benzaldehyde and inserted in the vial, upon which the relative distribution of flies over time is measured.

flies is counted that cross the 3 cm line (the first time point follows immediately after the 15 s of recovery).

The avoidance score of the flies is calculated as the number of flies under the 3 cm line average over the ten measurements (144).

3.3.8. Learning and Memory Behavior

Learning and memory are required to allow organisms to adapt to environmental changes in an experience-dependent manner. They improve the ability to survive in novel situations as well as to avoid harmful stimuli. *Drosophila* makes use of different types of memory to learn from various visual, olfactory or tactile stimuli. These memory types can be subdivided into protein synthesis-dependent and non-protein synthesis-dependent processes (Fig. 16). Short-term memory, which lasts for less than an hour, as well as anesthesia-resistant memory, which lasts up to 24 h, are independent of protein synthesis. Middle-term memory on the other hand, lasting 1–4 h, relies on translation of preexisting mRNA while long term memory requires both de novo transcription and translation (145–148).

Different brain structures have been shown to be involved in different parts of the learning and memory process (147, 149–154). The mushroom bodies and the central complex are necessary for olfactory learning, and also the dorsal paired medial neurons, antennal lobes have been shown to contribute to this behavior. Courtship conditioning on the other hand requires the mushroom bodies, the central complex and the antennal lobes as well as parts of the lateral protocerebrum. Interestingly, while the mushroom bodies are indispensible for these two types of learning, they are not necessary for spatial learning in the heat box (see below) or

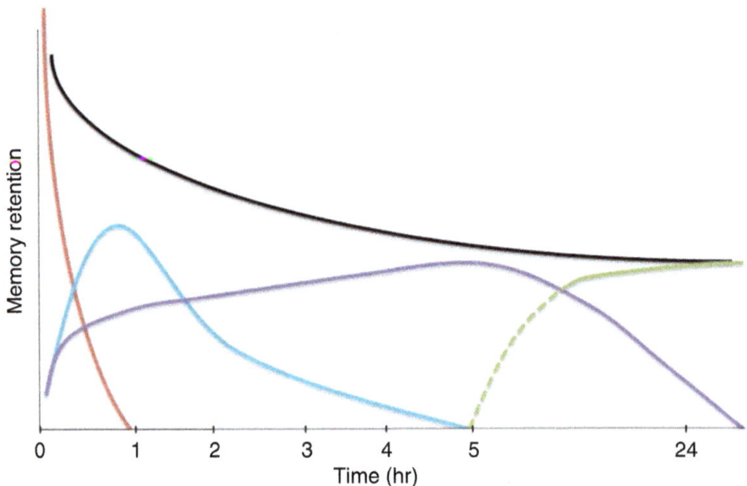

Fig. 16. Learning and memory model. *Red*: short-term memory; *blue*: middle-term memory; *purple*: anesthesia-resistant memory; *green*: long-term memory; *black*: observed memory decaying over time. From DeZazzo and Tully [146], with permission.

basic visual learning. However, context dependent visual learning does again involve this neuropil.

The best studied signaling cascade which is indispensable in the learning process is the cAMP/protein kinase A pathway of which the two first discovered learning and memory genes, *dunce* and *rutabaga*, are part (153, 155–157). However, multiple other pathways and mechanisms have been implicated in this behavior, such as pathways involved in neurotransmission or synaptic function, genes involved in RNA transport and translation or other kinases such as protein kinase C and calcium/calmodulin-dependent protein kinase II (153, 158–166).

Methods

Different learning assays have been developed to analyze learning and memory behavior in larvae or adults. Depending on the number of training sessions and whether this training is massed or spaced over multiple sessions, different types of memory can be tested.

Larval Learning and Memory

Drosophila larvae possess 21 pairs of olfactory sensory neurons, 80 pairs of gustatory sensory neurons, and only 12 neurons for vision. The adult fly in comparison has approximately 1,300, 650, and 6,000, respectively. This reduced complexity in larvae has led to the development of multiple assays to analyze learning and memory in this lifestage (167–169).

- Larval learning and memory assays consist of the association of olfaction or light to an aversive stimulus or a gustatory reinforcement (170–175).

Adult Learning and Memory

- Operant olfactory avoidance conditioning: Flies are attracted into tubes by phototaxis. These tubes contain odors associated with a shock or control odors without a shock. After training, the number of flies that avoids entering a tube with the "shock" odor is determined (176).
- Olfactory classical conditioning: Flies are placed into a chamber in which they receive the test odor with foot shocks and the control odor without shock. After training they are transferred to a T-maze where they are offered the choice between both odors. This assay gives a more robust result than the operant assay (155, 164, 177–179).
- Sucrose reward learning: similar to the classical conditioning, but, instead of foot shocks, flies receive a sucrose reward (164, 180–182).
- The proboscis extension reflex assay: makes use of the reflex of hungry flies to extend their proboscis if they taste sugar on their forelegs (183–185).
- Flight simulator: Aversive odors have also been used to train flies in the previously described flight simulator. In this paradigm,

either visual learning can be investigated by teaching the fly to fly to a certain landmark as motor learning, teaching the fly to turn in a particular direction. Alternatively, heat has also been used as an aversive stimulus in this paradigm (186–190).

- Buridan's paradigm: Spatial orientation memory can be analyzed using an adaptation of Buridan's paradigm. While the flies usually walk between two unreachable but fixed landmarks, these landmarks now disappear, while a landmark perpendicular to the position of the fly appears. If this novel landmark also disappears, the flies seem to remember the position of their first target and reorientate in that direction (191, 192).
- Spatial orientation—heat box: the fly learns to avoid one half of the test chamber using a heat stimulus (151, 193).
- Loser or winner mentality: Makes use of the previously described aggression assays. Flies seem to adapt their fighting strategy based on previous wins or losses. Furthermore, they tend not to fight with flies they previously encountered (60).
- Conditioned courtship suppression: This assay is an operant conditioning assay in which naive males are trained by pairing with mated females. If they learned from this training, they will spend less time courting virgin females than untrained males (105, 194–196).

3.3.9. Geo- or Gravitaxis Behavior

Drosophila has a negative geo- or gravitaxis instinct, which implies that flies move against the direction of gravity. This behavior is essential for the fly to navigate in its environment. Multiple genes have been shown to play a role in this behavior, including circadian clock genes such as *cry* and *Pdf* (197, 198).

Methods

- Hirsch classification maze: The most common method to analyze geotactic behaviors makes use of the Hirsch classification maze. Flies are introduced in this maze by a single entrance after which they encounter multiple junctions where they must choose to go up or down while they are attracted by a light source at the exit. Depending on the height of the exit where the fly leaves the maze, its geotaxis score is determined (199).
- Cylindrical assay: This assay makes use of a set of flies with clipped wings that are put into a cylindrical vial. The flies are tapped down after which the number of flies is counted that reaches a predetermined height during a fixed interval (42).

3.3.10. Phototaxis Behavior

Phototaxis behavior implies the movement of an organism in the direction of a light source or away from it. While adult flies are positively phototactic, larvae are photophobic. These light preferences can be crucial for survival.

Adult phototaxis behavior is polygenic and is influenced by different factors such as age or rhabdomere structure of the eye (200).

Light avoidance in larvae is regulated by the paired Bolwig organs, which form the primitive eye structures in this life stage and which are connected to the pigment dispersing factor expressing lateral neurons. Further, it has also been shown that light sensitive dendrites in the body wall contribute to this behavior as well as two pairs of isomorphic neurons in the central brain which seem to be involved in the decision between light and dark (201, 202).

Methods

Larval Assays

- Plate assay: Larvae are placed on an agar plate that is divided into four equal quadrants. Two quadrants block light; the other two allow the transmission of light. After the test period (usually 5 min), the number of larvae in each quadrant can be counted (203).
- Tube assay: larvae are placed into a tube with alternate dark or light sections. After 5 min, the number of larvae in the different sections can be counted (203).

Adult Assays

- Hirsch classification maze: An adaptation of the Hirsch classification maze used for the analysis of geotaxis behavior allows the investigation of adult phototaxis behavior (204).
- Tube assay: flies are allowed to choose between the dark or light side of two joined test tubes (38).

4. Resources

4.1. The Standard Brain

While hundreds of articles have been published about the *Drosophila* brain, the vast majority of these focus on a particular neuropil or a limited subset of neurons. However, insight in how the brain functions requires first that the overall architecture of the brain is understood, and second an understanding of how the approximately 100,000 neurons in the adult brain connect and interact with one another. In 2002, "The Standard Brain," a first effort in this direction was published by the group of Martin Heisenberg (205). Therein the size, location, and morphology of major neuropils in the adult brain were characterized in brains labeled with the antibody nc82, which recognizes the synaptic marker Bruchpilot. Heisenberg and colleagues were able to demonstrate that numerous examples of sexual dimorphism exist in the brain, but also that neuropils in different wild-type strains of *D. melanogaster* can differ significantly in size. Since then, a number of studies have been published that describe in great detail (1) the various types of neurons in the brain (1, 32–37), (2) the extent of sexual dimorphism in regions of the brain and in individual neurons (206), and (3) initial descriptions of the neural wiring of the brain (207, 208).

Indeed, the *Drosophila* connectome, the deciphering of every synapse made by every neuron in the brain, is the next goal. Several research groups and consortia are now busy with this challenging and ambitious endeavor, each using different approaches. Simultaneously, new tools are being developed to aid in these projects. Chiang and colleagues have created FlyCircuit, a database of tens of thousands of three-dimensional single neuron MARCM clones, which they used to reveal the connections between neuropils and develop a wiring diagram of the entire adult brain.

4.2. Online Resources

FlyBrain, an online atlas and database of the *Drosophila* nervous system: http://flybrain.neurobio.arizona.edu.

FlyCircuit, a database of *Drosophila* brain neurons: http://www.flycircuit.tw.

FlyView, a *Drosophila* image database: http://flyview.unimuenster.de.

Virtual Fly Brain (VFB): "An interactive tool for neurobiologists to explore the detailed neuroanatomy, gene expression, and associated phenotypes of the adult *Drosophila melanogaster* brain" (verbatim from Web site): http://www.virtualflybrain.org.

4.3. Antibodies

In the past decades, three major efforts have been undertaken to generate monoclonal antibodies against *Drosophila* proteins. Immunization was with complete head extracts (209), with head and brain extracts (210, 211), and with enriched membrane fractions of embryonic nervous system (212). From these studies, a number of monoclonal antibodies have been characterized and the corresponding epitopes identified (e.g., (213)). These monoclonal antibodies are now made available through the DSHB. The Würzburg hybridoma library is discussed separately as it contains additional monoclonal antibodies that have not been characterized, but that may well be important for the community and for future work.

4.3.1. DSHB

The Developmental Studies Hybridoma Bank at the University of Iowa is a repository of monoclonal antibodies against antigens from a variety of species. It currently has 196 monoclonal antibodies against *Drosophila* proteins plus a number of additional antibodies that cross-react with *Drosophila* antigens.

4.3.2. Würzburg Hybridoma Library

The Würzburg hybridoma library was generated by Alois Hofbauer in the group of Erich Buchner (210, 211). Some 1,000 hybridoma clones were generated by injection of homogenized *Drosophila* brains or heads into mice and fusion of their spleen cells with myeloma cells. Testing the mAbs secreted by these clones identified a library of about 200 mAbs, which selectively stain specific structures of the *Drosophila* brain. Using the approach "from antibody to gene," several genes coding for novel proteins of the presynaptic terminal were cloned and characterized (1).

4.4. Fly Stocks

FlyView stock collection: Enhancer trap lines from the FlyView project can be obtained at University of Muenster, Germany at http://flyview.uni-muenster.de/.

Flytrap, a database of P{GAL4} enhancer traps and their expression in brains:

Flytrap, University of Edinburgh, UK

FlyTrap, Yale University, USA

GETDB, a GAL4 enhancer trap database: GETDB, National Institute of Basic Biology, Okazaki, Japan

NIG-FLY: Fly stocks of the National Institute of Genetics: Stocks for RNA interference experiments may be obtained from the NIG RNAi fly unit at http://www.shigen.nig.ac.jp/fly/nigfly/.

Transgenic RNAi Project: TRiP at the Harvard Medical School plans to generate 6,250 transgenic RNAi lines targeted to attP2 on chromosome 3. Stocks are distributed through Bloomington *Drosophila* Stock Center (BDSC), http://www.flyrnai.org/TRiP-HOME.html.

Vienna Drosophila RNAi Center: The VDRC at IMP/IMBA in Vienna provides two genome-wide transgenic *Drosophila* RNAi libraries, http://www.vdrc.at.

Janelia Farm Research Campus: A major effort is under way to generate a collection of 5,000 transgenic lines that drive expression in patterns encompassing all neurons in the brain. This ongoing effort was described in a proof-of-principle paper (214). These transgenic lines will become available once this collection is complete and characterized.

Bloomington Stock Center: Information about stocks at the BDSC at Indiana University, USA can be obtained at http://flystocks.bio.indiana.edu.

Gene Disruption Project: Information about insertion lines produced by the Gene Disruption Project (GDP), Baylor College of Medicine, Texas, USA can be found at http://flypush.imgen.bcm.tmc.edu/pscreen/.

5. Perspectives

These are exciting times for the study of the *Drosophila* brain and of *Drosophila* behavior. The combination of the already available tools with the large efforts under way to describe in great detail the wiring of the fly brain will enable unprecedented insight into how a complex brain functions in vivo. Targeted disruption of gene function and manipulation of individual neurons as well as neural networks will allow to understand the genetic complexity that governs behavior and also how natural variation in behavior is affected by genetic variation and by gene–environment interaction. A number of important questions need to be answered in the context of the ongoing efforts. These include (1) the identification and description of pre-and postsynaptic sites in neural networks, (2) the description of the networks by means of transsynaptic markers, (3) the selective perturbation of limited numbers of neurons within networks by means of intersectional strategies and novel tools, and (4) in vivo imaging of a functioning brain when performing various tasks. Some new developments related to these are described below. Furthermore, new developments in the automated analysis of behavior can be expected. Finally, some of the challenges for the future include the development of methodologies to study synaptic plasticity in vivo in real-time and the description of the molecular identity of the types of neurons that are currently being described in great detail as a basis to understand their physiology.

5.1. Determining Pre- and Postsynaptic Sites

It is safe to assume that the vast majority of neurons in the *Drosophila* brain have been identified. However, in many of these cases their characterization is very superficial and leaves many unanswered questions. Unlike the mushroom bodies, for example, which have been thoroughly examined and in which axons and dendrites have been identified and well studied, the pre- and postsynaptic characteristics of many neurons remains speculative if not completely unknown. These questions can be clarified via the use of antibodies that recognize numerous pre- and postsynaptic markers (e.g., *synaptotagmin* and *discs large*), although the accuracy of these markers for the identification of pre- and postsynaptic compartments in the brain is not clear.

In vivo tools for the identification and labeling of axons and dendrites have also been developed. Fusion proteins with bovine tau (e.g., tau-GFP) preferentially label axons and have been used as axonal markers. Two GFP-tagged synaptic proteins, synaptobrevin-GFP (215) and synaptotagmin-GFP (216), have also been developed as presynaptic markers and can be used to label axons. A number of postsynaptic/dendritic markers have also been developed, such as nod-GFP (217), Homer-GFP (218), DSCAM17.1 (219), and DenMark (220).

5.2. Transsynaptic Markers

Identifying the synaptic partners for a given neuron remains a challenging goal. With the pre- and postsynaptic markers now available (e.g., synaptotagmin-GFP and DenMark), it is possible to demonstrate a physical interaction between one neuron's axons and another neuron's dendrites. However, this requires that the presynaptic marker can be expressed cleanly in one neuron and the postsynaptic marker in the other, a prerequisite that is often not obvious. A tool now being applied in other model systems such as mouse and zebrafish is the transsynaptic marker wheat germ agglutinin conjugated to HRP (WGA-HRP), which when expressed or injected into a neuron can be transported via the synapse to the synaptic partner, which also becomes labeled by the marker and thus identified. This technique, however, has only had very limited success in *Drosophila* (221).

5.3. Intersectional Strategies

The GAL4-UAS system has become an integral tool for *Drosophila* research. However, the desired expression of a transgene is sometimes more limited than the available GAL4 lines allow. In order to satisfy these demands a number of intersectional strategies have been developed to further limit the cells in which GAL4-mediated activation of a transgene occurs (222).

One technique to limit GAL4-mediated expression is with Gal80. For example, one might have a GAL4 line that expresses in the αβ lobes of the mushroom body as well as a few neurons elsewhere in the brain, and a Gal80 line that expresses specifically in all mushroom body neurons. When both the GAL4 and the Gal80 lines are present together with a UAS-transgene, the transgene would only be expressed in those few neurons outside the mushroom body.

A second system that limits the number of cells activated in the GAL4-UAS system is Split-GAL4 (223). In this system, the DNA-binding domain (DBD) and the activation domain (AD) of GAL4 are split and driven by two separate enhancers. Only in cells that express both the DBD and AD, which then reassociate via a leucine zipper to form an active GAL4 protein, will the UAS-transgene be expressed.

More recently, two additional binary systems have been described that can be used in combinations with the GAL4 system, namely, the LexA and the Q systems (224, 225).

5.4. Functional Interference

MARCM analysis has proven to be an important tool for gaining insights into the requirements of a given gene product in neuronal development (226). However, MARCM is limited to the observation of developmental phenotypes and can only be generated during mitosis of the developing neuron or one of its precursors. More recently, technologies have been developed that allow a protein of interest to be deleted, tagged, or inhibited conditionally.

IMAGO uses an integrase-based technique to generate knockout or tagged alleles of a gene of interest conditionally, thus providing more experimental freedom in comparison to MARCM (227).

FlAsH-FALI is a method allowing proteins expressing a tetracysteine tag to be inhibited upon illumination with light at 488 nm when in the presence of the fluorophore FlAsH (228). Furthermore, it is possible with a relatively straight-forward recombineering strategy to tag any protein of interest with a tetracysteine tag (or any other tag of interest), providing a method for the temporal and local inhibition of protein function, enabling its study in a given process (229).

5.5. In Vivo Imaging

In vivo imaging involves the direct visualization of neuronal activation by means of a fluorescent signal. This signal comes from a genetically encoded calcium indicator in which calmodulin is linked to GFP that has been circularly permuted, such as cameleons and the G-CaMPs with higher signal-to-noise (230, 231). Action potentials trigger a rise in intracellular Ca^{2+}, which in turn is bound by the calmodulin of the indicator causing a change in the conformation of the GFP and leading to increased fluorescence (232).

References

1. Otsuna H, Ito K (2006) Systematic analysis of the visual projection neurons of *Drosophila melanogaster*. I. Lobulu-specific pathways. J Comp Neurol 497:928–958
2. Shinomiya K, Matsudat K, Oishi T, Otsuna H, Ito K (2011) Flybrain neuron database: a comprehensive database system of the *Drosophila* brain neurons. J Comp Neurol 519:807–833
3. Heisenberg M, Böhl K (1979) Isolation of anatomical brain mutants of *Drosophila* by histological means. Z Naturforsch 34:143–147
4. Blest AD (1961) Some modifications of Holme's silver method for insect central nervous system. Q J Microsc Sci 102:413–417
5. Bodian D (1937) The staining of paraffin sections with activated protargol. The role of fixatives. Anat Rec 69:153–162
6. Protocol at: http://flybrain.neurobio.arizona.edu/Flybrain/html/atlas/golgi/index.html
7. Fischbach KF, Götz C (1981) Das Experiment: ein Blick ins Fliegenhirn: Golgi-gefärbte Nervenzellen bei *Drosophila*. Biol Z 11:183–187
8. Colonnier M (1964) The tangential organization of the visual cortex. J Anat 98:327–344
9. Nässel D (1996) Advances in the immunocytochemical localization of neuroactive substances in the insect nervous system. J Neurosci Methods 69:3–23
10. Hamasaka Y, Nässel D (2006) Mapping of serotonin, dopamine, and histamine in relation to different clock neurons in the brain of *Drosophila*. J Comp Neurol 494:314–330
11. Kolodziejczyk A, Sun X, Meinertzhagen IA, Nässel DR (2008) Glutamate, GABA and acetylcholine signaling components in the lamina of the *Drosophila* visual system. PLoS One 3:e2110
12. Nässel D, Winther A (2010) *Drosophila* neuropeptides in regulation of physiology and behavior. Prog Neurobiol 92:42–104
13. Goossens T, Kang YY, Wuytens G, Zimmermann P, Callaerts-Vegh Z, Pollarolo G, Islam R, Hortsch M, Callaerts P (2011) The *Drosophila* L1CAM homolog neuroglian signals through distinct pathways to control different aspects of mushroom body axon development. Development 138:1595–1605
14. Mardon G, Solomon NM, Rubin GM (2004) dachshund encodes a nuclear protein required for normal eye and leg development in *Drosophila*. Development 120(12):3473–3486
15. Clements J, Hens K, Francis C, Schellens A, Callaerts P (2008) Conserved role for the *Drosophila* Pax6 homolog eyeless in differentiation and function of insulin-producing neurons. Proc Natl Acad Sci U S A 105(42):16183–16188
16. O'Kane CJ, Gehring WJ (1987) Detection in situ of genomic regulatory elements in *Drosophila*. Proc Natl Acad Sci U S A 84:9123–9127
17. Bellen HJ, O'Kane CJ, Wilson C, Grossniklaus U, Pearson RK, Gehring WJ (1989) P-element-mediated enhancer detection: a versatile method to study development in *Drosophila*. Genes Dev 3:1288–1300
18. Yang MY, Armstrong JD, Vilinsky I, Strausfeld NJ, Kaiser K (1995) Subdivision of the *Drosophila* mushroom bodies by enhancer-trap expression patterns. Neuron 15:45–54
19. Han P-L, Meller V, Davis RL (1996) The *Drosophila* brain revisited by enhancer detection. J Neurobiol 31:88–102

20. Ito K, Suzuki K, Estes P, Ramaswami M, Yamamoto D, Strausfeld NJ (1998) The organization of extrinsic neurons and their implications in the functional roles of the mushroom bodies in *Drosophila melanogaster* Meigen. Learn Mem 5:52–77
21. Renn SCP, Armstrong JD, Yang M, Wang Z, An X, Kaiser K, Taghert PH (1999) Genetic analysis of the *Drosophila* ellipsoid body neuropil: organization and development of the central complex. J Neurobiol 41:189–207
22. Brand AH, Perrimon N (1993) Targeted gene expression as a means of altering cell fates and generating dominant phenotypes. Development 118:401–415
23. Duffy JB (2002) GAL4 system in *Drosophila*: a fly geneticist's Swiss army knife. Genesis 34:1–15
24. Elliott DA, Brand AH (2008) The GAL4 system: a versatile system for the expression of genes. Methods Mol Biol 420:79–95
25. Salvaterra PM, Kitamoto T (2001) *Drosophila* cholinergic neurons and processes visualized with GAL4/UAS-GFP. Brain Res Gene Expr Patterns 1:73–82
26. Ng M, Roorda RD, Lima SQ, Zemelman BV, Morcillo P, Miesenböck G (2002) Transmission of olfactory information between three populations of neurons in the antennal lobe of the fly. Neuron 36:463–474
27. Daniels RW, Gelfand MV, Collins CA, DiAntonio A (2008) Visualizing glutamatergic cell bodies and synapses in *Drosophila* larval and adult CNS. J Comp Neurol 508:131–152
28. Alekseyenko OV, Lee C, Kravitz EA (2010) Targeted manipulation of serotonergic neurotransmission affects the escalation of aggression in adult male *Drosophila melanogaster*. PLoS One 5:e10806
29. Friggi-Grelin F, Coulom H, Meller M, Gomez D, Hirsh J, Birman S (2003) Targeted gene expression in *Drosophila* dopaminergic cells using regulatory sequences from tyrosine hydroxylase. J Neurobiol 54:618–627
30. Cole SH, Carney GE, McClung CA, Willard SS, Taylor BJ, Hirsh J (2005) Two functional but noncomplementing *Drosophila* tyrosine decarboxylase genes: distinct roles for neural tyramine and octopamine in female fertility. J Biol Chem 280:14948–14955
31. Li H, Chaney S, Roberts IJ, Forte M, Hirsh J (2000) Ectopic G-protein expression in dopamine and serotonin neurons blocks cocaine sensitization in *Drosophila melanogaster*. Curr Biol 10:211–214
32. Tanaka NK, Tanimoto H, Ito K (2008) Neuronal assemblies of the *Drosophila* mushroom body. J Comp Neurol 508:711–755
33. Aso Y, Grübel K, Busch S, Friedrich AB, Siwanowicz I, Tanimoto H (2009) The mushroom body of adult *Drosophila* characterized by GAL4 drivers. J Neurogenet 23:156–172
34. Miyazaki T, Ito K (2010) Neural architecture of the primary gustatory center of *Drosophila melanogaster* visualized with GAL4 and LexA enhancer-trap systems. J Comp Neurol 518:4147–4181
35. Kamikouchi A, Shimada T, Ito K (2006) Comprehensive classification of the auditory sensory projections in the brain of the fruit fly *Drosophila melanogaster*. J Comp Neurol 499:317–356
36. Tanaka NK, Awasaki T, Shimada T, Ito K (2004) Integration of chemosensory pathways in the *Drosophila* second-order olfactory centers. Curr Biol 14:449–457
37. Young JM, Armstrong JD (2010) Structure of the adult central complex in *Drosophila*: organization of distinct neuronal subsets. J Comp Neurol 518:1500–1524
38. Benzer S (1967) Behavioral mutants of *Drosophila* isolated by countercurrent distribution. Proc Natl Acad Sci U S A 58: 1112–1119
39. Konopka RJ, Benzer S (1971) Clock mutants of *Drosophila melanogaster*. Proc Natl Acad Sci U S A 68:2112–2116
40. de Belle JS, Heisenberg M (1994) Associative odor learning in *Drosophila* abolished by chemical ablation of mushroom bodies. Science 263:692–695
41. Strauss R, Hanesch U, Kinkelin M, Wolf R, Heisenberg M (1992) No-bridge of *Drosophila melanogaster*: portrait of a structural brain mutant of the central complex. J Neurogenet 8:125–155
42. Strauss R, Heisenberg M (1993) A higher control center of locomotor behavior in the *Drosophila* brain. J Neurosci 13: 1852–1861
43. Coombe PE, Heisenberg M (1986) The structural brain mutant vacuolar medulla of *Drosophila melanogaster* with specific behavioral defects and cell degeneration in the adult. J Neurogenet 3:135–158
44. Perrimon N, Ni JQ, Perkins L (2010) In vivo RNAi: today and tomorrow. Cold Spring Harb Perspect Biol 2:a003640
45. Kitamoto T (2001) Conditional modification of behavior in *Drosophila* by targeted expression of a temperature-sensitive shibire allele in defined neurons. J Neurobiol 47:81–92
46. Parisky KM, Agosto J, Pulver SR, Shang Y, Kuklin E, Hodge JJ, Kang K, Liu X, Garrity PA, Rosbash M, Griffith LC (2008) PDF cells are a GABA-responsive wake-promoting

component of the *Drosophila* sleep circuit. Neuron 60:672–682
47. Shang Y, Griffith LC, Rosbash M (2008) Light-arousal and circadian photoreception circuits intersect at the large PDF cells of the *Drosophila* brain. Proc Natl Acad Sci U S A 105:19587–19594
48. Hodge JJ (2009) Ion channels to inactivate neurons in *Drosophila*. Front Mol Neurosci 2:13
49. Hamada FN, Rosenzweig M, Kang K, Pulver SR, Ghezzi A, Jegla TJ, Garrity PA (2008) An internal thermal sensor controlling temperature preference in *Drosophila*. Nature 454:217–220
50. Pulver SR, Pashkovski SL, Hornstein NJ, Garrity PA, Griffith LC (2009) Temporal dynamics of neuronal activation by channelrhodopsin-2 and TRPA1 determine behavioral output in *Drosophila* larvae. J Neurophysiol 101:3075–3088
51. Lima SQ, Miesenböck G (2005) Remote control of behavior through genetically targeted photostimulation of neurons. Cell 121:141–152
52. Schroll C, Riemensperger T, Bucher D, Ehmer J, Völler T, Erbguth K, Gerber B, Hendel T, Nagel G, Buchner E, Fiala A (2006) Light-induced activation of distinct modulatory neurons triggers appetitive or aversive learning in *Drosophila* larvae. Curr Biol 16:1741–1747
53. Fiala A, Suska A, Schlüter OM (2010) Optogenetic approaches in neuroscience. Curr Biol 20:R897–R903
54. Hoffman AA (1987) A laboratory study of male territoriality in the sibling species *Drosophila melanogaster* and *D. simulans*. Anim Behav 35:807–818
55. Hoffman AA (1987) Territorial encounters between *Drosophila* males of different sizes. Anim Behav 35:1899–1901
56. Hoffman AA (1989) Geographic variation in the territorial success of *Drosophila melanogaster* males. Behav Genet 19:241–255
57. Hoffman AA (1990) The influence of age and experience with conspecifics on territorial behaviour in *Drosophila melanogaster*. J Insect Behav 3:1–12
58. Hoffmann AA (1988) Heritable variation for territorial success in two *Drosophila melanogaster* populations. Anim Behav 36: 1180–1189
59. Nilsen SP, Chan YB, Huber R, Kravitz EA (2004) Gender-selective patterns of aggressive behavior in *Drosophila melanogaster*. Proc Natl Acad Sci U S A 101:12342–12347
60. Yurkovic A, Wang O, Basu AC, Kravitz EA (2006) Learning and memory associated with aggression in *Drosophila melanogaster*. Proc Natl Acad Sci U S A 103:17519–17524
61. Penn JK, Zito MF, Kravitz EA (2010) A single social defeat reduces aggression in a highly aggressive strain of *Drosophila*. Proc Natl Acad Sci U S A 107:12682–12686
62. Baier A, Wittek B, Brembs B (2002) *Drosophila* as a new model organism for the neurobiology of aggression? J Exp Biol 205:1233–1240
63. Certel SJ, Leung A, Lin CY, Perez P, Chiang AS, Kravitz EA (2010) Octopamine neuromodulatory effects on a social behavior decision-making network in *Drosophila* males. PLoS One 5:e13248
64. Certel SJ, Savella MG, Schlegel DC, Kravitz EA (2007) Modulation of *Drosophila* male behavioral choice. Proc Natl Acad Sci U S A 104:4706–4711
65. Chan YB, Kravitz EA (2007) Specific subgroups of FruM neurons control sexually dimorphic patterns of aggression in *Drosophila melanogaster*. Proc Natl Acad Sci U S A 104:19577–19582
66. Dierick HA, Greenspan RJ (2007) Serotonin and neuropeptide F have opposite modulatory effects on fly aggression. Nat Genet 39:678–682
67. Hoyer SC, Eckart A, Herrel A, Zars T, Fischer SA, Hardie SL, Heisenberg M (2008) Octopamine in male aggression of *Drosophila*. Curr Biol 18:159–167
68. Johnson O, Becnel J, Nichols CD (2009) Serotonin 5-HT(2) and 5-HT(1A)-like receptors differentially modulate aggressive behaviors in *Drosophila melanogaster*. Neuroscience 158:1292–1300
69. Lee G, Hall JC (2000) A newly uncovered phenotype associated with the *fruitless* gene of *Drosophila melanogaster*: aggression-like head interactions between mutant males. Behav Genet 30:263–275
70. Mundiyanapurath S, Chan YB, Leung AK, Kravitz EA (2009) Feminizing cholinergic neurons in a male *Drosophila* nervous system enhances aggression. Fly (Austin) 3:179–184
71. Potter CJ, Luo L (2008) Octopamine fuels fighting flies. Nat Neurosci 11:989–990
72. Rollmann SM, Zwarts L, Edwards AC, Yamamoto A, Callaerts P, Norga K, Mackay TF, Anholt RR (2008) Pleiotropic effects of *Drosophila neuralized* on complex behaviors and brain structure. Genetics 179:1327–1336
73. Zhou C, Rao Y (2008) A subset of octopaminergic neurons are important for

Drosophila aggression. Nat Neurosci 11: 1059–1067

74. Edwards AC, Rollmann SM, Morgan TJ, Mackay TF (2006) Quantitative genomics of aggressive behavior in *Drosophila melanogaster*. PLoS Genet 2:e154
75. Edwards AC, Ayroles JF, Stone EA, Carbone MA, Lyman RF, Mackay TF (2009) A transcriptional network associated with natural variation in *Drosophila* aggressive behavior. Genome Biol 10:R76
76. Edwards AC, Zwarts L, Yamamoto A, Callaerts P, Mackay TF (2009) Mutations in many genes affect aggressive behavior in *Drosophila melanogaster*. BMC Biol 7:29
77. Dierick HA, Greenspan RJ (2006) Molecular analysis of flies selected for aggressive behavior. Nat Genet 38:1023–1031
78. Ueda A, Wu CF (2009) Effects of social isolation on neuromuscular excitability and aggressive behaviors in *Drosophila*: altered responses by Hk and gstsl, two mutations implicated in redox regulation. J Neurogenet 23:378–394
79. Mundiyanapurath S, Certel S, Kravitz EA (2007) Studying aggression in *Drosophila* (fruit flies). J Vis Exp 2:155
80. Dankert H, Wang L, Hoopfer ED, Anderson DJ, Perona P (2009) Automated monitoring and analysis of social behavior in *Drosophila*. Nat Methods 6:297–303
81. Dierick HA (2007) A method for quantifying aggression in male *Drosophila melanogaster*. Nat Protoc 2:2712–2718
82. Hardin PE (2005) The circadian timekeeping system of *Drosophila*. Curr Biol 15:R714–R722
83. Allada R, Chung BY (2010) Circadian organization of behavior and physiology in *Drosophila*. Annu Rev Physiol 72:605–624
84. Nitabach MN, Taghert PH (2008) Organization of the *Drosophila* circadian control circuit. Curr Biol 18:R84–R93
85. Shaw P (2003) Awakening to the behavioral analysis of sleep in *Drosophila*. J Biol Rhythms 18:4–11
86. Harbison ST, Mackay TF, Anholt RR (2009) Understanding the neurogenetics of sleep: progress from *Drosophila*. Trends Genet 25: 262–269
87. Crocker A, Sehgal A (2010) Genetic analysis of sleep. Genes Dev 24:1220–1235
88. Shaw PJ, Cirelli C, Greenspan RJ, Tononi G (2000) Correlates of sleep and waking in *Drosophila melanogaster*. Science 287: 1834–1837
89. Hendricks JC, Finn SM, Panckeri KA, Chavkin J, Williams JA, Sehgal A, Pack AI (2000) Rest in *Drosophila* is a sleep-like state. Neuron 25:129–138
90. Shaw PJ, Tononi G, Greenspan RJ, Robinson DF (2002) Stress response genes protect against lethal effects of sleep deprivation in *Drosophila*. Nature 417:287–291
91. Hamblen M, Zehring WA, Kyriacou CP, Reddy P, Yu Q, Wheeler DA, Zwiebel LJ, Konopka RJ, Rosbash M, Hall JC (1986) Germ-line transformation involving DNA from the period locus in *Drosophila melanogaster*: overlapping genomic fragments that restore circadian and ultradian rhythmicity to per0 and per-mutants. J Neurogenet 3:249–291
92. Nitz DA, van Swinderen B, Tononi G, Greenspan RJ (2002) Electrophysiological correlates of rest and activity in *Drosophila melanogaster*. Curr Biol 12:1934–1940
93. Van Swinderen B, Nitz DA, Greenspan RJ (2004) Uncoupling of brain activity from movement defines arousal states in *Drosophila*. Curr Biol 14:81–87
94. Kume K, Kume S, Park SK, Hirsh J, Jackson FR (2005) Dopamine is a regulator of arousal in the fruit fly. J Neurosci 25:7377–7384
95. Lebestky T, Chang JS, Dankert H, Zelnik L, Kim YC, Han KA, Wolf FW, Perona P, Anderson DJ (2009) Two different forms of arousal in *Drosophila* are oppositely regulated by the dopamine D1 receptor ortholog DopR via distinct neural circuits. Neuron 64:522–536
96. Yamamoto A, Zwarts L, Callaerts P, Norga K, Mackay TF, Anholt RR (2008) Neurogenetic networks for startle-induced locomotion in *Drosophila melanogaster*. Proc Natl Acad Sci U S A 105:12393–12398
97. Wu Z, Gong Z, Feng C, Guo A (2000) An emergent mechanism of selective visual attention in *Drosophila*. Biol Cybern 82:61–68
98. van Swinderen B, Flores KA (2007) Attention-like processes underlying optomotor performance in a *Drosophila* choice maze. Dev Neurobiol 67:129–145
99. O'Dell KM (2003) The voyeurs' guide to *Drosophila melanogaster* courtship. Behav Processes 64:211–223
100. Ferveur JF (2010) *Drosophila* female courtship and mating behaviors: sensory signals, genes, neural structures and evolution. Curr Opin Neurobiol 20:764–769
101. Siwicki KK, Kravitz EA (2009) Fruitless, doublesex and the genetics of social behavior in *Drosophila melanogaster*. Curr Opin Neurobiol 19:200–206

102. Villella A, Gailey DA, Berwald B, Ohshima S, Barnes PT, Hall JC (1997) Extended reproductive roles of the fruitless gene in *Drosophila melanogaster* revealed by behavioural analysis of new fru mutants. Genetics 147:1107–1130
103. Rybak F, Sureau G, Aubin T (2002) Functional coupling of acoustic and chemical signals in the courtship behaviour of the male *Drosophila melanogaster*. Proc Biol Sci 269:695–701
104. MacBean IT, Parsons PA (1967) Directional selection for duration of copulation in *Drosophila melanogaster*. Genetics 56: 233–239
105. Siegel RW, Hall JC (1979) Conditioned responses in courtship behavior of normal and mutant *Drosophila*. Proc Natl Acad Sci U S A 76:3430–3434
106. Jallon JM, Hotta Y (1979) Genetic and behavioral studies of female sex appeal in *Drosophila*. Behav Genet 9:257–275
107. Ejima A, Griffith LC (2007) Measurement of courtship behavior in *Drosophila melanogaster*. Cold Spring Harb Protoc. doi:10.1101/pdb.prot4847
108. Melcher C, Bader R, Pankratz MJ (2007) Amino acids, taste circuits, and feeding behavior in *Drosophila*: towards understanding the psychology of feeding in flies and man. J Endocrinol 192:467–472
109. Britton JS, Lockwood WK, Li L, Cohen SM, Edgar BA (2002) *Drosophila*'s insulin/PI3-kinase pathway coordinates cellular metabolism with nutritional conditions. Dev Cell 2:239–249
110. Junger MA, Rintelen F, Stocker H, Wasserman D, Végh M, Radimerski T, Greenberg ME, Hafen E (2003) The *Drosophila* forkhead transcription factor FOXO mediates the reduction in cell number associated with reduced insulin signaling. J Biol 2:20
111. Kramer JM, Davidge JT, Lockyer JM, Staveley BE (2003) Expression of *Drosophila* FOXO regulates growth and can phenocopy starvation. BMC Dev Biol 3:5
112. Melcher C, Bader R, Walther S, Simakov O, Pankratz MJ (2006) Neuromedin U and its putative *Drosophila* homolog hugin. PLoS Biol 4:e68
113. Melcher C, Pankratz MJ (2005) Candidate gustatory interneurons modulating feeding behavior in the *Drosophila* brain. PLoS Biol 3:e305
114. Meng X, Wahlström G, Immonen T, Kolmer M, Tirronen M, Predel R, Kalkkinen N, Heino TI, Sariola H, Roos C (2002) The *Drosophila* hugin gene codes for myostimulatory and ecdysis-modifying neuropeptides. Mech Dev 117:5–13
115. Zinke I, Kirchner C, Chao LC, Tetzlaff MT, Pankratz MJ (1999) Suppression of food intake and growth by amino acids in *Drosophila*: the role of pumpless, a fat body expressed gene with homology to vertebrate glycine cleavage system. Development 126:5275–5284
116. Brown MR, Crim JW, Arata RC, Cai HN, Chun C, Shen P (1999) Identification of a *Drosophila* brain-gut peptide related to the neuropeptide Y family. Peptides 20: 1035–1042
117. Shen P, Cai HN (2001) *Drosophila* neuropeptide F mediates integration of chemosensory stimulation and conditioning of the nervous system by food. J Neurobiol 47:16–25
118. Wu Q, Zhao Z, Shen P (2005) Regulation of aversion to noxious food by *Drosophila* neuropeptide Y- and insulin-like systems. Nat Neurosci 8:1350–1355
119. Wu Q, Zhang Y, Xu J, Shen P (2005) Regulation of hunger-driven behaviors by neural ribosomal S6 kinase in *Drosophila*. Proc Natl Acad Sci U S A 102:13289–13294
120. Vosshall LB, Stocker RF (2007) Molecular architecture of smell and taste in *Drosophila*. Annu Rev Neurosci 30:505–533
121. Gerber B, Stocker RF, Tanimura T, Thum AS (2009) Smelling, tasting, learning: *Drosophila* as a study case. Results Probl Cell Differ 47:139–185
122. Carvalho GB, Kapahi P, Anderson DJ, Benzer S (2006) Allocrine modulation of feeding behavior by the sex peptide of *Drosophila*. Curr Biol 16:692–696
123. Prasad NG, Shakarad M, Anitha D, Rajamani M, Joshi A (2001) Correlated responses to selection for faster development and early reproduction in *Drosophila*: the evolution of larval traits. Evolution 55:1363–1372
124. Amrein H, Thorne N (2005) Gustatory perception and behavior in *Drosophila melanogaster*. Curr Biol 15:R673–R684
125. Wong R, Piper MD, Wertheim B, Partridge L (2009) Quantification of food intake in *Drosophila*. PLoS One 4:e6063
126. Shiraiwa T, Carlson JR (2007) Proboscis extension response (PER) assay in *Drosophila*. J Vis Exp 3:193
127. Marella S, Fischler M, Kong P, Asgarian S, Reukhert E, Scott K (2006) Imaging taste response in the fly brain reveals a functional map of taste category and behavior. Neuron 49:285–295
128. Serway CN, Kaufman RR, Strauss R, de Belle JS (2009) Mushroom bodies enhance initial

motor activity in *Drosophila*. J Neurogenet 23:173–184
129. Strauss R (2002) The central complex and the genetic dissection of locomotor behaviour. Curr Opin Neurobiol 12:633–638
130. Fry SN, Rohrseitz N, Straw AD, Dickinson MH (2008) TrackFly: virtual reality for a behavioral system analysis in free-flying fruit flies. J Neurosci Methods 171:110–117
131. Straw AD, Lee S, Dickinson MH (2010) Visual control of altitude in flying *Drosophila*. Curr Biol 20:1550–1556
132. Straw AD, Branson K, Neumann TR, Dickinson MH (2011) Multi-camera real-time three-dimensional tracking of multiple flying animals. J R Soc Interface 8:395–409
133. Frye MA (2007) Behavioral neurobiology: a vibrating gyroscope controls fly steering maneuvers. Curr Biol 17:R134–R136
134. Borst A, Haag J, Reiff DF (2010) Fly motion vision. Annu Rev Neurosci 33:49–70
135. Gotz KG (1980) Visual guidance in *Drosophila*. Basic Life Sci 16:391–407
136. Martin JR (2004) A portrait of locomotor behaviour in *Drosophila* determined by a video-tracking paradigm. Behav Processes 67: 207–219
137. Robie AA, Straw AD, Dickinson MH (2010) Object preference by walking fruit flies, *Drosophila melanogaster*, is mediated by vision and graviperception. J Exp Biol 213: 2494–2506
138. Valente D, Golani I, Mitra PP (2007) Analysis of the trajectory of *Drosophila melanogaster* in a circular open field arena. PLoS One 2:e1083
139. Strauss R (1998) Automatische Diagnose genetisch bedingter Laufanomalien der Fliege *Drosophila* bei freier Bewegung in realer und virtueller Umgebung. Forsch Wiss Rechnen GWDG-Bericht Nr 51:53–78
140. Pick S, Strauss R (2005) Goal-driven behavioral adaptations in gap-climbing *Drosophila*. Curr Biol 15:1473–1478
141. Larsson MC, Domingos AI, Jones WD, Chiappe ME, Amrein H, Vosshall LB (2004) Or83b encodes a broadly expressed odorant receptor essential for *Drosophila* olfaction. Neuron 43:703–714
142. Benton R, Sachse S, Michnick SW, Vosshall LB (2006) Atypical membrane topology and heteromeric function of *Drosophila* odorant receptors in vivo. PLoS Biol 4:e20
143. Wang JW, Wong AM, Flores J, Vosshall LB, Axel R (2003) Two-photon calcium imaging reveals an odor-evoked map of activity in the fly brain. Cell 112:271–282
144. Anholt RR, Lyman RF, Mackay TF (1996) Effects of single P-element insertions on olfactory behavior in *Drosophila melanogaster*. Genetics 143:293–301
145. Tully T, Preat T, Boynton SC, Del Vecchio M (1994) Genetic dissection of consolidated memory in *Drosophila*. Cell 79:35–47
146. DeZazzo J, Tully T (1995) Dissection of memory formation: from behavioral pharmacology to molecular genetics. Trends Neurosci 18:212–218
147. Dubnau J, Chiang AS, Tully T (2003) Neural substrates of memory: from synapse to system. J Neurobiol 54:238–253
148. Isabel G, Pascual A, Preat T (2004) Exclusive consolidated memory phases in *Drosophila*. Science 304:1024–1027
149. Joiner MA, Griffith LC (1999) Mapping of the anatomical circuit of CaM kinase-dependent courtship conditioning in *Drosophila*. Learn Mem 6:177–192
150. Liu L, Wolf R, Ernst R, Heisenberg M (1999) Context generalization in *Drosophila* visual learning requires the mushroom bodies. Nature 400:753–756
151. Putz G, Heisenberg M (2002) Memories in *Drosophila* heat-box learning. Learn Mem 9:349–359
152. Chiang AS, Blum A, Barditch J, Chen YH, Chiu SL, Regulski M, Armstrong JD, Tully T, Dubnau J (2004) radish encodes a phospholipase-A2 and defines a neural circuit involved in anesthesia-resistant memory. Curr Biol 14:263–272
153. Xia S, Miyashita T, Fu TF, Lin WY, Wu CL, Pyzocha L, Lin IR, Saitoe M, Tully T, Chiang AS (2005) NMDA receptors mediate olfactory learning and memory in *Drosophila*. Curr Biol 15:603–615
154. Skoulakis EM, Grammenoudi S (2006) Dunces and da Vincis: the genetics of learning and memory in *Drosophila*. Cell Mol Life Sci 63:975–988
155. Dudai Y, Jan YN, Byers D, Quinn WG, Benzer S (1976) Dunce, a mutant of *Drosophila* deficient in learning. Proc Natl Acad Sci U S A 73:1684–1688
156. Livingstone MS, Sziber PP, Quinn WG (1984) Loss of calcium/calmodulin responsiveness in adenylate cyclase of rutabaga, a *Drosophila* learning mutant. Cell 37:205–215
157. Levin LR, Han PL, Hwang PM, Feinstein PG, Davis RL, Reed RR (1992) The *Drosophila* learning and memory gene rutabaga encodes a Ca2+/calmodulin-responsive adenylyl cyclase. Cell 68:479–489

158. Tempel BL, Livingstone MS, Quinn WG (1984) Mutations in the dopa decarboxylase gene affect learning in *Drosophila*. Proc Natl Acad Sci U S A 81:3577–3581
159. Choi KW, Smith RF, Buratowski RM, Quinn WG (1991) Deficient protein kinase C activity in turnip, a *Drosophila* learning mutant. J Biol Chem 266:15999–16006
160. Chapman PF, Frenguelli BG, Smith A, Chen CM, Silva AJ (1995) The alpha-Ca2+/calmodulin kinase II: a bidirectional modulator of presynaptic plasticity. Neuron 14:591–597
161. Grotewiel MS, Beck CD, Wu KH, Zhu XR, Davis RL (1998) Integrin-mediated short-term memory in *Drosophila*. Nature 391:455–460
162. Cheng Y, Endo K, Wu K, Rodan AR, Heberlein U, Davis RL (2001) *Drosophila* fasciclinII is required for the formation of odor memories and for normal sensitivity to alcohol. Cell 105:757–768
163. Dubnau J, Chiang AS, Grady L, Barditch J, Gossweiler S, McNeil J, Smith P, Buldoc F, Scott R, Certa U, Broger C, Tully T (2003) The staufen/pumilio pathway is involved in *Drosophila* long-term memory. Curr Biol 13:286–296
164. Schwaerzel M, Monstirioti M, Scholz H, Friggi-Grelin F, Birman S, Heisenberg M (2003) Dopamine and octopamine differentiate between aversive and appetitive olfactory memories in *Drosophila*. J Neurosci 23: 10495–10502
165. Godenschwege TA, Reisch D, Diegelmann S, Eberle K, Funk N, Heisenberg M, Hoppe V, Hoppe J, Klagges BR, Martin JR, Nikitina EA, Putz G, Reifegerste R, Reisch N, Rister J, Schaupp M, Scholz H, Schwärzel M, Werner U, Zars TD, Buchner S, Buchner E (2004) Flies lacking all synapsins are unexpectedly healthy but are impaired in complex behaviour. Eur J Neurosci 20:611–622
166. Presente A, Boyles RS, Serway CN, de Belle JS, Andres AJ (2004) Notch is required for long-term memory in *Drosophila*. Proc Natl Acad Sci U S A 101:1764–1768
167. Stocker RF (1994) The organization of the chemosensory system in *Drosophila melanogaster*: a review. Cell Tissue Res 275:3–26
168. Stocker RF (2001) *Drosophila* as a focus in olfactory research: mapping of olfactory sensilla by fine structure, odor specificity, odorant receptor expression, and central connectivity. Microsc Res Tech 55:284–296
169. Gerber B, Stocker RF (2007) The *Drosophila* larva as a model for studying chemosensation and chemosensory learning: a review. Chem Senses 32:65–89
170. Aceves-Pina EO, Quinn WG (1979) Learning in normal and mutant *Drosophila* larvae. Science 206:93–96
171. Scherer S, Stocker RF, Gerber B (2003) Olfactory learning in individually assayed *Drosophila* larvae. Learn Mem 10:217–225
172. Gerber B, Scherer S, Neuser K, Michels B, Hendel T, Stocker RF, Heisenberg M (2004) Visual learning in individually assayed *Drosophila* larvae. J Exp Biol 207:179–188
173. Hendel T, Michels B, Neuser K, Schipanski A, Kaun K, Sokolowski MB, Marohn F, Michel R, Heisenberg M, Gerber B (2005) The carrot, not the stick: appetitive rather than aversive gustatory stimuli support associative olfactory learning in individually assayed *Drosophila* larvae. J Comp Physiol A 191:265–279
174. Michels B, Diegelmann S, Tanimoto H, Schwenkert I, Buchner E, Gerber B (2005) A role for synapsin in associative learning: the *Drosophila* larva as a study case. Learn Mem 12:224–231
175. Yarali A, Hendel T, Gerber B (2006) Olfactory learning and behaviour are 'insulated' against visual processing in larval *Drosophila*. J Comp Physiol A 192:1133–1145
176. Quinn WG, Harris WA, Benzer S (1974) Conditioned behavior in *Drosophila melanogaster*. Proc Natl Acad Sci U S A 71:708–712
177. Jellies JA (1981) Associative olfactory conditioning in *Drosophila melanogaster* and memory retention through metamorphosis. Master thesis, Illinois State University, Illinois
178. Tully T, Quinn WG (1985) Classical conditioning and retention in normal and mutant *Drosophila melanogaster*. J Comp Physiol A 157:263–277
179. Quinn WG, Sziber PP, Booker R (1979) The *Drosophila* memory mutant amnesiac. Nature 277:212–214
180. Tempel BL, Bonini N, Dawson DR, Quinn WG (1983) Reward learning in normal and mutant *Drosophila*. Proc Natl Acad Sci U S A 80:1482–1486
181. Keene AC, Krashes MJ, Leung B, Bernard JA, Waddell S (2006) *Drosophila* dorsal paired medial neurons provide a general mechanism for memory consolidation. Curr Biol 16:1524–1530
182. Krashes MJ, Waddell S (2008) Rapid consolidation to a radish and protein synthesis-dependent long-term memory after single-session appetitive olfactory conditioning in *Drosophila*. J Neurosci 28:3103–3113
183. Medioni J, Vaysse G (1975) [Conditional suppression of a reflex in *Drosophila*

melanogaster: acquisition and extinction]. C R Seances Soc Biol Fil 169:1386–1391

184. DeJianne D, McGuire TR, Pruzan-Hotchkiss A (1985) Conditioned suppression of proboscis extension in *Drosophila melanogaster*. J Comp Psychol 99:74–80
185. Chabaud MA, Devaud JM, Pham-Delegue MH, Preat T, Kaiser L (2006) Olfactory conditioning of proboscis activity in *Drosophila melanogaster*. J Comp Physiol A 192:1335–1348
186. Dill M, Wolf R, Heisenberg M (1993) Visual pattern recognition in *Drosophila* involves retinotopic matching. Nature 365:751–753
187. Guo A, Gotz KG (1997) Association of visual objects and olfactory cues in *Drosophila*. Learn Mem 4:192–204
188. Xia S, Liu L, Feng C, Guo A (1997) Memory consolidation in *Drosophila* operant visual learning. Learn Mem 4:205–218
189. Wolf R, Wittig T, Liu L, Wustmann G, Eyding D, Heisenberg M (1998) *Drosophila* mushroom bodies are dispensable for visual, tactile, and motor learning. Learn Mem 5:166–178
190. Brembs B, Heisenberg M (2000) The operant and the classical in conditioned orientation of *Drosophila melanogaster* at the flight simulator. Learn Mem 7:104–115
191. Strauss R, Pichler J (1998) Persistence of orientation toward a temporarily invisible landmark in *Drosophila melanogaster*. J Comp Physiol A 182:411–423
192. Neuser K, Triphan T, Mronz M, Poeck B, Strauss R (2008) Analysis of a spatial orientation memory in *Drosophila*. Nature 453:1244–1247
193. Wustmann G, Heisenberg M (1997) Behavioral manipulation of retrieval in a spatial memory task for *Drosophila melanogaster*. Learn Mem 4:328–336
194. Gailey DA, Jackson FR, Siegel RW (1984) Conditioning mutations in *Drosophila melanogaster* affect an experience-dependent behavioral modification in courting males. Genetics 106:613–623
195. Ejima A, Smith BP, Lucas C, Levine JD, Griffith LC (2005) Sequential learning of pheromonal cues modulates memory consolidation in trainer-specific associative courtship conditioning. Curr Biol 15:194–206
196. McBride SM, Giuliani G, Choi C, Krause P, Correale D, Watson K, Baker G, Siwicki KK (1999) Mushroom body ablation impairs short-term memory and long-term memory of courtship conditioning in *Drosophila melanogaster*. Neuron 24:967–977
197. Toma DP, White KP, Hirsch J, Greenspan RJ (2002) Identification of genes involved in *Drosophila melanogaster* geotaxis, a complex behavioral trait. Nat Genet 31:349–353
198. Mertens I, Vandingenen A, Johnson EC, Shafer OT, Li W, Trigg JS, De Loof A, Schoofs L, Taghert PH (2005) PDF receptor signaling in *Drosophila* contributes to both circadian and geotactic behaviors. Neuron 48:213–219
199. Hirsch J (1959) Studies in experimental behavior genetics. II. Individual differences in geotaxis as a function of chromosome variations in synthesized *Drosophila* populations. J Comp Physiol Psychol 52:304–308
200. Gong Z (2009) Behavioral dissection of *Drosophila* larval phototaxis. Biochem Biophys Res Commun 382:395–399
201. Gong Z, Liu J, Guo C, Zhou Y, Teng Y, Liu L (2010) Two pairs of neurons in the central brain control *Drosophila* innate light preference. Science 330:499–502
202. Xiang Y, Yuan Q, Vogt N, Looger LL, Jan LY, Jan YN (2010) Light-avoidance-mediating photoreceptors tile the *Drosophila* larval body wall. Nature 468:921–926
203. Sawin-McCormack EP, Sokolowski MB, Campos AR (1995) Characterization and genetic analysis of *Drosophila melanogaster* photobehavior during larval development. J Neurogenet 10:119–135
204. Hadlern NM (1964) Genetic influence on phototaxis in *Drosophila melanogaster*. Biol Bull 126:264–273
205. Rein K, Zöckler M, Mader MT, Grübel C, Heisenberg M (2002) The *Drosophila* standard brain. Curr Biol 12:227–231
206. Cachero S, Ostrovsky AD, Yu JY, Dickson BJ, Jefferis GSXE (2010) Sexual dimorphism in the fly brain. Curr Biol 20:1589–1601
207. Chiang AS, Lin CY, Chuang CC, Chang HM, Hsieh CH, Yeh CW, Shih CT, Wu JJ, Wang GT, Chen YC, Wu CC, Chen GY, Ching YT, Lee PC, Lin CY, Lin HH, Wu CC, Hsu HW, Huang YA, Chen JY, Chian HJ, Lu CF, Ni RF, Yeh CY, Hwang JK (2011) Three-dimensional reconstruction of brain-wide wiring networks in *Drosophila* at single-cell resolution. Curr Biol 21:1–11
208. Yu JY, Kanai MI, Demir E, Jefferis GSXE, Dickson BJ (2010) Cellular organization of the neural circuit that drives *Drosophila* courtship behavior. Curr Biol 20:1602–1614
209. Fujita SC, Zipursky SL, Benzer S, Ferrus A, Shotwell SL (1982) Monoclonal antibodies against the *Drosophila* nervous system. Proc Natl Acad Sci U S A 79:7929–7933
210. Hofbauer A (1991) A library of monoclonal antibodies against the brain of *Drosophila*

melanogaster. Professorial Dissertation, University of Würzburg, Germany

211. Hofbauer A, Ebel T, Waltenspiel B, Oswald P, Y-c C, Halder P, Biskup S, Lewandrowski U, Winkler C, Sickmann A, Buchner S, Buchner E (2009) The Wuerzburg hybridoma library against *Drosophila* brain. J Neurogenet 23:78–91
212. Goodman CS, Bastiani MJ, Doe CQ, du Lac S, Helfand SL, Kuwada JY, Thomas JB (1984) Cell recognition during neuronal development. Science 225:1271–1279
213. Zipursky SL, Venkatesh TR, Benzer S (1985) From monoclonal antibody to gene for a neuron-specific glycoprotein in *Drosophila*. Proc Natl Acad Sci U S A 82:1855–1859
214. Pfeiffer BD, Jenett A, Hammonds AS, Ngo T-TB, Misra S, Murphy C, Scully A, Carlson JW, Wan KH, Laverty TR, Mungall C, Svirskas R, Kadonaga JT, Doe CQ, Eisen MB, Celniker SE, Rubin GM (2008) Tools for neuroanatomy and neurogenetics in *Drosophila*. Proc Natl Acad Sci U S A 105:9715–9720
215. Estes PS, Ho GL, Narayanan R, Ramaswami M (2000) Synaptic localization and restricted diffusion of a *Drosophila* neuronal synaptobrevin-green fluorescent protein chimera in vivo. J Neurogenet 13:233–255
216. Zhang YQ, Rodesch CK, Broadie K (2002) Living synaptic vesicle marker: synaptotagmin-GFP. Genesis 34:142–145
217. Andersen R, Li Y, Resseguie M, Brenman JE (2005) Calcium/calmodulin-dependent protein kinase II alters structural plasticity and cytoskeletal dynamics in *Drosophila*. J Neurosci 25:8878–8888
218. Diagana TT, Thomas U, Prokopenko SN, Xiao B, Worley PF, Thomas JB (2002) Mutation of *Drosophila* homer disrupts control of locomotor activity and behavioral plasticity. J Neurosci 22:428–436
219. Wang J, Ma X, Yang JS, Zheng X, Zugates CT, Lee CH, Lee T (2004) Transmembrane/juxtamembrane domain-dependent Dscam distribution and function during mushroom body neuronal morphogenesis. Neuron 43:663–672
220. Nicolaï LJ, Ramaekers A, Raemaekers T, Drozdzecki A, Mauss AS, Yan J, Landgraf M, Annaert W, Hassan BA (2010) Genetically encoded dendritic marker sheds light on neuronal connectivity in *Drosophila*. Proc Natl Acad Sci U S A 107:20553–20558
221. Yoshihara Y, Mizuno T, Nakahira M, Kawasaki M, Watanabe Y, Kagamiyama H, Jishage K, Ueda O, Suzuki H, Tabuchi K, Sawamoto K, Okano H, Noda T, Mori K (1999) A genetic approach to visualization of multisynaptic neural pathways using plant lectin transgene. Neuron 22:33–41
222. Pfeiffer BD, Ngo TT, Hibbard KL, Murphy C, Jenett A, Truman JW, Rubin GM (2010) Refinement of tools for targeted gene expression in *Drosophila*. Genetics 186:735–755
223. Luan H, Peabody NC, Vinson CR, White BH (2006) Refined spatial manipulation of neuronal function by combinatorial restriction of transgene expression. Neuron 52:425–436
224. Lai SL, Lee T (2006) Genetic mosaic with dual binary transcriptional systems in *Drosophila*. Nat Neurosci 9:703–709
225. Potter CJ, Tasic B, Russler EV, Liang L, Luo L (2010) The Q system: a repressible system for transgene expression, lineage tracing, and mosaic analysis. Cell 141:536–548
226. Lee T, Luo L (1999) Mosaic analysis with a repressible cell marker for studies of gene function in neuronal morphogenesis. Neuron 22:451–461
227. Choi CM, Vilain S, Langen M, Van Kelst S, De Geest N, Yan J, Verstreken P, Hassan BA (2009) Conditional mutagenesis in *Drosophila*. Science 324:54
228. Marek KW, Davis GW (2002) Transgenically encoded protein photoinactivation (FlAsH-FALI): acute inactivation of synaptotagmin I. Neuron 36:805–813
229. Venken KJ, Kasprowicz J, Kuenen S, Yan J, Hassan BA, Verstreken P (2008) Recombineering-mediated tagging of *Drosophila* genomic constructs for in vivo localization and acute protein inactivation. Nucleic Acids Res 36:e114
230. Miyawaki A, Llopis J, Heim R, McCaffery JM, Adams JA, Ikura M, Tsien RY (1997) Fluorescent indicators for Ca2+ based on green fluorescent proteins and calmodulin. Nature 388:882–887
231. Nakai J, Ohkura M, Imoto K (2001) A high signal-to-noise Ca(2+) probe composed of a single green fluorescent protein. Nat Biotechnol 19:137–141
232. Tian L, Hires SA, Mao T, Huber D, Chiappe ME, Chalasani SH, Petreanu L, Akerboom J, McKinney SA, Schreiter ER, Bargmann CI, Jayaraman V, Svoboda K, Looger LL (2009) Imaging neural activity in worms, flies and mice with improved GCaMP calcium indicators. Nat Methods 6:875–881

Chapter 2

Genetically Encoded Markers for *Drosophila* Neuroanatomy

Ariane Ramaekers, Xiao-jiang Quan, and Bassem A. Hassan

Abstract

The description of the anatomy of neural circuits provides a framework for predictions about their functions. During the last 2 decades, the explosion of genetically encoded tools for manipulating and visualizing the neural circuits in the fruit fly allowed important advances in correlating neural circuits and behavior. In this chapter, we review the properties of the main genetically encoded markers that are used to study *Drosophila* neuroanatomy, including data on toxicity when available.

Key words: Genetically encoded markers, Binary systems, Neuroanatomy, Subcellular localization

1. Introduction

The brain of *Drosophila* is organized into structural and functional units, or compartments, that include neurons deriving from a single neuronal lineage (reviewed in (1)). In order to limit the space for reasonable predictions about their function, an accurate and comprehensive description of the connectivity between and within compartments is required. Ideally, this would include information about the directionality of the information flow between neurons.

Development of versatile and genetically encoded tools to study fruit fly neuroanatomy was originally based on P element technology and the GAL4/UAS binary system (2). The use of green and red fluorescent proteins, GFP and RFP (isolated respectively from the jellyfish *Aequorea* and from the reef coral *Discosoma*) (3), and the subsequent engineering of brighter and more photostable variants characterized by a palette of different spectra (4, 5) further enriched the toolbox. Recent innovations including the development of alternative binary systems, LexA/LexAOp (6) and

Bassem A. Hassan (ed.), *The Making and Un-Making of Neuronal Circuits in Drosophila*, Neuromethods, vol. 69,
DOI 10.1007/978-1-61779-830-6_2, © Springer Science+Business Media, LLC 2012

the latest—and therefore still less characterized QF/QUAS (7) as well as mosaic approaches taking advantage of several recombinase variants allowed the development of even more highly versatile and extremely sophisticated tools (8, 9).

In this introductory chapter, we would like to provide an overview of the principal features of genetically encoded markers available to visualize *Drosophila* neurons, including data on toxicity when available. All these markers are available as UAS constructs and transgenic stocks listed in FlyBase. In addition, some of them also exist as LexA-responsive (LexAOp) reporter lines (6). It is likely that all markers will be available for all expression systems in the near future. Tables displaying FlyBase references for the cited markers and company reference numbers for commercial antibodies can be found at the end of the chapter.

2. Labeling the Entire Neuron

Genetically encoded neuronal reporters can be distinguished based on their subcellular localization—nuclear, cytoplasmic, or associated to the membrane, to the cytoskeleton, or to specific organelles. Differences in labeling properties between reporter molecules also depend on their structural properties. In particular, their molecular weight and whether they are active or not as monomers influence their diffusion in thin cellular processes. Importantly, markers also differ regarding their innocuity or toxicity to neurons. For instance, the expression of most cytoskeleton-bound reporters was shown to perturb neuronal morphology and even induce lethality under some circumstances (10).

2.1. Cytoplasmic markers

The most common cytoplasmic markers consist of β-galactosidase (β-Gal), an *Escherichia coli* enzyme—encoded by the gene *lacZ* (11) —and GFP (3). The two proteins differ in size: GFP is small (27 kDa) and acts mostly as a monomer, while β-Gal acts as a large 464kDa tetramer (12, 13). As a consequence, GFP diffuses better in neurites as compared to β-Gal. In addition, in contrast to β-Gal, whose expression is visualized through immunostaining or by providing a chromogenic substrate (X-Gal), GFP emits fluorescence in a spontaneous manner, allowing for live imaging studies. Low levels of GFP expression can also be detected using specific monoclonal or polyclonal commercially available antibodies.

Both UAS-GFP (3, 14, 15) and UAS-lacZ (16) reporter lines were generated more than a decade ago and are generally considered to be innocuous for neurons. However, a recent publication indicates that, in long-term assays, GFP expression could, unlike β-gal, affect fly health and result in a decrease in longevity and in locomotor activity (17).

2.2. Nuclear Markers

Though nuclear markers cannot provide information about circuitry itself, they can be useful in a quantitative approach—counting the number of cells belonging to a certain population, or for developmental studies. They consist of the fusion of fluorescent as well as nonfluorescent reporters together with a nuclear localization signal (*nls* sequence). Most commonly used nuclear UAS lines include UAS-lacZ.NZ (18), UAS-GFP^{nls} (19) and UAS-RedStinger (consisting of the fast-maturing RFP variant DsRed.T4 associated to a *nls* sequence) (20). In addition, the chromosome-associated fluorescent markers H2Av-GFP and H2Av-RFP produced by the fusion of the fluorescent markers with the Histone H2A variant protein can also be used to visualize all nuclei (21, 22).

2.3. Membrane-Bound Markers

The most widely used membrane-associated reporter is the product of the fusion of the mammalian lymphocyte marker CD8 and GFP (23). Another lymphocyte membrane-bound protein, CD2, is also used and visualized thanks to a commercially available monoclonal antibody (24). Expression of one or both constructs does not seem to provoke any deleterious effects in neurons. This may be due to the fact that both CD8 and CD2 have no homologues in the fly and therefore are less likely to interfere with endogenous cellular processes (23–25). Though cytoplasmic and membrane-bound GFP do label the entire neuron, the two reporters should not be viewed as equivalent (25). Indeed, thin and thick processes differ in their cytoplasm versus membrane ratio. Therefore, membrane-bound reporters should be preferred when focus is directed towards thin fibers, whereas cytoplasmic markers better label thick processes and larger structures such as varicosities (23, 25).

CD2-HRP, a fusion between CD2 and horseradish peroxidase (HRP), constitutes another membrane-bound marker designed for electron microscopy usage. Indeed, HRP catalyses the formation of an electron-dense product from diaminobenzidine (DAB) that can be visualized by EM (26). The use of this tool was reported for ultrastructural studies in pupal and adult brains (27–29).

2.4. Cytoskeleton-Bound Markers

In order to target reporter genes all along the neurites, several cytoskeleton-bound reporter genes were generated by the fusion of lacZ or GFP together with tau or kinesin proteins. Also, the tau protein itself can be used as a probe and labeled using an antibody (30). However, both kinesin and tau fusion constructs proved to be toxic to neurons and to induce lethality in combination with several GAL4 lines (30). In addition, tau-GFP, tau-lacZ reporters as well as a human form of tau alone were shown to induce morphological defects in axonal arborization (10).

2.5. Secretory Pathway and Mitochondrial Markers

Genetically encoded markers for the secretory pathway, Endoplasmic Reticulum (ER) and Golgi Apparatus (GA) include the ER marker Lys-GFP-KDEL (31) and the GA markers ManII-GFP

(32) and GalT-GFP (31). Outposts of both ER and GA were detected in *Drosophila* neurites though their precise subcellular localization varies depending on neuronal type. For example, the ER marker GFP-KDEL, as well as the GA marker ManII-GFP, were observed in the entire axons when expressed in developing photoreceptor neurons (33). In contrast, in larval md neurons, ManII-GFP is specifically localized in the soma and at dendrites branching points (32). Therefore, it seems that the presence of secretory outposts is actually not restricted either to dendrites or axons but rather depend on particular developmental or physiological states of the neurites.

A genetically encoded mitochondrial marker, mito-GFP, was also established and reported to label all neuronal mitochondria when expressed in *Drosophila* neurons (34, 35).

3. Presynaptic Markers

3.1. Syt 1-GFP

Synaptotagmin 1 (Syt 1; also simply named Syt) is the founding member of a large family of vesicular Ca^{2+} sensor proteins that regulate membrane traffic in neurons and other cell types (reviewed in (36)). Among the seven *Drosophila* synaptotagmins, only two, Syt 1 and Syt 4, are present at most if not all synapses (37, 38). In particular, Sytl is associated to synaptic vesicles in probably all *Drosophila* presynaptic terminals (37). Accordingly, Sytl::GFP became a commonly used presynaptic marker (39) and a valuable tool for mapping neuronal circuits (see for example (40, 41)). Concerning toxicity of the fusion protein for the neurons, the only published data reveal that, in contrast to N-Syb-GFP (see below), expression of Sytl-GFP in a subset of CNS neurons throughout development (using the Gal4 driver G4) is fully viable (39). However, controls for more subtle defects, in particular on synaptic morphology, were not reported.

3.2. N-Syb-GFP

In *Drosophila*, two Synaptobrevins were isolated, a ubiquitous form, Syb, and a neuronal form N-Syb (42–44). N-Syb is associated to the membrane of synaptic vesicles and is required for evoked neurotransmitter release (45).

Similar to Sytl-GFP, the fusion construct N-Syb-GFP is often used as a reporter of the presynaptic compartment of *Drosophila* neurons, in particular in studies aiming at describing neuronal circuits (see for instance (46–48)). A first fusion construct of N-Syb with the GFP variant S65T was published (49, 50), followed by a similar fusion to the next generation GFP variant eGFP (39). More recently, UAS lines bearing a N-Syb::RFP fusion (monomeric DsRed) have also been established (51). The only information about the toxicity of N-Syb::eGFP fusion protein comes from

Zhang et al. (39) and indicates that expression of the construct in a large subset of neurons throughout development induces lethality at the pupal stage.

3.3. Brp-GFP

The *Drosophila* coiled-coil domain protein Bruchpilot (BRP) was first identified as the epitope recognized by the presynaptic active zone specific monoclonal antibody NC82 (52, 53). BRP is present at the electron dense T-bars where it is crucial for the structure of the active zone and the release of neurotransmitter, in particular by interacting directly with Ca^{2+} channels (reviewed in (54)). A UAS-BRP-GFP reporter was constructed by fusing GFP to the N terminus of BRP, and its subcellular localization was shown to match endogenous BRP expression (52). A UAS-BRP-mRFP line was also recently generated (55). However, no data is available concerning the innocuity/toxicity of these two transgenes.

4. Somatodendritic and Postsynaptic Markers

4.1. DenMark

The recently published DenMark is a fluorescent protein resulting from the fusion of the red fluorescent protein mCherry (5) with the mouse ICAM5/telencephalin (41). In the mammalian telencephalon, ICAM5 is specifically expressed in dendrites (56). Thanks to ICAM5, expression of DenMark is highly enriched at membranes of the somatodendritic compartment and is also detected at postsynaptic sites. We showed that DenMark expression does not affect neuronal physiology or dendritic morphology. Similar to CD8-GFP, this could be explained by the fact that ICAM5 has no invertebrate homologue and is therefore less likely to interfere with endogenous processes. Importantly, we found that in immature neurons, DenMark expression is not polarized and is detected in axonal as well as in somatodendritic compartments. Since the same was observed for another somatodendritic marker, DsCam 17.1—GFP (see below), we assume that this finding may reflect the developmental dynamic of neuronal polarization rather than a lack of specificity of the marker. Finally, an antibody against the mCherry progenitor DsRed (Clontech #) also labels DenMark and can be used to compensate for low levels of expression with some GAL4 driver lines (41).

4.2. Dscam 17.1-GFP

The Dscam 17.1 protein corresponds to one of the putative 38016 membrane-bound cell adhesion molecules potentially encoded by the *Dscam 1* (*Drosophila* Down syndrome cell adhesion molecule) locus. In *Drosophila*, the multiple roles of Dscams in the formation of neural circuits are achieved through homophilic or heterophilic interactions (57). Subcellular localization of the Dscam proteins is controlled by alternative splicing of the transmembrane domain encoded by exon 17. Isoforms containing the alternative exon 17.2 are localized to the axons, while the exon 17.1 targets the proteins at the somatodendritic compartment (58). Accordingly, Dscam 17.1-GFP fusion proteins present a somatodendritic localization and have already been adopted as a tool for mapping neuronal circuits (1, 40, 59, 60). As mentioned above, similar to DenMark, Dscam17.1-GFP is detected in axons of immature neurons (41). However, unlike DenMark, Dscam17.1-GFP is toxic at high dose—overexpression using the panneural elav-Gal4 driver is fully lethal (41). In addition, given the multiple functions of Dscams in the development of neuronal circuits (57), further investigations concerning the effects of Dscam17.1-GFP expression on fine dendritic morphology would be useful.

5. Cautionary Considerations

Genetically encoded markers used in combination with the GAL/UAS constitute powerful and versatile tools for the *Drosophila* neuroanatomist. However, like any tool, they have their limitations and pitfalls. For instance, as already discussed, they can be toxic for the neurons, provoking either cell death or more problematically, affecting fine morphology. As already discussed, markers deriving from exogenous proteins seem to be less prone to affect neuronal development and physiology, perhaps because they are less likely to interfere with too many endogenous cellular functions. In contrast, markers derived from endogenous important cell components, like cytoskeleton-bound probes or Dscam-GFP—are highly toxic (10, 30, 41). In this context, it is important to keep in mind that control experiments indicating that markers do not affect the morphology of neurons are still lacking for most of them.

Another important feature of certain genetically encoded markers is the specificity of their subcellular localization. As described above, many markers were designed by fusing a fluorescent (in most of the cases GFP) or a nonfluorescent marker (usually β-gal) together with full-length proteins, protein domains, or signal peptides controlling their subcellular localization. However, depending on the amount of fusion protein expressed in neurons, the cellular machinery governing protein trafficking and/or localization can become saturated resulting in the loss of the specificity of the fusion protein localization. Obviously, saturating concentration of the marker will depend on the nature of the marker itself, the "strength" of the GAL4 driver, the number of copies of UAS and GAL4 driver constructs and—due to the thermosensitivity of the GAL/UAS system (61)—the temperature at which flies are kept. For instance, significant reduction in subcellular specificity of the somatodendritic marker DenMark was observed in flies homozygous both for the mushroom body driver 201Y-GAL4 and UAS-DenMark constructs (41).

Similarly, antibodies recognizing markers such as GFP or DsRed are very useful for detecting low expression levels, especially in single or sparse neurons. However, they can also draw misleading interpretation when used with specific subcellular fusion constructs. Indeed, specific subcellular localization often consists of an enrichment of a marker at a specific compartment rather than its complete absence from the "nonlabeled" compartment. Thus, immunolabeling can lead to a saturation of the signal and thereby mask the difference between enriched and nonenriched structures.

6. What's Missing?

Despite the wealth of tools already available, one can think about several extra ones that could enrich the toolbox and potentially lead to qualitative improvements.

First, concerning the directionality of neuronal information flow, available tools encompass presynaptic and somatodendritic markers. However, we still lack a specific marker for postsynaptic densities. Second, a tool to visualize synapses was recently developed in *Caenorhabditis elegans* (62). The GRASP (GFP Reconstitution Across Synaptic Partners) method is based on the expression of two complementary fragments of GFP at the membrane of two distinct cells. When the two fragments come in contact—for instance when both cells form a synapse—the GFP is reconstituted and emits fluorescence. GRASP was then rapidly implemented in the fruit fly by taking advantage of the two binary systems, GAL/UAS and LexA/LexAOp (63). However, since both GFP fragments are expressed at membranes, GFP reconstitution occurs not only at synapses but also at any contact point between the cells. Therefore, a synapse-specific GRASP system, possibly based on the fusion of both GFP fragments with proteins respectively enriched at pre- and postsynaptic sites, still needs to be developed. Third, a limitation inherent to the GRASP system is linked to the stability of the GFP. Indeed, once reconstituted, the fluorescent protein will remain attached between the pre- and postsynaptic sites preventing its internalization. Besides possible toxic consequences, this will "fix" the synapse at the stage where the GFP was reconstituted. In contrast, a transsynaptic marker which, unlike GRASP, could be internalized and recycled would in principle allow monitoring plasticity of synaptic contacts between cells in living animals.

7. Products and Tools

7.1. Constructs

We provide here the Flybase references for constructs corresponding to the genetically encoded markers described in this chapter. We restricted the list to the constructs implemented in the major binary system, GAL4/UAS. However, new LexA-responsive LexAOp reporter lines are regularly generated and rendered available in stock centers.

	Construct	# Flybase
Cytoplasmic	UAS-GFP	FBtp0001403
	UAS-LacZ	FBtp0000355
Nuclear	UAS-GFP^{nls}	FBtp0001204
	UAS-RedStinger	FBtp0018199
	UAS-LacZ^{nls}	FBtp0001611
Membrane-bound	UAS-CD8::GFP	FBtp0002652
	UAS-CD2	FBtp0000378
	UAS-CD2::HRP	FBtp0019068
Organites	UAS-GFP::KDEL	FBtp0021899
	UAS-ManII::GFP	FBtp0039245
	UAS-GAlT::GFP	FBtp0041224
	UAS-mito::GFP	FBtp0041285
Presynaptic	UAS-syt1::GFP	FBtp0016185
	UAS-Nsyb::GFP	FBtp0013062
	UAS-BRP::GFP	FBtp0022731
Postsynaptic	UAS-DenMark	FBtp0056788
	UAS-DsCam17.1::GFP	FBtp0021447

7.2. Antibodies

The following list provides references for the antibodies mentioned in the text.

	Company	Host species
Anti-β gal	Cappel	Rabbit
Anti-GFP	mAB 3E6 Invitrogen #A11120	Mouse
	Invitrogen #A11122	Rabbit
Anti-CD2	Serotec clone OX-34	Rat
Anti-dsRed	Clontech #632496	Rabbit
Anti-bruchpilot	DHSB NC82	Mouse
Anti-synapsin	DHSB 3C11	Mouse
Anti-Ncadherin	DHSB DN-EX#8	Rat

References

1. Spindler SR, Hartenstein V (2010) The *Drosophila* neural lineages: a model system to study brain development and circuitry. Dev Genes Evol 220:1–10
2. Brand AH, Dormand EL (1995) The GAL4 system as a tool for unravelling the mysteries of the *Drosophila* nervous system. Curr Opin Neurobiol 5:572–578
3. Brand A (1995) GFP in Drosophila. Trends Genet 11:324–325
4. Tsien RY (1998) The green fluorescent protein. Annu Rev Biochem 67:509–544
5. Shaner NC et al (2004) Improved monomeric red, orange and yellow fluorescent proteins derived from *Discosoma* sp. red fluorescent protein. Nat Biotechnol 22:1567–1572
6. Lai SL, Lee T (2006) Genetic mosaic with dual binary transcriptional systems in *Drosophila*. Nat Neurosci 9:703–709
7. Potter CJ, Tasic B, Russler EV, Liang L, Luo L (2010) The Q system: a repressible binary system for transgene expression, lineage tracing, and mosaic analysis. Cell 141:536–548
8. Hadjieconomou D et al (2011) Flybow: genetic multicolor cell labeling for neural circuit analysis in *Drosophila melanogaster*. Nat Methods 8(3):260–6
9. Hampel S et al (2011) *Drosophila* Brainbow: a recombinase-based fluorescence labeling technique to subdivide neural expression patterns. Nat Methods 8(3):253–9
10. Williams DW, Tyrer M, Shepherd D (2000) Tau and tau reporters disrupt central projections of sensory neurons in *Drosophila*. J Comp Neurol 428:630–640
11. O'Kane CJ, Gehring WJ (1987) Detection in situ of genomic regulatory elements in *Drosophila*. Proc Natl Acad Sci U S A 84: 9123–9127
12. Prasher DC, Eckenrode VK, Ward WW, Prendergast FG, Cormier MJ (1992) Primary structure of the Aequorea victoria green-fluorescent protein. Gene 111:229–233
13. Matthews BW (2005) The structure of *E. coli* beta-galactosidase. Cr Biol 328:549–556
14. Yeh E, Gustafson K, Boulianne GL (1995) Green fluorescent protein as a vital marker and reporter of gene expression in *Drosophila*. Proc Natl Acad Sci U S A 92:7036–7040
15. Dickson BJ (1996) Transgenic lines 1010T2 and 1010T10. Personal communication to FlyBase FBrf0086268
16. Brand AH, Perrimon N (1993) Targeted gene expression as a means of altering cell fates and generating dominant phenotypes. Development 118:401–415
17. Mawhinney RM, Staveley BE (2011) Expression of GFP can influence aging and climbing ability in *Drosophila*. Genet Mol Res 10:494–505
18. Bloomington *Drosophila* Stock (1998) C. Alleles, transposons and insertions not in FlyBase. Personal communication to FlyBase FBrf0104719
19. Shiga Y, TanakaMatakatsu M, Hayashi S (1996) A nuclear GFP beta-galactosidase fusion protein as a marker for morphogenesis in living *Drosophila*. Dev Growth Differ 38:99–106
20. Barolo S, Castro B, Posakony JW (2004) New *Drosophila* transgenic reporters: insulated P-element vectors expressing fast-maturing RFP. Biotechniques 36(3):436–440
21. Clarkson M, Saint R (1999) A His2AvDGFP fusion gene complements a lethal His2AvD mutant allele and provides an in vivo marker for *Drosophila* chromosome behavior. DNA Cell Biol 18:457–462
22. Heidmann SP (2007) {His2Av-mRFP1} insertions. Personal communication to FlyBase FBrf0200083
23. Lee T, Luo L (1999) Mosaic analysis with a repressible cell marker for studies of gene function in neuronal morphogenesis. Neuron 22:451–461
24. Dunin-Borkowski OM, Brown NH (1995) Mammalian CD2 is an effective heterologous marker of the cell surface in *Drosophila*. Dev Biol 168:689–693
25. Ito K, Okada R, Tanaka NK, Awasaki T (2003) Cautionary observations on preparing and interpreting brain images using molecular biology-based staining techniques. Microsc Res Tech 62:170–186
26. Larsen CW, Hirst E, Alexandre C, Vincent JP (2003) Segment boundary formation in *Drosophila* embryos. Development 130: 5625–5635
27. Watts RJ, Schuldiner O, Perrino J, Larsen C, Luo L (2004) Glia engulf degenerating axons during developmental axon pruning. Curr Biol 14:678–684
28. Lin HH, Lai JS, Chin AL, Chen YC, Chiang AS (2007) A map of olfactory representation in the *Drosophila* mushroom body. Cell 128:1205–1217
29. Edwards TN, Meinertzhagen IA (2009) Photoreceptor neurons find new synaptic targets when misdirected by overexpressing runt in *Drosophila*. J Neurosci 29:828–841

30. Ito K, Sass H, Urban J, Hofbauer A, Schneuwly S (1997) GAL4-responsive UAS-tau as a tool for studying the anatomy and development of the *Drosophila* central nervous system. Cell Tissue Res 290:1–10
31. Snapp EL, Iida T, Frescas D, Lippincott-Schwartz J, Lilly MA (2004) The fusome mediates intercellular endoplasmic reticulum connectivity in *Drosophila* ovarian cysts. Mol Biol Cell 15:4512–4521
32. Ye B et al (2007) Growing dendrites and axons differ in their reliance on the secretory pathway. Cell 130:717–729
33. Yogev S, Schejter ED, Shilo BZ (2010) Polarized secretion of *Drosophila* EGFR ligand from photoreceptor neurons is controlled by ER localization of the ligand-processing machinery. PLoS Biol 8(10):e1000505 doi: 10.1371/journal.pbio.1000505
34. Pilling A, Saxton WP(2004) {MitoGFP.AP} construct and insertions. Personal communication to FlyBase FBrf0178877
35. Rikhy R, Ramaswami M, Krishnan KS (2003) A temperature-sensitive allele of *Drosophila* sesB reveals acute functions for the mitochondrial adenine nucleotide translocase in synaptic transmission and dynamin regulation. Genetics 165:1243–1253
36. Pang ZP, Sudhof TC (2010) Cell biology of Ca2+-triggered exocytosis. Curr Opin Cell Biol 22:496–505
37. Littleton JT, Bellen HJ, Perin MS (1993) Expression of synaptotagmin in *Drosophila* reveals transport and localization of synaptic vesicles to the synapse. Development 118: 1077–1088
38. Adolfsen B, Saraswati S, Yoshihara M, Littleton JT (2004) Synaptotagmins are trafficked to distinct subcellular domains including the postsynaptic compartment. J Cell Biol 166: 249–260
39. Zhang YQ, Rodesch CK, Broadie K (2002) Living synaptic vesicle marker: synaptotagmin-GFP. Genesis 34:142–145
40. Zhang K, Guo JZ, Peng Y, Xi W, Guo A (2007) Dopamine-mushroom body circuit regulates saliency-based decision-making in *Drosophila*. Science 316:1901–1904
41. Nicolai LJ et al (2010) Genetically encoded dendritic marker sheds light on neuronal connectivity in *Drosophila*. Proc Natl Acad Sci U S A 107:20553–20558
42. DiAntonio A et al (1993) Identification and characterization of *Drosophila* genes for synaptic vesicle proteins. J Neurosci 13:4924–4935
43. Sudhof TC, Baumert M, Perin MS, Jahn R (1989) A synaptic vesicle membrane protein is conserved from mammals to *Drosophila*. Neuron 2:1475–1481
44. Chin AC, Burgess RW, Wong BR, Schwarz TL, Scheller RH (1993) Differential expression of transcripts from syb, a *Drosophila melanogaster* gene encoding VAMP (synaptobrevin) that is abundant in non-neuronal cells. Gene 131:175–181
45. Deitcher DL et al (1998) Distinct requirements for evoked and spontaneous release of neurotransmitter are revealed by mutations in the *Drosophila* gene neuronal-synaptobrevin. J Neurosci 18:2028–2039
46. Ramaekers A et al (2005) Glomerular maps without cellular redundancy at successive levels of the *Drosophila* larval olfactory circuit. Curr Biol 15:982–992
47. Otsuna H, Ito K (2006) Systematic analysis of the visual projection neurons of *Drosophila melanogaster* I. Lobula-specific pathways. J Comp Neurol 497:928–958
48. Helfrich-Forster C et al (2007) Development and morphology of the clock-gene-expressing lateral neurons of *Drosophila melanogaster*. J Comp Neurol 500:47–70
49. Ito K et al (1998) The organization of extrinsic neurons and their implications in the functional roles of the mushroom bodies in *Drosophila melanogaster* Meigen. Learn Mem 5:52–77
50. Estes PS, Ho GL, Narayanan R, Ramaswami M (2000) Synaptic localization and restricted diffusion of a *Drosophila* neuronal synaptobrevin–green fluorescent protein chimera in vivo. J Neurogenet 13:233–255
51. Raghu SV, Joesch M, Borst A, Reiff DF (2007) Synaptic organization of lobula plate tangential cells in *Drosophila*: gamma-aminobutyric acid receptors and chemical release sites. J Comp Neurol 502:598–610
52. Wagh DA et al (2006) Bruchpilot, a protein with homology to ELKS/CAST, is required for structural integrity and function of synaptic active zones in *Drosophila*. Neuron 49: 833–844
53. Hofbauer A et al (2009) The Wuerzburg hybridoma library against *Drosophila* brain. J Neurogenet 23:78–91
54. Wichmann C, Sigrist SJ (2010) The active zone T-bar–a plasticity module? J Neurogenet 24:133–145
55. Mauss A, Tripodi M, Evers JF, Landgraf M (2009) Midline signalling systems direct the formation of a neural map by dendritic targeting in the *Drosophila* motor system. PLoS Biol 7:e1000200
56. Mori K, Fujita SC, Watanabe Y, Obata K, Hayaishi O (1987) Telencephalon-specific

antigen identified by monoclonal antibody. Proc Natl Acad Sci U S A 84:3921–3925

57. Hattori D, Millard SS, Wojtowicz WM, Zipursky SL (2008) Dscam-mediated cell recognition regulates neural circuit formation. Annu Rev Cell Dev Biol 24:597–620
58. Wang J et al (2004) Transmembrane/juxtamembrane domain-dependent Dscam distribution and function during mushroom body neuronal morphogenesis. Neuron 43:663–672
59. Vomel M, Wegener C (2008) Neuroarchitecture of aminergic systems in the larval ventral ganglion of *Drosophila melanogaster*. PLoS One 3:e1848
60. Young JM, Armstrong JD (2010) Structure of the adult central complex in *Drosophila*: organization of distinct neuronal subsets. J Comp Neurol 518:1500–1524
61. Duffy JB (2002) GAL4 system in *Drosophila*: a fly geneticist's Swiss army knife. Genesis 34:1–15
62. Feinberg EH et al (2008) GFP Reconstitution Across Synaptic Partners (GRASP) defines cell contacts and synapses in living nervous systems. Neuron 57:353–363
63. Gordon MD, Scott K (2009) Motor control in a *Drosophila* taste circuit. Neuron 61: 373–384

Chapter 3

Subcellular Resolution Imaging in Neural Circuits

W. Ryan Williamson, Chih-Chiang Chan, and P. Robin Hiesinger

Abstract

Drosophila combines advanced genetics with a brain of ideal size for high-resolution imaging in toto. However, imaging of intracellular compartments pushes the limits of light microscopy in every system, and at the subcellular level the small size of fly neurons presents a challenge. In this chapter, we review recent imaging advances that, often for the first time, allow the visualization of intracellular biology of neurons in the context of their neuronal circuits. We discuss the different preparations that keep neural circuit architectures intact for live and fixed imaging. Finally, we review advances in light microscopy and imaging probes in combination with these preparations and provide a guide to which high-resolution microscopy techniques are applicable to the different *Drosophila* preparations. We focus on the imaging of intracellular membrane trafficking dynamics. However, since any imaging of intracellular trafficking constitutes an example of imaging at subcellular resolution, many approaches discussed here will be useful for the study of neuronal cell biology in *Drosophila* in general.

Key words: Fluorescent microscopy, High-resolution imaging, Brain dissection, Immunohistochemistry, Live imaging

1. Imaging Approaches in Neural Circuit Preparations

The principle requirements for the visualization of subcellular compartments are the same in all systems: The goal is to visualize distinguishable structures at the highest resolution possible. In addition, live imaging demands minimal phototoxicity. Limits are imposed by both the markers for intracellular proteins and the microscopy technique itself. Recent years have seen the development of many new approaches to high-resolution fluorescence microscopy as well as many new fluorescent probes. The following two sections present recent advances in fluorescence imaging technologies and explain how these technologies can be applied to study subcellular biology in neural circuits in fly preparations in vivo.

Bassem A. Hassan (ed.), *The Making and Un-Making of Neuronal Circuits in Drosophila*, Neuromethods, vol. 69,
DOI 10.1007/978-1-61779-830-6_3, © Springer Science+Business Media, LLC 2012

1.1. Overview of High-Resolution Fluorescence Microscopy Imaging Approaches

With a steadily increasing number of high-resolution microscope types, the educated choice of which microscope to choose for a specific preparation has become more difficult. For subcellular resolution imaging in living *Drosophila* preparations, we only consider fluorescence microscopy techniques based on far-field high-resolution fluorescence microscopes. Near-field microscopy techniques, like total internal reflection fluorescence (TIRF), are not likely to be useful for imaging neural circuits in vivo due to their lack of working distance (1). There are some applications where conventional light microscopy approaches will be sufficient, but probably none where they would be superior to all high-resolution fluorescence techniques. The basic concept and different incarnations of far-field high-resolution microscopy are described in detail elsewhere (1–3). We only briefly summarize the most applicable approaches for *Drosophila*. In addition, we focus on what microscopy techniques work best with the different preparations.

1.1.1. Conventional Confocal Laser Scanning and Deconvolution

The traditional point-scanning confocal microscope has been the most widely used high-resolution fluorescence microscopy technique for over 15 years. With excellent optics, the resolution of today's laser scanning microscopes indeed closely approaches the diffraction barrier with 250 nm in the x/y plane and 600 nm in the z axis (light path through the lens). The theoretical resolution limit in the x/y plane is given by half the excitation wavelength, i.e., 244 nm for a 488 Argon laser line (GFP illumination), and 316.5 nm for a 633-nm Helium–Neon laser line (far-red, e.g., Cy5, illumination). The resolution limit in the z axis is more than twice that of x/y and described by the *point spread function* (*PSF*). The PSF is a mathematically defined description of the detectable light spread from a point light source (4). In three-dimensional (3D) space, this spread has the shape of a rocket along the z-axis (Fig. 1). If the point from which the spread originates and the shape of the PSF are known, then the PSF can be transformed back into a point for the purpose of removing light scatter and out-of-focus light in confocal datasets. This is the principle of *deconvolution* (5, 6). *Nonblind deconvolution* utilizes a measured PSF (as shown in Fig. 1a), whereas blind deconvolution assumes a simple PSF (e.g., Gaussian) and then tests every voxel in a 3D dataset for whether this PSF is indeed applicable—altering the image data and PSF in turn in an iterative computation-intensive process (e.g., (7)). As shown in Fig. 1, the "point spread" inside a *Drosophila* brain is inhomogeneous and therefore limits the applicability of the same PSF for every point in a 3D or 4D dataset (7). Deconvolution brings the data closer to the resolution limit by removing artificial light scattering introduced by the optics and the preparation itself. However, deconvolution does not remove the diffraction barrier. The principle of deconvolution will become more important in the

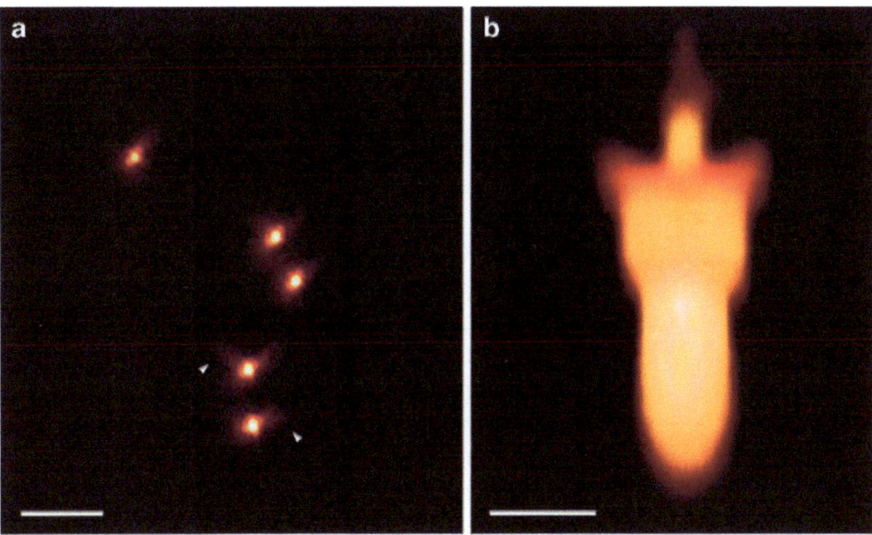

Fig. 1. A point in the confocal microscope. (**a**) 500-nm fluorescent beads that were injected into a fluorescently immunolabeled adult *Drosophila* brain and scanned with a conventional confocal microscope at 20 μm depth (7). All five beads show different light scattering due to unpredictable tissue-dependent distortions of the point spread function (PSF). (**b**) Volume rendering of a three-dimensional (3D) conventional confocal scan of one of the beads shown in (**a**). The light path to the lens is up. Such scans of a perfectly round object show the principle shape of the PSF for fluorescent point minus further tissue-dependent distortions as shown in (**a**). Scale bars in (**a**) 5 μm, in (**b**) 1 μm.

section on superresolution imaging techniques that effectively remove the diffraction barrier for visible light (1). For conventional confocal microscopy datasets numerous powerful software implementations exist that can be used independently of the microscope with which the data was obtained.

Confocal microscopes have the very useful ability to optimize the imaging and resolution setting for each lens and specimen size by adjusting several parameters. In practice, one wants to use the lens with the highest numerical aperture possible, adjust the confocal pinhole to an airy disc of 1 or just slightly above 1, and finally digitally zoom to adjust the scan area to match the pixel size to the resolution limit. The principle of the confocal pinhole is discussed elsewhere (2); here, it shall suffice to note that an airy disc of 1 defines the point of maximum confocality; below this value light is further lost without gaining resolution in the *z* axis. In contrast, an airy disc above 1 increases light detection by opening the pinhole and losing "confocality," i.e., resolution in the *z* axis. In the *x/y* plane, it is rarely useful to increase digital zoom such that the pixel size falls below 100 nm (e.g., 50 μm^2 for a 512×512 pixel scan). Note that any scan resolution below the diffraction limits of 200–250 nm (i.e., any scan of less than 100 μm^2 at 512×512) is theoretically empty resolution and only helpful as a means of "spatial averaging," which is an application-dependent alternative to temporal averaging, i.e., rescanning the same points several times (see below).

For live imaging purposes, the second major set of parameters to consider are the scan speed and dwell times. Conventional confocal microscopes operate by point-scanning and digitally integrating pictures from scanned lines. This is a slow process and the speed depends on how fast the point-scanner moves. The standard galvanometer scanning mirror in confocal microscopes operates between 400 and 1,000 Hz. Hence, a single-channel 512×512 image (without averaging) will take between 0.5 and 1 s. If timing permits, averaging should always be used as it dramatically improves image quality by averaging out random noise. Using ×4 line or frame averaging of a 512×512 pixel scan in a single channel will therefore take around 2 s to obtain. A 3D dataset of 512×512×64 voxels will therefore take 128 s per channel. Importantly, the slow video rate is not the only problem of slow scan speed. Slower point-scanning also means longer dwell times of the laser on fluorophores which greatly increases photobleaching and phototoxicity (3, 8).

1.1.2. Resonant Scanning and Spinning Disc Confocal

Both Resonant Scanning and Spinning Disc Confocal Microscopes overcome the slow scan speed of conventional confocal microscopes. The resonant scanning technique is based on the ability of a galvanometer mirror to operate at a resonant frequency that is approximately ten times higher than its normal scan speed (e.g., 8,000 Hz). Technically, this is achieved by sending a sine wave control function to the galvanometer motor. The ×10 acceleration is sufficient to turn a point-scanning confocal into a real live imaging microscope with substantially reduced phototoxicity and scan speed at video rates. A 512×512 pixel scan takes 0.05–0.1 s, i.e., up to 20 frames per second (fps) at 512×512, 512×256 or 256×256 scans are twice as fast. Averaging can be applied to increase image quality at the expense of speed. Dramatic increases of averaging (e.g., ×32 or ×64) yield the same quality high-resolution images as conventional confocal microscopy at the same slow speed—but with one major difference: photobleaching and phototoxicity are substantially reduced due to much shorter dwell times of the point scanner at the time of excitation (8). The disadvantage of the resonant scanner over the conventional microscope is that the ability to adjust the scan speed and regions of interest are lost. In almost all other aspects, a resonant confocal is identical to the conventional confocal. Bimodal microscopes have been available for several years.

The spinning disc confocal microscope is based on an architecture different from that of conventional point scanners. The principle is reviewed in many excellent references (9, 10). We focus on the differences with respect to typical *Drosophila* imaging preparations. In brief, Spinning Disc microscopes use a quickly rotating so-called Nipkow-Disc with defined pinhole size (the pinhole can only be changed by exchanging the Nipkow Disc). Illumination does not require a laser and photon detection is done with a fast charge-coupled device (CCD) camera. Despite these major differences, spinning disc microscopes are true confocal microscopes in that out-of-focus light is largely blocked from reaching the detector through, in this case, many confocal pinholes. However, the image is not "scanned" and does not need to be digitally integrated as would be required with laser scanning confocal microscopes; the CCD camera indeed "sees" the whole visible field at a fast video rate. The major limitation of this technique is that some out-of-focus light still reaches the detector through adjacent pinholes (so-called pinhole cross talk). The problem increases with the depth of the tissue under investigation and causes depth-dependent loss of confocality. In addition, illumination that does not pass through the pinholes can get reflected by the disc resulting in higher background noise. Finally, spinning discs do not offer any of the advantages that come with laser point-scanners, like photobleaching or photoactivation in small regions of interest.

1.1.3. Multiphoton Microscopy

A key parameter that we have so far not discussed is the working distance: How deep can I scan? Multiphoton microscopy addresses this issue by using far-red excitation lasers that penetrate deeper into tissue with less light-scatter. The basic concept is that photon density reaches a threshold for excitation only in the focal plane, thereby completely eliminating out-of-focus excitation. Since only one point is illuminated, no confocal pinhole is required to eliminate out-of-focus light. The focused excitation also dramatically reduces phototoxicity due to unproductive excitation. Multiphoton microscopy techniques are reviewed elsewhere (2, 11). It should be noted that multiphoton microscopes are diffraction-limited similar to other light microscopes; indeed, they have a theoretically reduced maximal resolution due to the long excitation wavelength. In practice, however, the ability to exclude out-of-focus excitation, especially in deep tissue, often yields higher resolution data than confocal microscopes in similar circumstances. Examples of multiphoton microscopy are discussed in the following sections.

1.1.4. Far-Field Superresolution-Imaging with STED and PALM/STORM

The diffraction barrier described by the formula of Ernst Abbe in 1873 dictates that two simultaneously illuminated points (or fluorescent molecules) must be separated by at least a distance of half the wavelength of light in order to be resolved. The shorter wavelength of an electron similarly defines the resolution limit of electron microscopy. Only within the last few years have widely applicable fluorescent microscopy techniques become available that effectively break the diffraction barrier (3, 12). All techniques are based on the idea that two fluorescent molecules that are closer together than 200 nm can be excited sequentially. In other words, they are not separated spatially, because diffraction cannot be removed per se, but temporally. However, the technical hurdles of exciting two fluorophores separately within 200 nm are substantial, simply because no lens-focused beam of light illuminates a spot smaller than 200 nm. Different techniques have emerged in recent years that achieve superresolution imaging with different approaches. Two approaches have become available for practical usage in the last few years: stimulated emission depletion (STED) microscopy and photoactivatable localization microscopy (PALM) or stochastic optical reconstruction microscopy (STORM). The technical details of these techniques are discussed elsewhere (1, 3). Here, we focus on the basic principles and differences that serve as a foundation for the choice of application in *Drosophila*.

STED is based on confocal point-scanning microscopy; indeed, the first commercially available STED setup can be obtained as an upgrade to an existing confocal laser scanning microscope. The basic trick is a red-shifted so-called STED laser that illuminates in a donut shape around the standard excitation point laser. Wherever the STED laser provides sufficient energy fluorescence is suppressed due to the photophysical property of stimulated emission. The "hole" in the middle of the STED laser is characterized by a gradient of decreasing STED laser energy. With increasing STED laser intensity the hole becomes smaller and only fluorophores that are below a photophysical threshold become excited by the excitation laser. By point-scanning with the excitation/STED laser pair in an otherwise conventional confocal setup, neighboring fluorophores that are closer together than 200 nm can be sequentially activated and thus resolved.

PALM/STORM microscopy uses a radically different approach to temporally separate two fluorophores within 200 nm: sparse illumination (13, 14). In effect, both PALM and STORM use threshold illumination that randomly illuminates fluorophores in a specimen such that typically no fluorophores within a 200 nm radius are illuminated at the same time. If a sparsely illuminated fluorophore emits enough photons, its location can be determined using the same principle described above for deconvolution. The key difference between STED and PALM/STORM therefore is this: With STED the microscope "knows" where every single photon comes from, whereas in PALM/STORM its location needs to be determined. Few photons from a fluorescent molecule suffice to determine its localization in STED, but it must be capable of repeating many on/off cycles through stimulated emission. In contrast, in PALM/STORM any given fluorophore may theoretically only have to be excited once, as long as it emits enough photons to deconvolve its localization. In practical terms, this has led to a greater applicability of PALM/STORM for more fluorescent molecules. STED, on the other hand, is currently the faster method for live imaging. The development of these approaches is very fast-paced, and increasing speed and applicability can be expected for both types of systems at the time of publication.

1.2. What Microscopy Technique Should I Use for My Preparation?

In the following section, we review the applicability of the above described microscopy techniques for *Drosophila* preparations that are useful for imaging of neural circuitry. The following considerations apply:

1.2.1. Considerations for Choosing the Right Microscopy Technique

Consideration 1: Live or Fixed?

Conventional confocal laser-scanning microscopy still offers the greatest versatility for imaging fixed preparations for the size of the *Drosophila* brain. However, at high resolution and with weak fluorescence (and consequently high laser intensities) photobleaching becomes a serious issue. In such cases, resonant-scanning helps by decreasing photobleaching at the cost of losing some versatility (mainly a smaller minimal field of view and no scanning of asymmetric regions of interest).

Spinning Disc, Resonant Scanning, and Multiphoton confocals are all suitable for live imaging due to their reduced phototoxicity. However, they achieve this by three different means—weak nonlaser illumination for the spinning disc, reduced point-laser dwell time for the resonant confocal, and reduced out-of-focus excitation in the case of multiphoton. We are not aware of a direct comparison of the three for the same preparation and we do not have high-end spinning disc or multiphoton microscopes available to perform this comparison. However, information from successful experiments in *Drosophila* preparations together with knowledge of the different architectures of these microscopes allow several conclusions to be drawn regarding what should and should not work. As outlined above, if the tissue depth is small and imaging close to the diffraction barrier is not required, spinning disc confocal microscopy is still a good choice. Recent work on imaging the development of the *Drosophila* wing imaginal disc offers some details that should be applicable to neural circuit preparations (15). In addition, a video protocol is available for another preparation that only requires a small working distance using spinning disc microscopy (16).

Resonant scanning has become more popular in recent years as an extension of the applicability of conventional confocal laser scanning microscopes. We have recently used resonant scanning for live imaging in developing eye disc–brain culture (17, 18). For high-resolution far-field imaging, we use a ×63 (NA 1.3) glycerine lens that increases the working distance by more than 10 μm compared to oil immersion lenses. However, light scattering deeper than 20 μm in the *Drosophila* tissue in water precludes high-resolution imaging. Furthermore, even strong fluorophores require averaging between ×8 and ×48 to reduce noise. More details are available in a video protocol (18).

Multiphoton microscopy has been very successfully applied in a number of *Drosophila* brain preparations, including live imaging of the olfactory lobe (19, 20), and is discussed below in the context of the adult brain preparation.

Consideration 2: How Deep Do You Need to Image?

The fluorescent microscopy techniques discussed above allow for quantitative imaging in deep tissue in the following order: Multiphoton > Resonant Confocal & STED > PALM/STORM > Spinning Disc. Successful application of spinning disc confocals is largely restricted to cell culture and thin preparations where tissue within only a few micrometers of the surface is imaged (e.g., (16)). In the case of *Drosophila* preparation, subcellular high-resolution imaging of structure deeper than 10 μm is most likely better performed using a resonant scanner. We routinely use resonant scanning for live pupal and adult brain preparations up to depths of 20–30 μm (18). Below a tissue depth of 20–30 μm multiphoton approaches provide significantly higher quality data.

Consideration 3: How Weak Are Your Fluorescent Probes?

Both the CCDs in spinning disc confocals and the photomultiplier tubes (PMTs) in line-scanning confocals are single photon detection devices. The biggest differences are speed and light-sensitivity: The PMT is the fastest available photon detector, generating an electrical output after photon detection within a few nanoseconds (providing the key reason for its use in line-scanning microscopes). With respect to light-sensitivity (and almost all other parameters), today's newest CCDs outperform PMTs. The quantum efficiency (i.e., probability of a single photon to cause a detectable charge) is 5–20% for PMT and 25–95% for CCDs. In addition, CCDs have a dynamic range ten times as large as those of PMTs as well as less dark signal and noise.

Consideration 4: How Fast Do You Need to Image?

CCD-based imaging is faster, simply because the CCD sees the whole image at any time point whereas line-scanning is very time costly. Hence, most imaging purposes that require an imaging rate of more than 25 frames per second are CCD-based (including the spinning disc confocal). For image rates below 25 frames per second many other considerations start to play a role for the choice of microscope. While a resonant confocal offers most of the advantages and flexibility of a full confocal microscope, the image quality increases only with lower speed. Where the speed/quality curves intersect depends on many parameters, including the brightness of the fluorophores. In practical terms, a high-resolution resonant scan at 25 frames per second will in most cases be too noisy for subcellular imaging. We typically need to average at least ×8 to discern subcellular structures in *Drosophila* brain or filet preparations below 500 nm in *x*/*y*, making the fastest reasonable speed for a single 512 × 512 scan 3 frames per second.

Consideration 5: How Long Do You Need to Image?

Long live imaging sessions (hours and longer) suffer from two main problems: drift and phototoxicity/photobleaching. Phototoxicity is discussed in Consideration 1. Drift originates from both movements of the preparation as well as the microscope. High-quality stages guarantee focus drift of less than 1 μm in the *z* direction per hour. Key to reducing microscope/stage drift on any system is to keep the temperature constant. Ideally, the microscope should be running for 2 h at least to reach an even operating temperature of all parts, especially for *Drosophila* preparations which do not require a heated stage or chamber. Drift is negligible in *x*/*y* for high-performance microscope stages and less in *z* for 2D imaging over time. For 3D imaging over time, a galvanotable with nanometer accuracy inside the stage is highly recommended. Drift of the preparation within the imaging chamber can be very difficult to control and are inherent for a moving (e.g., some filet preparations) or developing specimen (e.g., the eye–brain complex). Where some drift is unavoidable the simplest solution is to choose a generous bounding box, i.e., imaging a region with sufficient space around the region of interest to allow for a certain amount of drift. Several software solutions exist for realignment of 3D datasets. In addition, several smart integrated software/hardware solutions have recently been developed that can track and correct for preparation drift by automatically correcting stage positioning during long live imaging sessions. All of these options can be explored thoroughly with the major microscope and imaging software manufacturers.

With these considerations in mind, the following *Drosophila* "neural circuit" preparations are available for imaging:

1.2.2. Embryo

The embryonic central nervous system develops within a few hours and becomes functional only an hour before the embryo hatches. The embryonic CNS has therefore mostly been studied as a model for early nervous system development. The early brain and ventral ganglion are too deep inside the intact embryo (20–60 μm) for live imaging with spinning disc microscopy. Resonant scanning allows deeper imaging, but no high-resolution scanning deeper than 20 μm is possible. Beyond this depth, multiphoton is recommended. For fixed tissue, conventional confocal works very well in cleared tissue, as it allows high-resolution scanning through approximately half the thickness of the intact embryo (50–75 μm).

An alternative to the intact embryo is the embryo filet in which the CNS as well as all neuromuscular junctions (NMJs) are directly exposed (Fig. 2a). The preparation of embryo filets is made possible by water-polymerizing surgical glues (21). Imaging can be performed either using a water-dipping lens or inside a perfusion chamber (18). A high-resolution water-dipping lens can greatly improve the quality of live imaging of both the CNS or the NMJs using spinning disc or resonant confocal microscopes.

1.2.3. Larval Eye Disc–Brain Complexes

The larval brain-imaginal discs complex is easily dissected from third instar wandering larvae (18). The developing imaginal disc sends hundreds of photoreceptor axons into the larval brain hemispheres through the optic stalk; the leg discs are innervated by axons from the larval brain. Hence, the larval brain-imaginal disc

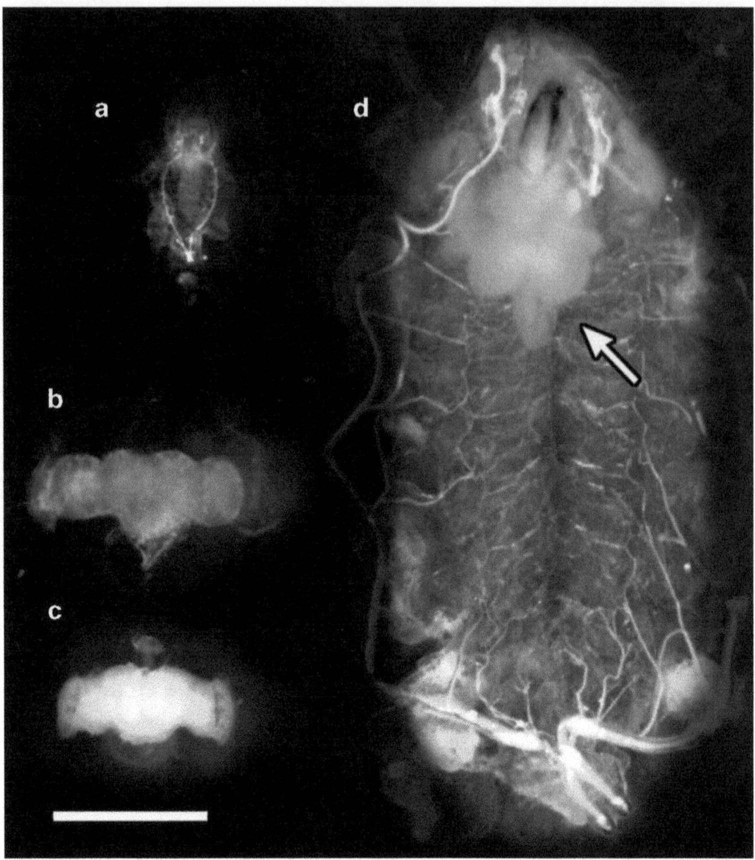

Fig. 2. *Drosophila* live neural circuit preparations. (**a**) Embryo filet preparation. (**b**) Pupal P+30% eye–brain preparation. (**c**) Adult brain. (**d**) L3 larval filet preparation. The larval brain–eye disc complex is marked with an *arrow*. Scale bar: 500 μm.

complex is an excellent model for neural circuit formation in vivo. Live imaging can be performed in a perfusion chamber using resonant confocal microscopy (18). The size and thickness of the larval brain are comparable to those of a whole-mount embryo (comp. Fig. 2a and arrow in 2D). Hence, all considerations discussed above for the intact embryo preparation apply.

1.2.4. Larval Filet

The larval filet preparation (Fig. 2d) has served as the work horse of synapse function and plasticity studies in *Drosophila* for more than 20 years. The larval filet presents an in vivo setting that is amenable to electrophysiological and imaging studies of the NMJ. The NMJs are large (up to 5 μm) bouton-like synaptic contacts between the motor neurons that originate in the ventral ganglion and the body wall musculature of the larva. The larval filet can be handled very similar to neuronal cell culture systems in terms of accessibility to bath solutions, dyes, and electrodes. Since the complete nervous system remains intact, the larval filet represents an ideal preparation for the study of motor neuron circuitry in vivo. Like the embryo filet, the larval filet can be prepared either using water-polymerizing surgical glue and imaged in a closed perfusion chamber, or (more commonly) with a water-dipping lens. In the latter case, the preparation is typical performed using small metal pins to immobilize the living preparation. The larval filet is amenable to high-resolution imaging with any live imaging technique (including spinning disc and resonant scanning) due to the direct exposure of the nerves and synaptic boutons in the preparation. Finally, superresolution imaging has been applied very successfully with this preparation (22).

1.2.5. Pupal Brain

The pupal brain (including the developing eyes) is an excellent preparation for the study of neural circuit development (Fig. 2b). During 20–40% of pupal development, the eye–brain complex is largely detached from other structures in the pupa as it undergoes metamorphosis. The dissection is easy and live imaging can be performed in a closed perfusion chamber (18). The first successful brain culture of the developing pupal brain was performed by Gibbs and Truman (23). Live imaging of development is difficult and few examples are available. Recent developments for the imaging of the wing disc may be helpful (15) as are optimized techniques for culture media and perfusion (18, 24).

1.2.6. Adult Brain

The adult *Drosophila* brain (Fig. 2c) is currently at the forefront of the quest to unravel neural circuitry in vivo. The adult fly brain is only a little bigger than the whole-mount embryo; hence, most imaging techniques for the embryo are applicable to the adult brain. There are a plethora of genetic driver and expression probes available to image the circuit function at the cellular and subcellular level. The main focus of the last few years has been on the olfactory and visual systems as well as central brain structures implicated in learning and memory, especially the mushroom bodies. With the possible exception of the first optic neuropil, the lamina and the glomeruli of the olfactory lobes, most adult brain structures of interest require some working distance and capability to perform high-resolution scanning at depths greater than 20 μm. Indeed, successful live imaging of neuronal activity in the olfactory lobe has been made possible by multiphoton microscopy (19, 20, 25). Although resonant scanning should theoretically perform similarly in this system, we are not aware of a comparable study. In addition, it is likely that STED and PALM/STORM will prove very useful in the study of neural circuitry in the adult brain, as both allow superresolution imaging with an increased working distance compared to conventional confocal microscopy (3). The visual system has proven especially useful with respect to subcellular resolution imaging of neurons in vivo. This is mostly because photoreceptor neurons have large and easily accessible cell bodies in the developing eye epithelium and require little working distance. Similarly, photoreceptor synaptic terminals are comparably large (cylindric shape of 1 μm diameter and more than 10 μm length). These synaptic terminals require imaging between 10 and 30 μm deep inside tissue in an intact brain preparation. However, a special eye preparation with the first optic neuropil (but not the remaining optic lobe) attached allows imaging of live photoreceptor terminals with less than 5 μm tissue depth (17, 18). In addition, numerous genetic tools are available for the photoreceptor-specific expression of fluorescent subcellular probes. Finally, both the photoreceptors and their postsynaptic targets can easily be genetically manipulated (26–28). Similar tools are available for the manipulation of the olfactory lobes and other brain structures.

2. Working with Fluorescent Reporters in Neural Circuit Preparations

Fluorescent reporters can be used to assay the size and location of subcellular compartments, the spatial and temporal dynamics of a compartment and the characteristics of the subcellular environment such as pH. This section includes reporters that have either proven useful in *Drosophila* preparations or suggest themselves for experiments in neural circuit preparations based on experiments in other systems (Table 1). We do not provide detailed protocols for the precise methods for implementing each technology. Instead, we focus on the key features and practical information for the application of the different fluorescent reporters in imaging neural circuit preparations.

2.1. Targeted Labeling

The following fluorophores have a single absorption/emission spectrum (i.e., nonphotoconvertible, nonphotoactivatable) and are used to tag proteins of interest with the primary purpose of determining protein localization. We focus on relatively new or otherwise special probes and do not discuss commonly used xFP-type fluorescent proteins.

2.1.1. mKate2

Most standard confocal microscope setups include a far-red laser (e.g., HeNe 633 nm) for the visualization of far-red fluorescent probes. Commonly used fluorescent probes in the far-red spectrum include Cy5™-conjugated antibodies and the nuclear dye Toto-3™. Together with blue laser excitation (GFP range) and green laser excitation (RFP range) the far-red spectrum is the most common choice for simultaneous imaging of a third channel. Nonetheless, the development of genetically encoded far-red fluorescent tags has been slow. For several years mPlum served as far-red fluorophore, although its excitation maximum of 590 nm (emission max. at 649 nm) is red-shifted by a large amount and its quantum yield and photostability are inferior to those of most xFPs.

Table 1
Selected Fluorescent Reporters for the Analysis of Neural Circuit Preparations

Sections	Class	Fluorophore	Activation wavelength	Reversible	Before activation Excitation max (nm)	Before activation Emission max (nm)	After activation Excitation max (nm)	After activation Emission max (nm)	Primary mode of introduction	Useful imaging technologies	References
2.1		mKate2	n/a	n/a	n/a	n/a	588	633	Gene	Conv, STED	(29)
		EBFP2	n/a	n/a	n/a	n/a	383	448	Gene	Conv	(30)
	Biarsenical	FlAsH	n/a	n/a	n/a	n/a	510	535	M-P tag	Conv	Invitrogen.com
		ReAsH	n/a	n/a	n/a	n/a	593	607	M-P tag	Conv	
		KillerRed	n/a	n/a	n/a	n/a	585	610	Gene	Conv	(36)
2.2	Photoconvertable	Phamret	405	No	458	480	458	520	Gene	Conv	(38)
		Dendra2	405, 488	No	490	507	553	573	Gene	Conv	(39)
	Photoactivatable	PAmCherry	405	No	n/a	n/a	564	595	Gene	PALM/STORM	(41)
		Dronpa	405	Yes	n/a	n/a	503	518	Gene	PALM/STORM	(42)
		bsDronpa	405	Yes	n/a	n/a	460	504	Gene	PALM/STORM	(45)
	Cyanine dyes	Cy5	350–570	Yes	n/a	n/a	647	665	IHC	PALM/STORM	(46)
		Cy7	350–570	Yes	n/a	n/a	746	773	IHC	PALM/STORM	
	Rhodamine amides	SRA545	375	Yes	n/a	n/a	Green	545	IHC	PALM/STORM	(47)
		SRA552	375	Yes	n/a	n/a	Green	552	IHC	PALM/STORM	
		SRA577	375	Yes	n/a	n/a	Green	577	IHC	PALM/STORM	
		SRA617	375	Yes	n/a	n/a	Green	617	IHC	PALM/STORM	
2.3	Quantum dots	Various	n/a	n/a	n/a	n/a	Various	Various	Endo	Conv, 2-Photon, STED	Invitrogen.com
		pHrodo	n/a	n/a	n/a	n/a	560	585	M-P dye	Conv	Invitrogen.com
	FM dyes	FM 1-43	n/a	n/a	n/a	n/a	510	626	M-P dye	Conv, 2-Photon	Invitrogen.com
		FM 4-64	n/a	n/a	n/a	n/a	558	734	M-P dye	Conv	

(continued)

Table 1
(continued)

					Before activation		After activation				
Sections	Class	Fluorophore	Activation wavelength	Reversible	Excitation max (nm)	Emission max (nm)	Excitation max (nm)	Emission max (nm)	Primary mode of introduction	Useful imaging technologies	References
2.4	Lysotracker	Blue DND-22	n/a	n/a	n/a	n/a	373	422	M-P dye	Conv	Invitrogen.com
		Green DND-26	n/a	n/a	n/a	n/a	504	511	M-P dye	Conv, 2-Photon	
		Red DND-99	n/a	n/a	n/a	n/a	577	590	M-P dye	Conv	
	Lysosensor	Blue DND-167	n/a	n/a	n/a	n/a	373	425	M-P dye	Conv	Invitrogen.com
		Green DND-189	n/a	n/a	n/a	n/a	443	505	M-P dye	Conv, 2-Photon	
		pHluorin	n/a	n/a	n/a	n/a	475	508	Gene	Conv	(66)
		HyPer	n/a	n/a	n/a	n/a	500	516	Gene	Conv, 2-Photon	(67)
		GCaMP3	n/a	n/a	n/a	n/a	490	520	Gene	Conv, 2-Photon	(85)
		CuFL	n/a	n/a	n/a	n/a	503	525	Gene	Conv, 2-Photon	(73)

Gene genetically encoded; *M-P tag* membrane-permeable tag; *IHC* immunohistochemistry; *Endo* endocytosis; *M-P dye* membrane permeable dye mKate2 is a monomeric, bright, and very photostable genetically encoded far-red fluorophore that represents an improved version of the previously developed TagFP635. It is reportedly threefold brighter than TagFP635 and tenfold brighter than mPlum (29). Expression of this protein has been demonstrated in *Xenopus* embryos and in mammalian cell lines but not to our knowledge in *Drosophila*. We have previously generated transgenic flies expressing proteins tagged with TagFP635. In our hands, this probe is sufficiently bright in the far-red spectrum, but exhibited significant overlap with probes in the green laser/red emission channel. The reported excitation maximum for mKate2 is 588 nm, with an emission maximum at 633 nm. Like mPlum, this probe is therefore best excited with an orange laser, but not a far-red (633 nm) laser. At this point, the development of a true far-red fluorescent tag is still outstanding.

2.1.2. EBFP2

Fluorophores that can be excited with ultraviolet lasers and fluoresce in the blue spectrum allow to add channels using shorter (blue-shifted) wavelengths in conjunction with fluorophores in the main visible spectrum (GFP-RFP range). The blue fluorescent protein EBFP2 has an excitation peak at 383 nm and maximum emission at 448 nm. It is therefore ideal for simultaneous imaging with GFP and higher wavelength fluorophores (30). However, EBFP2 forms weak dimers and is therefore of only limited use as a protein tag. Expression of myr-EBFP2 has been demonstrated in *Drosophila* neurons (31).

2.1.3. FlAsH and ReAsH

Bulky genetically encoded fluorophores can interfere with the function or localization of proteins. A possible solution is the use of small-molecule dyes that associate with high affinity with a short, nonbulky genetically encoded tetracysteine motif Cys–Cys–Pro–Gly–Cys–Cys. FlAsH and ReAsH are biarsenical compounds that must be added exogenously to the preparation (32). These small molecules are easily dissolved in culture media and diffuse freely across membranes. This allows the imaging of protein localization in vivo using small tags and a small fluorescent molecule. However, this technique relies on diffusion of the fluorescent molecule into cells that are directly exposed to the culture medium. FlAsH maximally excites at 508 nm and maximally emits at 528 nm. ReAsH maximally excites at 593 nm and emits at 608 nm.

An additional key use for FlAsH labeling is acute inactivation of the protein associated with the FlAsH molecule, a technology termed FlAsH-FALI (fluorophore-assisted light inactivation). At the *Drosophila* NMJ, the technique has been used with success to assay an endocytic function of Synaptotagmin during the synaptic vesicle cycle (33, 34). This technology was recently applied using recombineering-mediated insertion of the tetracysteine motif into a gene locus within a large genomic fragment, thereby eliminating the problems associated with over-expression of a construct via the Gal4/UAS system (33, 35). For more information about FlAsH-FALI, see Chap. 6.

2.1.4. KillerRed

KillerRed is a genetically encoded red fluorescent fluorophore that has been selected for maximal production of reactive oxygen species (ROS) (36). Light-induced reactive oxygen production leads to a dosage-dependent inactivation of neighboring proteins and subsequently the cell death. KillerRed forms dimers, limiting its use as a protein tag. Successful KillerRed-induced cell ablation has recently been reported in neural circuits of zebrafish (37). We have generated *Drosophila* strains for the expression of KillerRed, including expression of cytosolic KillerRed (UAS-KillerRed) as well as tagged neuronal intracellular trafficking proteins neuronal Synaptobrevin and V100 (unpublished data and (17)). All probes serve as excellent red fluorescent probes in *Drosophila* preparation. However, we have so far not succeeded in effecting any phototoxic effect with either green light or laser activation in *Drosophila*.

2.2. Photoactivation/ Photoconversion

Photoactivatable fluorescent proteins (PAFPs) are nonfluorescent until stimulated by an activating wavelength of defined intensity. After activation, PAFPs exhibit specific excitation/emission spectra. Photoconvertible FPs (PCFPs) exhibit a particular excitation/emission spectrum until excited by a specific wavelength of light that results in a shift of the emission maximum. The following PAFPs and PCFPs are applicable for live imaging and superresolution fluorescent microscopy.

2.2.1. Phamret

Phamret is an acronym for Photoactivation-mediated resonance energy transfer. This probe couples PA-GFP to a high-performance ECFP variant through a two amino acid linker. It is a PCFP that can be excited at the pre- and postphotoconverted state with the same excitation maximum at 458 nm, resulting in cyan fluorescence before photoconversion and PA-GFP emission after conversion. Photoactivation is effected with 405-nm illumination to evoke FRET between the ECFP moiety and activated PA-GFP. After photoactivation, Phamret exhibits green fluorescence with an emission maximum at 520 nm. One of the advantages of Phamret is the use of a single excitation wavelength before and after photoactivation. Since only one laser is required for imaging, this PCFP can be used to determine protein diffusion kinetics up to 100 $\mu m/s^2$ (38). Phamret has been successfully imaged in mammalian cell culture but not to our knowledge in *Drosophila*. Note that the photoconversion only effects a 15-fold change between the two fluorescent states (compare to ~4,000-fold for Dendra2, see below).

2.2.2. Dendra2

Dendra2 is a genetically encoded, monomeric green-to-red fluorophore. It is an improved variant of the original Dendra (39) with increased brightness both before and after photoconversion.

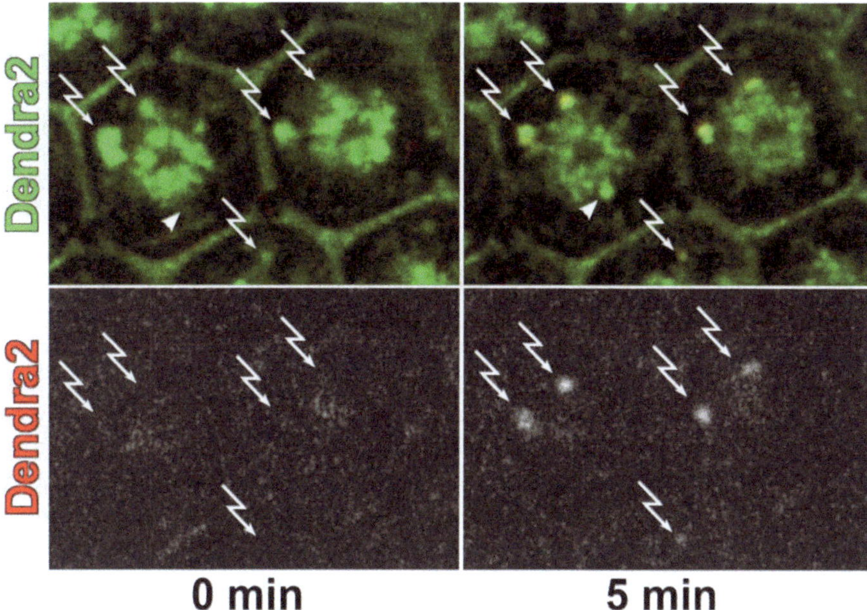

Fig. 3. Live imaging and photoconversion of Dendra2-marked intracellular compartments in the developing *Drosophila* visual system. A Dendra2-n-Syb fusion protein was expressed in the *Drosophila* visual system using GMR-Gal4. Shown is the live preparation of a *Drosophila* eye disc where the *green* and *red* fluorescent spectra are scanned simultaneously. Before photoactivation at 0 min no discernable signal is apparent in the red channel. Over the time course of 5 min, five individual *green* intracellular compartments are photoconverted using UV spot illumination of 10–50 ms (marked by *arrows*). A scan of the same section shows the live scan of these photoconverted compartments in the red channel. The *arrowhead* indicates a new compartment that formed/moved into the focal plane during the 5-min scanning period.

Before photoconversion, Dendra2 is a monomeric fluorescent protein with an excitation maximum at 490 nm and an emission maximum at 507 nm. Dendra2 is designed for photoconversion with both UV as well as normal blue laser (488 nm) illumination (39). After photoconversion, Dendra2 exhibits an excitation maximum of 553 nm and maximal emission at 573 nm. We have generated transgenic flies for the expression of Dendra2, both cytosolic (UAS-Dendra2) as well as tagged intracellular neuronal markers (UAS Dendra2-n-Syb and UAS Dendra2-v100). In our hands, photoconversion with 488 nm laser illumination using spot illumination and 400 Hz conventional or 8,000 Hz resonant laser illumination all lead to Dendra2 bleaching without significant photoconversion. In contrast, photoconversion using 405 nm spot illumination in the millisecond range yields robust Dendra2 photoconversion in *Drosophila* eye–brain preparations (Fig. 3). Notably, Dendra2 and Phamret (above) can be used to simultaneously assay the spatial dynamics of distinct intracellular compartments (40).

2.2.3. PAmCherry

PAmCherry is a recently developed genetically encoded, monomeric red PAFP. It is initially nonfluorescent and can be photoactivated by UV irradiation. PAmCherry is bright enough for use with PALM and has spectral characteristics that allow two-color PALM by simultaneously imaging with a green PAFP (41). The excitation/emission maxima are at 564/595 nm. So far, expression has only been demonstrated in mammalian cell culture.

2.2.4. Dronpa Variants

Dronpa is a genetically encoded, monomeric, PAFP with excitation and emission characteristics similar to those of GFP, but a quantum yield that is 40% higher than that of EGFP (42). Unlike GFP, Dronpa must be activated by irradiation at 405 nm, after which excitation at 488 nm both stimulates fluorescence and deactivates the fluorophore. Further, Dronpa can be reactivated/deactivated multiple times with minimal loss in fluorescence. Dronpa2 and Dronpa3 are half as bright as EGFP; however, they both have greatly enhanced kinetics for both activation and deactivation. This enhancement led to the advent of stroboscopic (S)-PALM, a form of PALM that dramatically shortens data acquisition time (43, 44). bsDronpa has blue-shifted excitation/emission characteristics. This Dronpa variant is activated with 405 nm laser light with excitation/emission maxima at 460 nm and 504 nm, respectively (45). Expression of Dronpa has been demonstrated in mammalian and *Drosophila* S2 cell cultures, but has yet to be demonstrated in vivo in the fly.

2.2.5. Cyanine Dyes and Rhodamine Amides

Although genetically encoded PA fluorophores have many advantages, several recently developed photoactivatable fluorescent small molecules have promising potential for use in superresolution fluorescent microscopy, especially in fixed preparations. These include the cyanine dyes (46) and rhodamine amides (47), both of which are photoactivatable and can be fused to a secondary antibody for use in immunohistochemistry. Applications may include antibody internalization, as described in the next section.

2.3. Extracellular Labeling and Endocytosis

2.3.1. Antibody Internalization

Pulse-chase experiments with specific antibodies are a common method employed to measure receptor endocytosis and intracellular trafficking. The basic idea is to add an antibody against a specific membrane protein or ligand to a live culture. Endocytosis rate and kinetics as well as downstream trafficking (chase) can be measured quantitatively, because the time point and the duration of antibody application (pulse) are defined. The tissue can be fixed after defined time periods to label the internalized antibody with a fluorophore-conjugated secondary antibody as well as other antibodies. For a live imaging variant the primary antibody needs to be conjugated to a fluorophore directly. A detailed protocol has been published for imaging receptor-mediated endocytosis in motor neurons at the *Drosophila* larval NMJ (48). This protocol has been used to study the trafficking of several receptors, including Fasciclin II, Frizzled-2, and Wntless (Wls/Evi) (49–52). Notch receptor endocytosis has been measured in wing discs (53).

2.3.2. Quantum Dots

Quantum dots are nanometer-sized crystals that function as semiconductors. Quantum dots have fluorescent properties that depend on the size and shape of the crystal. Since inorganic semiconductors are toxic and insoluble, the quantum dot core is coated with an amphiphilic material, commonly polyethylene glycol (PEG) (54). Additionally, the polymer coating can be conjugated with tags and proteins of interest. Quantum dots have unique advantages for intracellular labeling; compared to most organic dyes, quantum dots are roughly 20 times as bright, are 100 times as photostable, and have a much narrower emission spectrum, which improves fluorescent isolation in multichannel recordings (55). On the negative side, the bulkiness of quantum dots makes it difficult to label intracellular proteins without interfering with function. Recent technological developments have facilitated access to the cytoplasm (56–58). Despite this limitation, the spectral properties of quantum dots confer the ability to image single molecules on the surface of a cell using conventional fluorescence imaging techniques (59). Additionally, the fate of internalized compartments following a receptor-mediated endocytosis event can be monitored for long periods of time in live culture (60). Indeed, research using quantum dots has led to several recent advances in cell biology (61) and synaptic biology (62).

Quantum dots are currently available in a variety of colors. Commercially available Qdots® are available preconjugated with biological molecules designed for protein labeling. The following list includes some Qdots that can be applied to monitor the fate of cell-surface membrane proteins and endocytosed compartments in vivo.

- *Anti-GST*: This Qdot specifically binds the commonly used GST protein tag. In practice, anti-GST Qdots may be helpful for endocytosis experiments using extracellularly GST-tagged protein in live cultures.
- *Secondary antibody*: Qdots are available in a variety of colors and are conjugated with affinity-purified, highly cross-absorbed anti-mouse, rabbit, rat, chicken, or goat antibodies.
- *Amine-derivatized PEG*: This Qdot is sold with a kit that includes the materials and instructions necessary to covalently label this Qdot with primary antibodies. Live imaging of endocytosed Qdots fused to primary antibodies has been demonstrated in rat tumor cells (63).

2.3.3. pHrodo™ Dyes

pHrodo dyes are commercially available derivates of rhodamine and exhibit increasing fluorescence with decreasing pH. A variety of pHrodo dye conjugates are available; a 10-KDa dextran bead conjugate is useful for the in vivo tracking of endocytosed compartments along the endolysosomal pathway. These red fluorescent small molecules are nonfluorescent when added to tissue culture media at the manufacturer's recommended concentration but become increasingly fluorescent as they are endocytosed and trafficked to lysosomal compartments. Imaging can commence as soon as the dye is added to the tissue culture, since pHrodo dye remaining in the extracellular solution is invisible relative to endocytosed, acidified compartments. This dye could be used in conjunction with the internalization and trafficking assays described above for Qdots.

2.3.4. FM Dyes

FM dyes are lipophilic styryl compounds that are added exogenously to the medium of a live preparation where they quickly incorporate into membranes. Pulse-chase type experiments with FM dyes are widely used to assay membrane trafficking in a variety of tissues. In particular, FM (re-)uptake experiments have been critical in the study of synaptic vesicle cycling (64), including many studies at the *Drosophila* embryonic and larval NMJ (see Chap. 6 for details).

The most commonly used dyes are FM 1-43, FM 2-10, and FM 4-64, which differ mainly in their fluorescence characteristics, but have also been shown to exhibit different kinetics of membrane labeling. FM dyes are highly water soluble, and there they exhibit little fluorescence. FM dye fluorescence strongly increases upon membrane binding. In practice, the dye is dissolved at low concentrations in culture media, the media is added to a tissue culture, and the dye quickly associates with cell membranes where it can now be visualized by fluorescence microscopy. Vesicle cycling is then stimulated by one of the many available methods and the remaining extracellular dye is washed away. Wash efficiency is greatly improved by applying the compound ADVASEP-7, a small molecule added to the culture media that preferentially binds FM dye and quickly removes it from cell membranes. For further information see ref. (64) and Chap. 6.

2.4. Sensors of the Subcellular Environment

Subcellular conditions include pH, levels of nitric oxide (NO), ROS, and calcium concentration, among many others. Some of these features can be measured at the resolution limit of light for individual subcellular compartments in *Drosophila* neural circuit preparations.

2.4.1. Measuring Intracompartmental pH Using LysoTracker® and LysoSensor™

Most intracellular membrane compartments are acidified to varying degrees. For example, synaptic vesicles use the proton motive force resulting from acidification to load neurotransmitter, while acidification of endosomal compartments is directly implicated in signaling through receptor–ligand dissociation and endosomal maturation.

The two probes LysoTracker and LysoSensor use different and complementary strategies to visualize the pH of intracellular compartments. LysoTracker is a fluorescent probe that is added to the culture medium at such low concentrations that background fluorescence is negligible. We routinely use 100 nM for larval filet preparation as well as larval and pupal eye–brain culture. LysoTracker passively diffuses across membranes and selectively accumulates in highly acidified compartments, including lysosomes and autophagosomes, but not early endosomal structures. Hence, LysoTracker labeling of strongly acidified compartments is a function of time. We measure LysoTracker signal 5 min after application. At later time points LysoTracker becomes unreliable, as it may alter the subcellular environment itself. LysoTracker is available with blue, green, yellow, and red spectral properties. In our hands, the fixation of the live LysoTracker signal for subsequent immunolabeling leads to a substantial loss of signal.

In contrast to LysoTracker, LysoSensor is quenched at neutral pH and becomes increasingly fluorescent as pH decreases, becoming maximally fluorescent at pH 5. In other words, while LysoTracker does not change fluorescent properties while accumulating in acidified compartments, LysoSensor does not accumulate, but changes its fluorescent properties as a function of pH. In its practical application, LysoSensor is added to the culture medium similar to LysoTracker but at higher concentrations (≥1 μM). LysoSensor is available in either blue or green. While use of LysoSensor is a more quantitative way to measure intracompartmental pH, it yields weaker signals in *Drosophila* preparations in our hands.

Notes:

- When using these probes experimentally, results are easier to interpret in genetic mosaics where marked mutant and wild type cells can be used for a quantitative comparison of directly neighboring mutant and control cells. We have made extensive use of LysoTracker in 50/50 MARCM clones (65) in eye–brain cultures (17, 66).
- For whole-brain cultures at any developmental stage, the protective outer membrane must be marginally torn to provide the probes access to cortical cells (66).
- LysoSensor and LysoTracker are also available fused to dextran beads for experiments where an initial endocytosis event is desirable.
- Images should be recorded within the first 5 min after adding these probes to the culture due to a potentially confounding alkalizing effect.

2.4.2. Measuring Intracompartmental pH Using Synapto-pHluorin

pHluorin is a genetically encoded, pH-sensitive GFP variant that exhibits increasing fluorescence intensity with increasing lumenal pH, fluorescing minimally at pH 5 and maximally at pH 8 (67). In functional studies of *Drosophila* neurons, pHluorin fused to the lumenal end of synaptobrevin (synapto-pHluorin) has been used to image synaptic vesicle fusion, an event that results in the neutralization of an otherwise acidified synaptic vesicle and a consequential increase in synapto-pHluorin fluorescence (34). We have recently used synapto-pHluorin in *Drosophila* neurons to measure the lumenal pH of early endosomes in live eye–brain culture by acquiring confocal images before and after neutralization of the lumen (17). By targeting pHluorin-tagged molecules to specific intracellular compartments, lumenal pH can be measured in a targeted manner.

2.4.3. HyPer: Measuring Reactive Oxygen Species In Vivo

This ratiometric, genetically encoded hydrogen peroxide sensor has spectral characteristics similar to YFP and can be tagged to a gene of interest for targeted subcellular localization (68). HyPer effectively senses ROS in vivo without emitting ROS on its own. Expression of this FP has been demonstrated in mammalian cell culture (69) and Zebrafish (70), but not to our knowledge in *Drosophila*.

2.4.4. GCaMP3: The Latest in Calcium-Sensing

Calcium ion concentrations modulate a plethora of cell biological events; for a review of calcium signaling in cell biology, see ref. (71). Calcium influx upon synaptic activation has traditionally been the most powerful approach used to directly visualize and image synaptic activity. GCaMP3 is a green fluorescent, genetically encoded calcium sensor that exhibits increasing fluorescent intensity in direct proportion with increasing calcium ion concentrations. GCaMP3 has been used to report neural activity in model organisms including *Drosophila* (72). However, its application could theoretically extend into the realm of other Ca-dependent cell biological processes by targeting GCaMP3 to a specific subcellular region of interest.

2.4.5. CuFL: A Copper-Based Fluorescent Probe for Nitric Oxide

Nitric oxide (NO) has been implicated in the signaling program of *Drosophila* optic lobe development (23). CuFL, the first direct sensor for NO levels in living cells, was recently introduced and might prove useful in further cell biological studies involving NO (73). This fluorescein-based small molecule diffuses freely across membranes and fluoresces in the green spectrum only in the presence of NO.

3. Examples for Imaging of Intracellular Trafficking in Neural Circuits

Intracellular trafficking underlies many aspects of the development and function of neural circuits. In order to establish meaningful synaptic connections, neurons must present or interpret guidance cues at the right place and time. After connections are made, activity-dependent and -independent modifications require the regulated delivery or removal of channels, receptors, and many other signaling proteins. Finally, neuronal degeneration typically commences with diminished synaptic function and the accumulation of undegraded proteins. In short, intracellular trafficking affects all stages in the life of a neural circuit from development to maintenance to degeneration. In this section, we highlight a few examples that illuminate the importance and application of subcellular resolution imaging for the formation, function, and maintenance of neural circuits.

3.1. Intracellular Trafficking in Circuit Formation

Once regarded as merely passive transport machinery, intracellular vesicle trafficking is now known to play instructive roles in most aspects of developmental biology, including signal transduction, asymmetric cell division, cell fate specification and cell growth. Prominent examples include the regulation of cellular differentiation by endo-/exocytosis of the Notch ligand Delta (reviewed in refs. (74–76)) and the regulation of synaptic plasticity by AMPA receptor trafficking (reviewed in refs. (77, 78)). Comparatively little is known about the function of intracellular trafficking compartments during axon pathfinding, target selection, and synapse formation (79). This is surprising because a general feature of guidance receptors known to mediate synaptic targeting choices is precise spatiotemporal regulation, i.e., they must be presented at the right time and place on the membrane to convey meaningful synapse formation signals during brain wiring. This problem is amplified by the number of guidance receptors or their isoforms that need to be spatiotemporally regulated during brain development.

Perhaps the best characterized example of an instructive role of guidance receptor trafficking during axon targeting is the regulation of the guidance receptor Robo by the endosomal sorting receptor commissureless (80–82). Commissureless (Comm) is required cell-autonomously in ipsilateral pioneer neurons in the *Drosophila* embryo in order to allow midline crossing of these neurons. Robo is the receptor for the repellent Slit. In order to allow midline crossing, Comm temporarily diverts the Robo receptor from the Golgi to the endosomal/lysosomal pathway. Comm thus ensures a precise spatiotemporal developmental program to establish correct neuronal connectivity (82). Live imaging of the trafficking of a Robo-green fluorescent protein (GFP) fusion in living embryos demonstrated that Comm prevents the delivery of Robo-GFP to the growth cone (81).

Similarly, a study on intraaxonal patterning demonstrated a requirement for endocytosis in the spatiotemporal localization of the guidance receptors Robo3 and Derailed (31). Using a live imaging approach with fluorescently labeled receptors that include Fluorescence Recovery After Photobleaching (FRAP), the same study also showed that transmembrane proteins are mobile within their compartment but less mobile at intraaxonal compartment boundaries.

In more general terms, endosomal compartments can function as signaling hubs that control the activation and downregulation of guidance receptors (83, 84). We have recently shown that experimental control and simultaneous imaging of endolysosomal trafficking in eye–brain complexes in vivo provides a means to visualize what guidance receptors are actively "cycled" at a given time in a specific neuronal subcellular domain, e.g., the synapse or cell body (64).

3.2. Intracellular Trafficking in Circuit Function

Since synaptic function is reviewed elsewhere in this book (Chap. 6), we only briefly discuss the key intracellular trafficking events of relevance for subcellular resolution imaging. The key features of synaptic function with respect to imaging are: First, the conduction of an electrical potential, which can be imaged using voltage-dependent dyes and probes not discussed in this chapter. Second, calcium-influx at the synapse provides a fast and reliable readout for both the Ca-sensing machinery that triggers neurotransmitter release as well as the experimentalist imaging synaptic activity (see discussion of GCamP3 (85) in Sect. 2). Third, neurotransmitter release at chemical synapses is regulated by the synaptic vesicle cycle, a large-scale intracellular trafficking machinery that closely intersects with the secretory pathway and endosomal trafficking. Synaptic vesicle exocytosis and endocytosis can be imaged with a variety of powerful genetically encoded or exogenously applied probes. One of the most successfully applied tools for the study of *Drosophila* circuitry function is the genetically encoded exocytic probe synapto-pHlourin (66, 86). Since synapto-pHluorin can be expressed anywhere in the fly nervous system using the Gal4/UAS system, its applicability is limited mostly by the amount and detectability of synchronous fluorescence increase at active synapses in a given circuit. Of more limited use for circuit function are exogenously applied probes including FM dyes. However, any preparation that allows such probes to freely diffuse into the synaptic cleft can turn these probes into powerful assays for synaptic function and underlying intracellular trafficking.

3.3. Intracellular Trafficking in Circuit Degeneration

Numerous subcellular mechanisms that lead to neurodegeneration have been proposed, including axonal transport, protein aggregation, mitochondrial dysfunction, excitotoxicity, and intracellular transport (87–89). Many of these mechanisms have benefited greatly in recent years from the ability to visualize their dynamics at subcellular resolution in neuronal circuit preparations in vivo.

Axonal transport is critical for the cell body to communicate with the cell periphery. The highly polarized morphology and the differential requirement of membrane components in neurons represent a challenge for the trafficking machinery to correctly deliver cargos. Defects in axon transport have been implicated in the neurodegeneration in Alzheimer Disease, amyotrophic lateral sclerosis, and the polyglutamine diseases (90–93). In *Drosophila*, disruption of microtubule motor proteins kinesin1 or dynactin leads to accumulated cargos including vesicles, synaptic membrane proteins, and mitochondria. The axonal swellings ("organelle jams") are thought to block axon transport, resulting in neuromuscular defects and disruption of neuronal organization.

The removal of subcellular "debris" is essential for maintaining functional neurons. Neurons use several approaches for dispose of toxic protein aggregates and damaged organelles. These clearance mechanisms include targeting proteins for proteasomal degradation and transporting substrates such as protein complexes and organelles to lysosomes and autophagosomes for subsequent degradation (94). In recent years *Drosophila* has been employed as a model system to study the basic cell biological machinery underlying the subcellular defects observed in many neurodegenerative diseases. For example, defective lysosomal function in the mutant for the lysosomal sugar carrier *spinster/benchwarmer* (95) or the protective chaperone NMNAT (96) cause neurodegeneration and provide genetic inroads into lysosomal degradation and misfolded protein responses, respectively. We have recently reported a neuron-specific intracellular degradation pathway based on the function of the neuronal v-ATPase subunit V0a1 (17). Loss of v0a1 leads to adult-onset degeneration in photoreceptors and sensitizes neurons to neurotoxic insults, including human tau and Abeta proteins (97). The identification and characterization of this neuronal degradation mechanism was performed in eye–brain live culture and fixed preparations, using many of the techniques described above, including resonant confocal live imaging of Lysotracker, synapto-pHluorin, and other probes.

Autophagy plays a potentially protective role in neurodegeneration. In mammals, knockouts of autophagy-related genes result in intra-neuronal aggregates and neurodegeneration (98, 99). In *Drosophila*, *atg7* mutants display protein aggregation and neuronal degeneration in aged brains, indicating that autophagy plays a neuroprotective role in the CNS (100, 101). Numerous fluorescently tagged autophagy reporters exist, including Atg8-GFP (102).

Acknowledgments

We thank all members of the Hiesinger lab for discussion. We apologize to all of our colleagues whose work was not discussed because of space constraints or our shortcomings. Our work is supported by grants from the Welch Foundation (I-1657), a grant from the

Cancer Prevention Research Institute of Texas (CPRIT) and the National Institute of Health (NEI/NIH RO1018884). PRH is a Eugene McDermott Scholar in Biomedical Research at UT Southwestern Medical Center.

References

1. Schermelleh L, Heintzmann R, Leonhardt H (2010) A guide to super-resolution fluorescence microscopy. J Cell Biol 190:165
2. Pawley J (2006) Handbook of biological confocal microscopy, 3rd edn. Springer, Berlin
3. Hell SW (2007) Far-field optical nanoscopy. Science 316:1153–1158
4. Gibson SF, Lanni F (1992) Experimental test of an analytical model of aberration in an oil-immersion objective lens used in three-dimensional light microscopy. J Opt Soc Am A 9:154–166
5. Richardson WH (1972) Bayesian-based iterative method of image restoration. J Opt Soc Am 62:55–59
6. Holmes TJ (1988) Maximum-likelihood image restoration adapted for non-coherent optical imaging. J Opt Soc Am A5:666–673
7. Hiesinger PR, Scholz M, Meinertzhagen IA, Fischbach KF, Obermayer K (2001) Visualization of synaptic markers in the optic neuropils of *Drosophila* using a new constrained deconvolution method. J Comp Neurol 429:277–288
8. Borlinghaus RT (2006) MRT letter: high speed scanning has the potential to increase fluorescence yield and to reduce photobleaching. Microsc Res Tech 69:689–692
9. Graf R, Rietdorf J, Zimmermann T (2005) Live cell spinning disk microscopy. Adv Biochem Eng Biotechnol 95:57–75
10. Nakano A (2002) Spinning-disk confocal microscopy—a cutting-edge tool for imaging of membrane traffic. Cell Struct Funct 27: 349–355
11. Helmchen F, Denk W (2002) New developments in multiphoton microscopy. Curr Opin Neurobiol 12:593–601
12. Huang B, Babcock H, Zhuang X (2010) Breaking the diffraction barrier: super-resolution imaging of cells. Cell 143:1047–1058
13. Betzig E, Patterson GH, Sougrat R, Lindwasser OW, Olenych S, Bonifacino JS, Davidson MW, Lippincott-Schwartz J, Hess HF (2006) Imaging intracellular fluorescent proteins at nanometer resolution. Science 313:1642–1645
14. Zhuang X (2009) Nano-imaging with storm. Nat Photonics 3:365–367
15. Aldaz S, Escudero LM, Freeman M (2010) Live imaging of *Drosophila* imaginal disc development. Proc Natl Acad Sci U S A 107:14217–14222
16. Brink D, Gilbert M, Auld V (2009) Visualizing the live *Drosophila* glial-neuromuscular junction with fluorescent dyes. J Vis Exp 27:1154
17. Williamson W, Wang D, Haberman A, Hiesinger P (2010) A dual function of V0-ATPase a1 provides an endolysosomal degradation mechanism in *Drosophila melanogaster* photoreceptors. J Cell Biol 189:885
18. Williamson WR, Hiesinger PR (2010) Preparation of developing and adult *Drosophila* brains and retinae for live imaging. J Vis Exp 37:1936
19. Semmelhack JL, Wang JW (2009) Select *Drosophila* glomeruli mediate innate olfactory attraction and aversion. Nature 459:218–223
20. Wang JW, Wong AM, Flores J, Vosshall LB, Axel R (2003) Two-photon calcium imaging reveals an odor-evoked map of activity in the fly brain. Cell 112:271–282
21. Featherstone DE, Chen K, Broadie K (2009) Harvesting and preparing *Drosophila* embryos for electrophysiological recording and other procedures. J Vis Exp 27:1347
22. Kittel RJ, Wichmann C, Rasse TM, Fouquet W, Schmidt M, Schmid A, Wagh DA, Pawlu C, Kellner RR, Willig KI, Hell SW, Buchner E, Heckmann M, Sigrist SJ (2006) Bruchpilot promotes active zone assembly, Ca2+ channel clustering, and vesicle release. Science 312:1051–1054
23. Gibbs S, Truman J (1998) Nitric oxide and cyclic GMP regulate retinal patterning in the optic lobe of *Drosophila*. Neuron 20:83–93
24. Ayaz D, Leyssen M, Koch M, Yan J, Srahna M, Sheeba V, Fogle KJ, Holmes TC, Hassan BA (2008) Axonal injury and regeneration in the adult brain of *Drosophila*. J Neurosci 28:6010–6021
25. Suh GS, Wong AM, Hergarden AC, Wang JW, Simon AF, Benzer S, Axel R, Anderson DJ (2004) A single population of olfactory sensory neurons mediates an innate avoidance behaviour in *Drosophila*. Nature 431:854–859
26. Stowers RS, Schwarz TL (1999) A genetic method for generating *Drosophila* eyes

composed exclusively of mitotic clones of a single genotype. Genetics 152:1631–1639

27. Newsome TP, Asling B, Dickson BJ (2000) Analysis of *Drosophila* photoreceptor axon guidance in eye-specific mosaics. Development 127:851–860
28. Chotard C, Leung W, Salecker I (2005) glial cells missing and gcm2 cell autonomously regulate both glial and neuronal development in the visual system of *Drosophila*. Neuron 48:237–251
29. Shcherbo D, Murphy C, Ermakova G, Solovieva E, Chepurnykh T, Shcheglov A, Verkhusha V, Pletnev V, Hazelwood K, Roche P (2009) Far-red fluorescent tags for protein imaging in living tissues. Biochem J 418:567
30. Ai H, Shaner N, Cheng Z, Tsien R, Campbell R (2007) Exploration of new chromophore structures leads to the identification of improved blue fluorescent proteins. Biochemistry 46:5904–5910
31. Katsuki T, Ailani D, Hiramoto M, Hiromi Y (2009) Intra-axonal patterning: intrinsic compartmentalization of the axonal membrane in *Drosophila* neurons. Neuron 64:188–199
32. Marks K, Nolan G (2006) Chemical labeling strategies for cell biology. Nat Methods 3: 591–596
33. Marek K, Davis G (2002) Transgenically encoded protein photoinactivation (FlAsH-FALI): acute inactivation of synaptotagmin I. Neuron 36:805–813
34. Poskanzer K, Marek K, Sweeney S, Davis G (2003) Synaptotagmin I is necessary for compensatory synaptic vesicle endocytosis in vivo. Nature 426:559–563
35. Venken K, Kasprowicz J, Kuenen S, Yan J, Hassan B, Verstreken P (2008) Recombineering-mediated tagging of *Drosophila* genomic constructs for in vivo localization and acute protein inactivation. Nucleic Acids Res 36:e114
36. Bulina ME, Chudakov DM, Britanova OV, Yanushevich YG, Staroverov DB, Chepurnykh TV, Merzlyak EM, Shkrob MA, Lukyanov S, Lukyanov KA (2006) A genetically encoded photosensitizer. Nat Biotechnol 24:95–99
37. Del Bene F, Wyart C, Robles E, Tran A, Looger L, Scott EK, Isacoff EY, Baier H (2010) Filtering of visual information in the tectum by an identified neural circuit. Science 330:669–673
38. Matsuda T, Miyawaki A, Nagai T (2008) Direct measurement of protein dynamics inside cells using a rationally designed photoconvertible protein. Nat Methods 5:339–345
39. Gurskaya N, Verkhusha V, Shcheglov A, Staroverov D, Chepurnykh T, Fradkov A, Lukyanov S, Lukyanov K (2006) Engineering of a monomeric green-to-red photoactivatable fluorescent protein induced by blue light. Nat Biotechnol 24:461–465
40. Chudakov D, Lukyanov S, Lukyanov K (2007) Tracking intracellular protein movements using photoswitchable fluorescent proteins PS-CFP2 and Dendra2. Nat Protoc 2:2024–2032
41. Subach F, Patterson G, Manley S, Gillette J, Lippincott-Schwartz J, Verkhusha V (2009) Photoactivatable mCherry for high-resolution two-color fluorescence microscopy. Nat Methods 6:153–159
42. Ando R, Mizuno H, Miyawaki A (2004) Regulated fast nucleocytoplasmic shuttling observed by reversible protein highlighting. Science 306:1370
43. Ando R, Flors C, Mizuno H, Hofkens J, Miyawaki A (2007) Highlighted generation of fluorescence signals using simultaneous two-color irradiation on Dronpa mutants. Biophys J 92:L97–L99
44. Flors C, Hotta J, Uji-i H, Dedecker P, Ando R, Mizuno H, Miyawaki A, Hofkens J (2007) A stroboscopic approach for fast photoactivation-localization microscopy with Dronpa mutants. J Am Chem Soc 129:13970–13977
45. Andresen M, Stiel A, Fölling J, Wenzel D, Schönle A, Egner A, Eggeling C, Hell S, Jakobs S (2008) Photoswitchable fluorescent proteins enable monochromatic multilabel imaging and dual color fluorescence nanoscopy. Nat Biotechnol 26:1035–1040
46. Bates M, Huang B, Dempsey G, Zhuang X (2007) Multicolor super-resolution imaging with photo-switchable fluorescent probes. Science 317:1749
47. Bossi M, Folling J, Belov V, Boyarskiy V, Medda R, Egner A, Eggeling C, Schonle A, Hell S (2008) Multicolor far-field fluorescence nanoscopy through isolated detection of distinct molecular species. Nano Lett 8: 2463–2468
48. Ramachandran P, Budnik V (2010) Internalization and trafficking assay for *Drosophila* larvae. Cold Spring Harb Protoc 2010(8):pdb.prot5472
49. Ataman B, Ashley J, Gorczyca D, Gorczyca M, Mathew D, Wichmann C, Sigrist S, Budnik V (2006) Nuclear trafficking of *Drosophila* Frizzled-2 during synapse development requires the PDZ protein dGRIP. Proc Natl Acad Sci U S A 103(20):7841–7846
50. Korkut C, Ataman B, Ramachandran P, Ashley J, Barria R, Gherbesi N, Budnik V (2009) Trans-synaptic transmission of vesicular Wnt

signals through Evi/Wntless. Cell 139: 393–404

51. Mathew D, Popescu A, Budnik V (2003) *Drosophila* amphiphysin functions during synaptic Fasciclin II membrane cycling. J Neurosci 23:10710
52. Mathew D, Ataman B, Chen J, Zhang Y, Cumberledge S, Budnik V (2005) Wingless signaling at synapses is through cleavage and nuclear import of receptor DFrizzled2. Science 310:1344
53. Maitra S, Kulikauskas RM, Gavilan H, Fehon RG (2006) The tumor suppressors Merlin and Expanded function cooperatively to modulate receptor endocytosis and signaling. Curr Biol 16(7):702–709
54. Walling M, Novak J, Shepard J (2009) Quantum dots for live cell and in vivo imaging. Int J Mol Sci 10:441
55. Chan W, Nie S (1998) Quantum dot bioconjugates for ultrasensitive nonisotopic detection. Science 281:2016
56. Jablonski A, Humphries W IV, Payne C (2008) Pyrenebutyrate-mediated delivery of quantum dots across the plasma membrane of living cells. J Phys Chem B 113:405–408
57. Kim B, Jiang W, Oreopoulos J, Yip C, Rutka J, Chan W (2008) Biodegradable quantum dot nanocomposites enable live cell labeling and imaging of cytoplasmic targets. Nano Lett 8:3887–3892
58. Delehanty J, Mattoussi H, Medintz I (2009) Delivering quantum dots into cells: strategies, progress and remaining issues. Anal Bioanal Chem 393:1091–1105
59. Jaqaman K, Loerke D, Mettlen M, Kuwata H, Grinstein S, Schmid S, Danuser G (2008) Robust single-particle tracking in live-cell time-lapse sequences. Nat Methods 5: 695–702
60. Cui B, Wu C, Chen L, Ramirez A, Bearer E, Li W, Mobley W, Chu S (2007) One at a time, live tracking of NGF axonal transport using quantum dots. Proc Natl Acad Sci U S A 104:13666
61. Pinaud F, Clarke S, Sittner A, Dahan M (2010) Probing cellular events, one quantum dot at a time. Nat Methods 7:275–285
62. Triller A, Choquet D (2008) New concepts in synaptic biology derived from single-molecule imaging. Neuron 59:359–374
63. Tada H, Higuchi H, Wanatabe T, Ohuchi N (2007) In vivo real-time tracking of single quantum dots conjugated with monoclonal anti-HER2 antibody in tumors of mice. Cancer Res 67:1138
64. Gaffield M, Betz W (2007) Imaging synaptic vesicle exocytosis and endocytosis with FM dyes. Nat Protoc 1:2916–2921
65. Lee T, Luo L (1999) Mosaic analysis with a repressible cell marker for studies of gene function in neuronal morphogenesis. Neuron 22:451–461
66. Williamson WR, Yang T, Terman JR, Hiesinger PR (2010) Guidance receptor degradation is required for neuronal connectivity in the *Drosophila* nervous system. PLoS Biol 8:e1000553
67. Miesenböck G, De Angelis D, Rothman J (1998) Visualizing secretion and synaptic transmission with pH-sensitive green fluorescent proteins. Nature 394:192–195
68. Belousov V, Fradkov A, Lukyanov K, Staroverov D, Shakhbazov K, Terskikh A, Lukyanov S (2006) Genetically encoded fluorescent indicator for intracellular hydrogen peroxide. Nat Methods 3:281–286
69. Wu R, Ma Z, Liu Z, Terada L (2010) Nox4-derived H_2O_2 mediates endoplasmic reticulum signaling through local Ras activation. Mol Cell Biol 30:3553
70. Niethammer P, Grabher C, Look A, Mitchison T (2009) A tissue-scale gradient of hydrogen peroxide mediates rapid wound detection in zebrafish. Nature 459:996–999
71. Clapham D (2007) Calcium signaling. Cell 131:1047–1058
72. Seelig J, Chiappe M, Lott G, Dutta A, Osborne J, Reiser M, Jayaraman V (2010) Two-photon calcium imaging from head-fixed *Drosophila* during optomotor walking behavior. Nat Methods 7(7):535–540
73. Lim M, Xu D, Lippard S (2006) Visualization of nitric oxide in living cells by a copper-based fluorescent probe. Nat Chem Biol 2:375–380
74. Fischer JA, Eun SH, Doolan BT (2006) Endocytosis, endosome trafficking, and the regulation of *Drosophila* development. Annu Rev Cell Dev Biol 22:181–206
75. Chitnis A (2006) Why is delta endocytosis required for effective activation of notch? Dev Dyn 235:886–894
76. Le Borgne R, Bardin A, Schweisguth F (2005) The roles of receptor and ligand endocytosis in regulating Notch signaling. Development 132:1751–1762
77. Greger IH, Esteban JA (2007) AMPA receptor biogenesis and trafficking. Curr Opin Neurobiol 17(3):287–297
78. Nicoll RA, Tomita S, Bredt DS (2006) Auxiliary subunits assist AMPA-type glutamate receptors. Science 311:1253–1256

79. Chan CC, Epstein D, Hiesinger PR (2011) Intracellular trafficking in *Drosophila* visual system development: a basis for pattern formation through simple mechanisms. Dev Neurosci 71(12):1227–1245
80. Keleman K, Rajagopalan S, Cleppien D, Teis D, Paiha K, Huber LA, Technau GM, Dickson BJ (2002) Comm sorts robo to control axon guidance at the *Drosophila* midline. Cell 110:415–427
81. Keleman K, Ribeiro C, Dickson BJ (2005) Comm function in commissural axon guidance: cell-autonomous sorting of Robo in vivo. Nat Neurosci 8:156–163
82. Dickson BJ, Gilestro GF (2006) Regulation of commissural axon pathfinding by slit and its Robo receptors. Annu Rev Cell Dev Biol 22:651–675
83. O'Donnell M, Chance RK, Bashaw GJ (2009) Axon growth and guidance: receptor regulation and signal transduction. Annu Rev Neurosci 32:383–412
84. Seto ES, Bellen HJ, Lloyd TE (2002) When cell biology meets development: endocytic regulation of signaling pathways. Genes Dev 16:1314–1336
85. Tian L, Hires SA, Mao T, Huber D, Chiappe ME, Chalasani SH, Petreanu L, Akerboom J, McKinney SA, Schreiter ER, Bargmann CI, Jayaraman V, Svoboda K, Looger LL (2009) Imaging neural activity in worms, flies and mice with improved GCaMP calcium indicators. Nat Methods 6:875–881
86. Ng M, Roorda RD, Lima SQ, Zemelman BV, Morcillo P, Miesenbock G (2002) Transmission of olfactory information between three populations of neurons in the antennal lobe of the fly. Neuron 36:463–474
87. Lu B, Vogel H (2009) *Drosophila* models of neurodegenerative diseases. Ann Rev Pathol 4:315–342
88. Lessing D, Bonini N (2009) Maintaining the brain: insight into human neurodegeneration from *Drosophila melanogaster* mutants. Nat Rev Genet 10:359–370
89. De Vos K, Grierson A, Ackerley S, Miller C (2008) Role of axonal transport in neurodegenerative diseases. Neuroscience 31:151
90. Williamson TL, Cleveland DW (1999) Slowing of axonal transport is a very early event in the toxicity of ALS-linked SOD1 mutants to motor neurons. Nat Neurosci 2:50–56
91. LaMonte BH, Wallace KE, Holloway BA, Shelly SS, Ascano J, Tokito M, Van Winkle T, Howland DS, Holzbaur EL (2002) Disruption of dynein/dynactin inhibits axonal transport in motor neurons causing late-onset progressive degeneration. Neuron 34:715–727
92. Puls I, Jonnakuty C, LaMonte BH, Holzbaur EL, Tokito M, Mann E, Floeter MK, Bidus K, Drayna D, Oh SJ, Brown RH Jr, Ludlow CL, Fischbeck KH (2003) Mutant dynactin in motor neuron disease. Nat Genet 33:455–456
93. Gunawardena S, Goldstein LS (2005) Polyglutamine diseases and transport problems: deadly traffic jams on neuronal highways. Arch Neurol 62:46–51
94. Garcia-Arencibia M, Hochfeld WE, Toh PP, Rubinsztein DC (2010) Autophagy, a guardian against neurodegeneration. Semin Cell Dev Biol 21:691–698
95. Dermaut B, Norga KK, Kania A, Verstreken P, Pan H, Zhou Y, Callaerts P, Bellen HJ (2005) Aberrant lysosomal carbohydrate storage accompanies endocytic defects and neurodegeneration in *Drosophila* benchwarmer. J Cell Biol 170:127–139
96. Zhai RG, Cao Y, Hiesinger PR, Zhou Y, Mehta SQ, Schulze KL, Verstreken P, Bellen HJ (2006) *Drosophila* NMNAT maintains neural integrity independent of its NAD synthesis activity. PLoS Biol 4:e416
97. Williamson WR, Hiesinger PR (2010) On the role of v-ATPase V0a1-dependent degradation in Alzheimer disease. Commun Integr Biol 3:604–607
98. Komatsu M, Waguri S, Chiba T, Murata S, Iwata J, Tanida I, Ueno T, Koike M, Uchiyama Y, Kominami E, Tanaka K (2006) Loss of autophagy in the central nervous system causes neurodegeneration in mice. Nature 441: 880–884
99. Hara T, Nakamura K, Matsui M, Yamamoto A, Nakahara Y, Suzuki-Migishima R, Yokoyama M, Mishima K, Saito I, Okano H, Mizushima N (2006) Suppression of basal autophagy in neural cells causes neurodegenerative disease in mice. Nature 441:885–889
100. Juhasz G, Erdi B, Sass M, Neufeld TP (2007) Atg7-dependent autophagy promotes neuronal health, stress tolerance, and longevity but is dispensable for metamorphosis in *Drosophila*. Genes Dev 21:3061–3066
101. McPhee CK, Baehrecke EH (2009) Autophagy in *Drosophila melanogaster*. Biochim Biophys Acta 1793:1452–1460
102. Chang YY, Neufeld TP (2009) An Atg1/Atg13 complex with multiple roles in TOR-mediated autophagy regulation. Mol Biol Cell 20:2004–2014

Chapter 4

In Vivo Single Cell Labeling Techniques

Chih-Fei Kao and Tzumin Lee

Abstract

Single cell labeling allows identification of neuron types based on neurite trajectories, an essential step to understand brain anatomy and function. For years, various techniques have been developed to achieve in vivo single cell labeling. In *Drosophila*, several genetic mosaic approaches have not only allowed labeling of specific neuron subsets but also enabled manipulations of gene function or neural activity in or outside the marked cell(s). They collectively provide an unprecedented level of versatility for diverse targeted studies. In this chapter, we selectively cover the general methods of single neuron labeling mediated by flip-out or mitotic recombination.

Key words: Single cell labeling, Neuronal morphology, Genetic mosaic, FLP/FRT, Flip-out, MARCM

1. Background and Historical Overview

A functional brain consists of neurons of diverse types that are wired in complex patterns to process information through specific neuron ensembles. To visualize individual neurons and determine their connections is essential for revealing the physical map of the brain, laying the necessary foundation for studying both brain development and function. Single cell labeling in whole-mount brains has proven informative in the dissection of neural circuitry in the *Drosophila* brain. Labeling single neurons by genetic mosaics further permits manipulation of gene functions specifically in the labeled cell(s), not only allowing detailed phenotypic analysis but also enabling studies of pleiotropic genes in specific neurons within otherwise normal brains.

Prior to the era of transgenesis, single neurons were labeled by chemical or physical means. (1) Chemical approach: the classic Golgi staining method is achieved by impregnating a fixed tissue with chemicals that lead to the fill of a limited number of cells at

Bassem A. Hassan (ed.), *The Making and Un-Making of Neuronal Circuits in Drosophila*, Neuromethods, vol. 69,
DOI 10.1007/978-1-61779-830-6_4, © Springer Science+Business Media, LLC 2012

random in their entirety with visible silver chromate (1). Although the Golgi staining allows visualization of single neurons with remarkably fine structural information, this technique can be only used in well-fixed brains and researchers have no control over the labeling patterns. (2) Physical approach: one can transfer dyes or fluorescence reporters into neurons by physical means, such as injection via a sharp electrode (2). Given the physical limitations, deep-lying neurons might not be accessible, and even with direct injection, it is challenging to target specific neurons in unstained samples.

The ability to introduce transgenes into the fly genome subsequently fostered the development of more sophisticated tools for single neuron labeling in intact brains. Such tools are largely built upon some binary transgene induction system, like GAL4/UAS where transgenes of interests are placed under the control of the UAS promoter whose activity depends on the GAL4 transcription factor (3). Diverse GAL4 drivers, obtained through enhancer trap (jumping a minimal-promoter-GAL4 transgene randomly in the genome) or promoter bashing (introducing GAL4 transgenes already carrying diverse *cis*-regulatory elements), are available in the fly community (3–6). They allow one to express UAS-transgenes in specific subsets of brain cells, to unveil the morphologies of the GAL4-positive cells with reporters and/or modulate their gene functions or neural activities with effectors. One can further apply additional targeting techniques to the GAL4-positive cells for single-cell analysis/manipulation in specific neuron populations across different organisms.

Single cell labeling among GAL4-positive neurons can be achieved physically, optically or genetically. (1) Physical approach: visualizing GAL4-positive cells in live samples is possible with UAS-fluorescent reporters, such as the ones described in Chap. 1, which can then be targeted for electrophysiology followed by dye filling to reveal single neuron morphology. (2) Optical approach: the combination of targeted expression of photochromic molecules, such as photoactivatable GFP (PA GFP) (7), Dronpa (8), and KAEDE (9), with the in vitro photoactivation/photoconversion processes can allow visualization of single neurons (or specific protein molecules) in particular neuron ensembles. Given the reversibility of photochromic molecules, one can in theory repeat photoconversion in different cells to sequentially mark individual GAL4-positive neurons in a given brain. However, it could be labor-intensive and technically challenging to focus on deep neurons with single-cell precision. (3) Genetic approach: by sparse de-repression or activation of GAL4 drivers or UAS reporters, one can selectively label a small subset of the driver-promoter-active cells in a stochastic manner that might be regulated by the pattern of cell proliferation. This approach requires genomic rearrangements that allow coregulation of multiple transgenes and/or creation of genetic mosaics. One can thus visualize single neurons and further manipulate gene function or activities within these neurons in otherwise normal brains. This approach has no physical limitation, and can be cleverly applied to systematically determine single neurons and their lineage of development. The power and versatility of this approach are currently unparalleled and can be continuously enhanced as further sophisticated genetic/transgenic tools are being developed.

Diverse strategies of single neuron labeling by genetics may vary in: (1) the induction of genomic rearrangement, (2) the nature of site-specific recombination, (3) the control of transgene by de-repression or activation, (4) the control over GAL4 or UAS-reporter, and (5) the design of reporters (Fig. 1). First, sparse sporadic labeling can be realized by transient and/or weak induction of the recombinase (e.g., *hs*-FLP [flippase]) that may trigger genomic rearrangement in precursors (for clonal analysis) or post-mitotic neurons (for single cell labeling). Second, the FLP-mediated site-specific recombination may occur between FRTs (flippase recognition target) situated in *cis* on the same chromosome arm or in *trans* between homologous chromosomes, leading to flip-out or mitotic recombination. Third, a transgene can be activated following excision of a stop cassette in flip-out or become de-repressed after loss of its repressor during mitotic recombination. Fourth, GAL4 can be antagonized by GAL80, while distinct reporters can be independently silenced with sequence-specific microRNAs

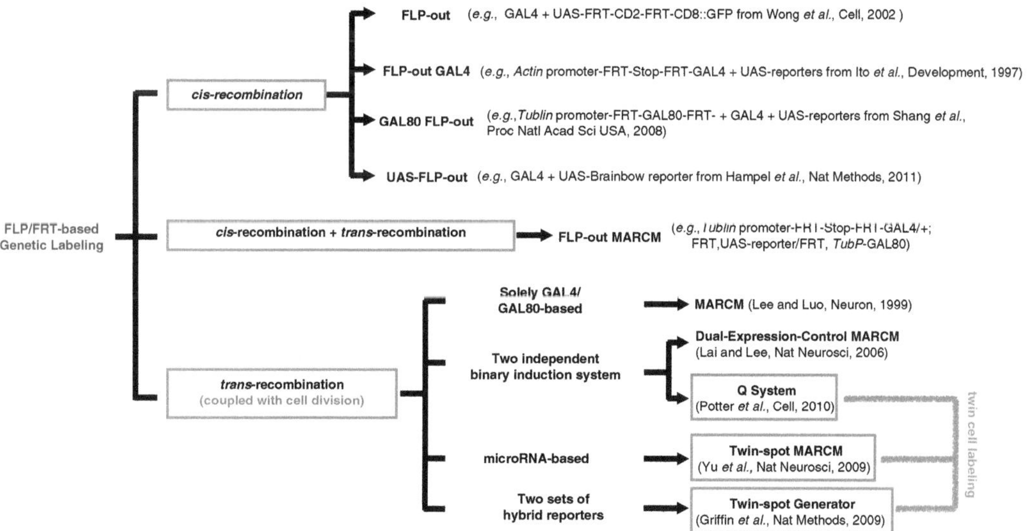

Fig. 1. Various genetic techniques for single cell labeling. One can flip in otherwise silenced markers or GAL4, or flip out GAL80 to label single cells via *cis*-recombination. By contrast, one can render loss of repressors, such as GAL80, QS, and microRNA, to activate reporters in single or twin cells via mitotic recombination between FRTs situated in *trans*. One can also reconstitute functional reporters through *trans*-recombination to achieve twin cell labeling. Flip-out (FLP-out) and Mosaic analysis with repressible cell markers (MARCM) can be further combined to derive FLP-out MARCM.

(10, 11). Fifth, one can significantly increase the throughput of single neuron analysis with BrainBow (12) or Flybow reporters (13) that carry multiple open reading frames flanked by pairs of recombination sites. These open reading frames encode distinct fluorescent proteins and compete for the sole promoter upon induction of the recombinase. One can thus express various reporters stochastically in different GAL4-positive neurons to label multiple single neurons simultaneously in distinct colors.

In this chapter, we selectively cover single neuron labeling by flip-out or following mitotic recombination. We discuss the principles and general methods of flip-out (FLP-out) labeling, conventional MARCM (Mosaic Analysis with Repressible Cell Markers), and various improvements on MARCM.

1.1. FLP-Out Labeling

To achieve single cell labeling, the FLP-out technique requires at least two genetic elements: FLP activity and a FLP-out transgene (14–18). A basic FLP-out transgene contains six DNA elements in the following order: (1) a promoter element to drive marker gene expression, (2) a FRT site, (3) a first marker gene, (4) a transcription termination sequence (Stop sequence), (5) a second FRT site (with the same direction as first FRT site), (6) a second marker gene (Fig. 2a). In the absence of FLP activity, expression of the first marker gene, which is driven by the selected promoter element, labels the cells with active promoter activity. And the second marker gene is off because of the Stop following the first marker gene. Upon induction of FLP, the DNA segment flanked by the direct FRT repeats (the first marker gene and transcription termination sequence) may be "flipped out" stochastically. As a result, the second marker can be selectively expressed in isolated single cells among those labeled by the first marker (Fig. 2a).

The pattern of labeling depends on the expression of FLP and the FLP-out promoter activity: only doubly positive cells may undergo "flipped out" reaction and become labeled by the second marker (Fig. 2b). The intersection allows single cell labeling in further restricted patterns. For temporal induction, a heat shock promoter can be used to regulate the timing of FLP expression. Clonal analysis is possible following stage-specific induction of FLP in precursors to unveil their patterns of proliferation.

1.2. Mosaic Analysis with Repressible Cell Markers

MARCM labels cells via mitotic recombination that allows coupling of cell division with the expression of GAL4-dependent UAS-marker through loss of the GAL4 repressor, GAL80 (10, 19). To achieve MARCM labeling requires at least at five genetic elements (FRTs, FLP, GAL80, GAL4, and UAS-marker) arranged in a specific configuration (Fig. 3a). In contrast to FLP-out somatic recombination, MARCM involves mitotic recombination between FRT sites located on the homologous chromosomes. After chromosomal segregation, one daughter cell inherits the maternal

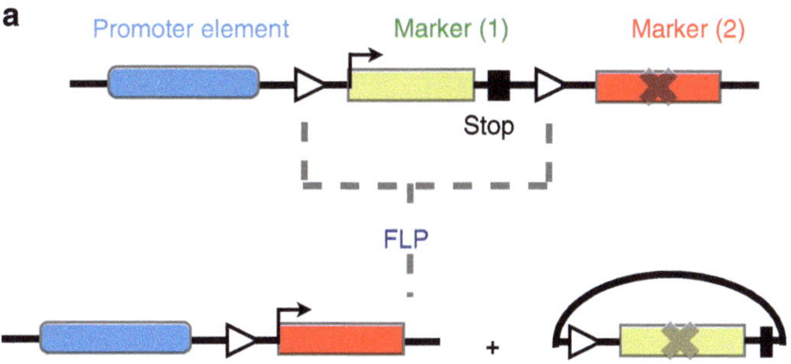

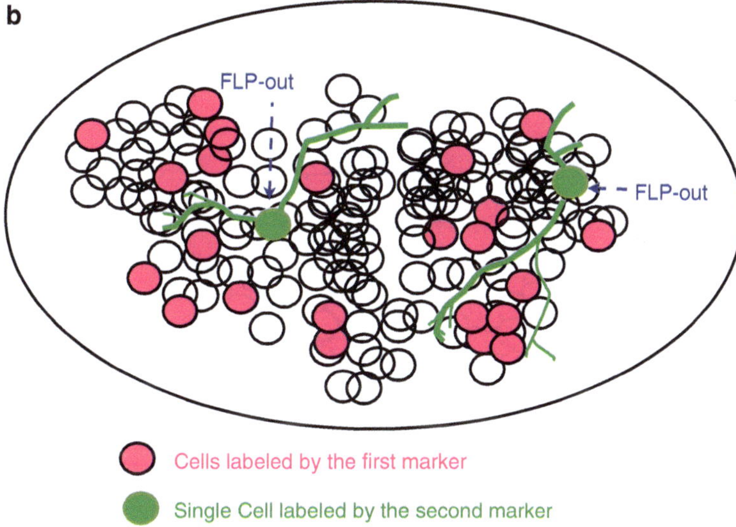

Fig. 2. Schematic and clonal patterns of FLP-out labeling technique. (**a**) FLP-out requires at least two genetic elements: FLP activity and a FLP-out cassette. A typical FLP-out cassette shown here contains the following six genetic elements: (1) a promoter element to initiate gene expression, (2) a FRT, (3) a first marker gene, (3) a transcription termination sequence (indicated as Stop), (4) a second FRT with same direction as the first FRT, (5) a second marker gene. Notably, the DNA fragment situated between two FRTs contains the first marker and Stop sequence, which, after FLP activity kicks in and somatic recombination occurs, would be flipped out. Only in the "flipped-out" cells, the second marker gene would become active. Given the flexibility in genetics, examples of distinct variations of FLP-out cassettes are shown in the bottom. (**b**) Patterns of FLP-out clones. Cells labeled by the first markers (indicated as *red* cells) represent the active expression pattern of selected promoter element. By contrast, the randomly "flipped-out" cells (indicated as *green* cells) are labeled by a second marker. If membrane-tagged markers are used, detailed neurite morphology can be revealed.

recombinant chromosome that carries a transposed paternal chromosome arm and the paternal nonrecombinant one, becoming homozygous for all the paternal alleles of the genes situated distal to the site of recombination. Conversely, the other daughter cell will carry the two maternal alleles as it inherits the transposed maternal chromosome arm. By placing a ubiquitously expressed

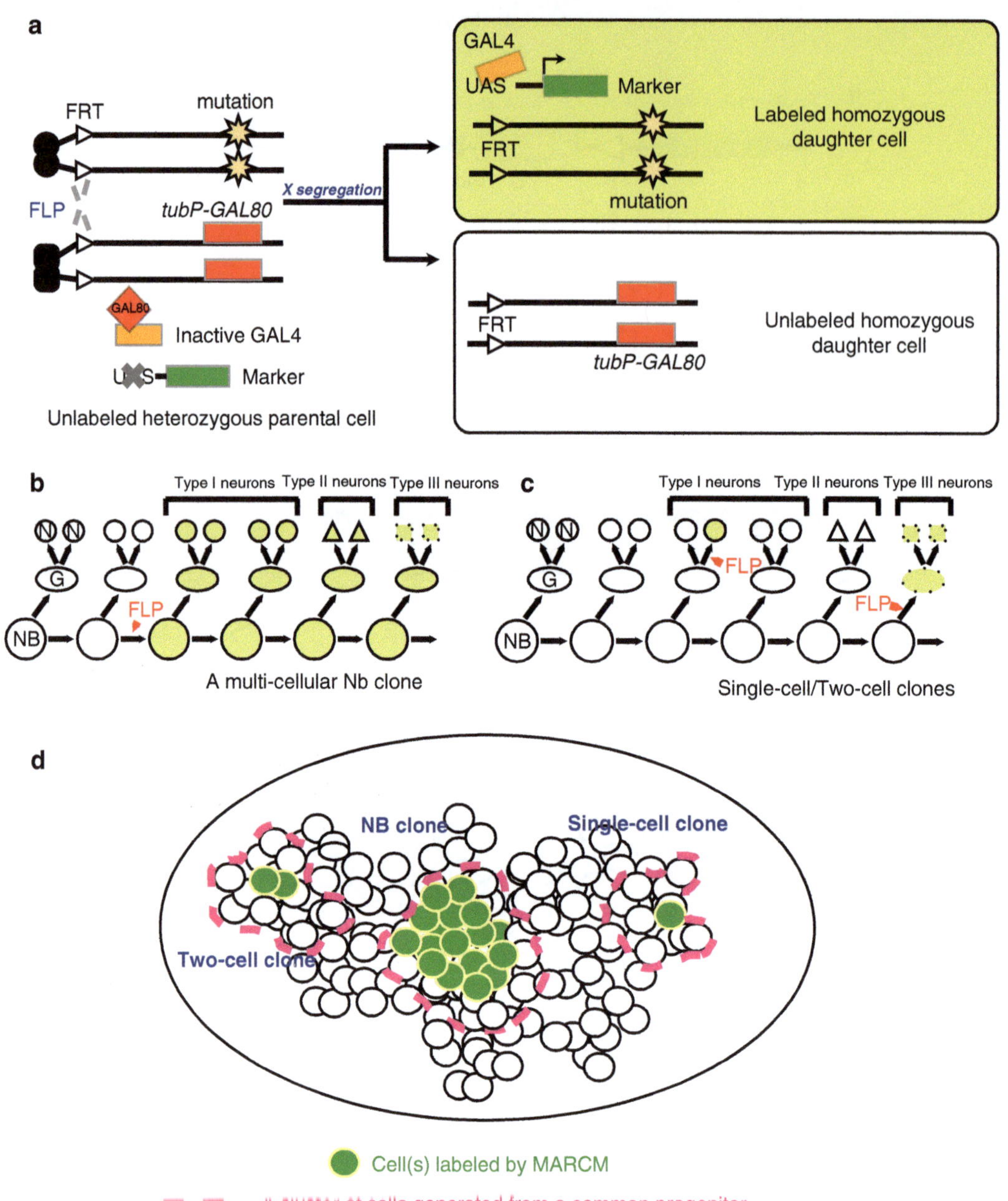

Fig. 3. Schematic and clonal patterns of MARCM labeling technique. (**a**) MARCM requires at least the following genetic elements: FRTs (*open triangles*), FLP, GAL4 (*orange box*), *tubP*-GAL80 (*red box*), and UAS-reporter (*green box*). After FLP/FRT-mediated mitotic recombination occurs in a given dividing cell, one of the daughter cell devoid of GAL80 activity is revealed by the GAL4-activated UAS-reporters. On the other hand, the other daughter cell would become homozygous of *tubP*-GAL80, suppressing GAL4 activity. Furthermore, if a mutation (*yellow star*) is introduced in *trans* to FRT, *tubP*-GAL80, the labeled MARCM clones will be homozygous for the mutation in otherwise heterozygous background. Consequently, this strategy is greatly useful to study the cell-autonomous functions of lethal mutations. (**b**) and (**c**) Schematic illustrations shows how FRT/FLP-mediated mitotic recombination in a developing neuronal lineage can lead to formation of two mutually exclusive classes of labeled clones. Only when mitotic recombination takes place in a dividing GMC, single-cell MARCM clones can be generated ((**b**); indicated as single type I *green* neuron). Accordingly, if GMC (shown as G in the figure) loses the repressor gene, its two daughter neurons will be labeled as a two-cell clone ((**b**); indicated as two type III *green* neurons). Notably, single- and two-cell MARCM clones can be generated independently in the same neuronal lineage at different developmental times. When mitotic recombination takes place in a regenerating neuroblast (NB), all the postmitotic neurons (N) subsequently generated will be labeled ((**c**); *green* cells), forming a NB clone. Cell type composition of a lineage will be preserved in a NB clone. (**d**) Patterns of three distinct types of MARCM clones: single-cell, two-cell, and Nb clones (*green* color cells). The *pink dot line* circles a neuronal lineage that is originated from a common progenitor.

GAL80 transgene (e.g., *tubP*-GAL80) distal to the site of mitotic recombination, one of the homozygous daughter cells will lose the GAL80 transgene. Clones of GAL80-minus cells can thus be generated among GAL80-containing cells, and those GAL4-positive cells in the GAL80-minus clones become uniquely labeled by the UAS-marker (Fig. 3a).

Given the genetic arrangement of MARCM, one of the daughter cells will become homozygous for alleles situated distal to the FRT site after mitotic recombination. This feature enables one to knock out an essential gene specifically in the GAL80-minus cells by introducing a recessive lethal mutation into the transposed chromosome arm that does not carry *tubP*-GAL80 (Fig. 3a, mutation). Moreover, depending on the purpose of the experiment, it is also possible to incorporate additional UAS transgenes to manipulate gene function in the uniquely labeled GAL80-minus cells of mosaic brains.

As in the FLP-out experiment, FLP activity in MARCM can be controlled under different regulatory mechanisms, such as a tissue-specific promoter for regional expression or a heat-shock-dependent promoter for temporal induction. However, distinct from FLP-out experiments, MARCM requires expression of FLP in precursors to achieve clonal labeling via mitotic recombination. A typical neuronal lineage originates from a progenitor cell (neuroblast; NB) that undergoes multiple rounds of self-renewing asymmetric cell divisions to deposit a series of intermediate precursors (ganglion mother cell; GMC). Each GMC divides once to produce two neurons that may acquire different fates through differential Notch signaling. Clones of different sizes may thus arise depending on which neural precursors have undergone mitotic recombination (Fig. 3d). Mitotic recombination in an asymmetric NB division would give rise to a multicellular NB clone consisting of the remaining lineage or a two-cell clone derived from the immediate GMC (Fig. 3b). And one can obtain isolated single-cell clones if mitotic recombination was induced during the neuron-producing GMC divisions (Fig. 3c). Since MARCM couples labeling with cell division, not only is neuronal morphology unveiled, analyses of single-cell clones generated systematically through lineage development will reveal how diverse neurons derive from a common progenitor (Fig. 3c). MARCM, therefore, permits cell lineage analysis and allows one to identify and target distinct neurons based on their developmental origin.

Various improved MARCMs and their parallels, including dual-expression-control MARCM (20), twin-spot MARCM (11), twin-spot generator (TSG) (21), and Q-MARCM (22), are available for further sophisticated mosaic studies (Fig. 1). Dual-expression-control MARCM employs two GAL80-repressible transcriptional activators (GAL4 vs. LexA::GAD) to label and manipulate different subsets of GAL80-minus cells independently in a given organism (20). Twin-spot MARCM utilizes microRNAs to silence individual reporters directly such that one can repress two markers independently and label the homozygous sister clones derived from a heterozygous precursor in different colors at the same time (11). By contrast, twin-spot labeling with TSG involves two complementary hybrid reporters whose reconstitution depends on recombination between FRTs across homologous chromosome arms (21). And Q-MARCM is built upon the Q repressible binary system where the QF transcriptional activator can be potently suppressed by the QS repressor. Double MARCM can achieve twin-spot labeling with GAL4/GAL80 and QF/QS acting in *trans* on homologous chromosome arms, or facilitate dual mosaic analyses with them assembled on non-homologous chromosome arms (22). Despite the transgene differences, these positive-labeling genetic mosaic systems all involve marker activation following FLP/FRT-mediated mitotic recombination and are thus carried out with similar protocols.

2. Equipments and Materials

2.1. Equipment

- 25°C Incubator to culture fly lines
- Water bath for heat shock-induced FLP expression (set at 32 or 37°C)
- Stereoscope and dissection apparatus for tissue retrieval
- Standard facility for tissue fixation and immunostaining
- Confocal microscope for imaging fluorescent signals
- Computers with image processing software

2.2. Materials

- Standard culturing media and containers for fruit fly cultivation
- Fly lines with required genetic elements that are appropriately arranged (Many of the FLP-out and MARCM ready fly lines have been deposited to Bloomington *Drosophila* Stock Center, Indiana University)
- Reagents for detection of markers by immunostaining: specific primary antibodies (Abs), various fluorophore-conjugated secondary Abs, slides, coverslips, mounting solution, and fingernail polish

3. Procedures

3.1. Singe Cell labeling Using FLP-Out Technique

1. Prepare FLP-out ready fly lines

   To get ready for FLP-out experiments, transgenes carrying a FLP-out reporter and FLP activity need to be introduced into the same organisms by genetics. An example of FLP-out reporter widely used contains the following elements: promoter-FRT-rCD2-stop-FRT-mCD8::GFP (Fig. 2a). Both rCD2 and mCD8::GFP are membrane-bound markers, adequate for tracing neurites and determining overall neuron morphology. Note that the activity of FLP can be induced by a variety of regulatory mechanisms, such as a heat-shock promoter to achieve temporal control of FLP expression. And the FLP-out events can be detected only in the cells where the promoter of FLP-out reporter is active.

2. Generate FLP-out clone
Prepare the FLP-out cross with appropriate numbers of virgin females and male flies (roughly 30 females and 10 males for collection of well synchronized embryos) in vials containing fresh medium and dry yeast extract. Maintain the cross for 1–2 days to allow fertilization of females and efficient egg laying. Keep transferring the parental flies to new vials until retrieval of enough progeny. For induction of FLP activity under the control of heat-shock promoter, conduct heat shock by submerging the vials into 37°C water bath for 10–15 min at desired developmental time. Reduce or prolong the duration of heat shock to adjust the FLP-out efficacy. After heat shock, return the samples back to 25°C incubator. If FLP activity is governed by a tissue-specific promoter, FLP will be expressed according to the native property of selected promoter element. In this case, just continue expanding the FLP-out progeny until enough samples are collected.

3. Sample dissection and staining
Depending on your experimental purposes, dissect the animals with right genotype at the developmental stage of interest. After tissue fixation, perform immunostaining with appropriate antibody sets for detection of marker expression. Finally, mount the samples on the glass slide with suitable mounting solution.

4. Image collection and data analysis
Images of the labeled cells can be captured by a regular fluorescence microscope with an attached camera or a confocal microscope. Follow the instructions provided by the manufacturer to operate the microscopes and collect images.

3.2. Singe Cell Labeling Using MARCM

1. Prepare MARCM-applicable fly lines
MARCM requires at least five genetic elements (FRTs, FLP, GAL80, GAL4, and UAS marker). According to the experimental design and genetic feasibility, these different genetic elements can be creatively assembled into two sexes and finally arranged together after crosses (Fig. 3a). It's crucial that GAL80 has to be put distal to one homologous FRT. And the

GAL4 and UAS-reporter cannot be put on the same chromosome arm as GAL80. A sample set of MARCM-applicable flies is *hs*-FLP; FRT G13, *tubP*-GAL80/Cyo and FRT G13, UAS-mCD8::GFP, GAL4.

2. Generate MARCM clone (analogous to step 2 in part A)
3. Sample dissection and staining (see step 3 in part A)
4. Image collection and data analysis (see step 4 in part A)

4. Experimental Variables

4.1. Choices of Makers/Effectors

A plethora of markers are available and can be readily incorporated into FLP-out or MARCM system. Fluorescent proteins with different excitation/emission spectra are particularly popular (23). They allow multicolor labeling in live samples, and the widely utilized GFP and RFP can be detected by immunostaining in fixed tissues as well. Other protein tags that can be reliably located by immunostaining in whole-mount fly brains include mCD8, rCD2, Myc, HA, and V5. In addition, various chimeras have been designed to highlight specific subcellular structures of the labeled cells. For example, a membrane-targeted fluorescent marker, such as mCD8::GFP and myr::RFP, can outline entire neuron morphology for tracing individual neurites. To reveal neuron polarity, one can mark dendrites selectively using Dscam [TM1]::GFP (24) or DenMark (25). And coinduction of multiple transgenes is possible, as the switch can be implemented on the driver (e.g., placing GAL4 after the flip-out cassette in Fig. 1). Furthermore, even with transient weak induction of FLP, it is common to obtain multiple FLP-out or MARCM clones. To trace multiple neurons unambiguously in the complex brain requires differential labeling of individual neurons, which was made possible recently with the development of BrainBow and Flybow that allow stochastic selection of multiple markers in the labeled cells.

If desired, various effectors can be coinduced in the labeled cells. To suppress gene function, one can silence specific genes by RNAi using various dsRNA or miRNA transgenes or block protein activities through expression of dominant-negative competitors. One can also express wild-type transgenes ectopically for gain-of-function studies. And please see other chapters for effectors that alter neuron activities.

4.2. Induction of FLP

FLP can be induced in various spatially and/or temporally controlled manners. For stochastic single-neuron labeling, a transient induction is preferred and this can be readily achieved with *hs*-FLP. Induction of FLP in adult flies is sufficient to elicit FLP-out in postmitotic neurons. But to obtain MARCM-labeled single neurons requires expression of FLP during the neuron-producing mitoses. Targeting specific neurons for single-cell MARCM labeling, thus, demands prior knowledge of when their immediate precursors are present in the developing nervous system. This can be empirically determined through systematic analysis of MARCM clones generated at different times of brain development. Such birthdating experiments require brief induction of FLP in well-synchronized developing organisms and use of *hs*-FLP and FRT pairs that yield no background clone in the absence of heat shock.

Tissue-specific induction of FLP driven by various endogenous promoters is mostly incompatible with single-neuron labeling given that most promoters are dynamic and may trigger recombination accumulatively in excessive cells. To target a particular population of cells for stochastic FLP-out or mitotic recombination requires not only spatial but also temporal regulation of FLP. This could be potentially achieved by binary induction of FLP, using an inducible driver (e.g., geneswitch) or via control of the driver with a temperature-sensitive repressor (e.g., GAL80[ts]) (26–29). In addition, a BrainBow or Flybow reporter allows multicolor labeling of single neurons and can be utilized for dense reconstruction of neural circuits with FLP driven by subtype-specific endogenous promoters.

4.3. Genetic Mosaic Analysis

MARCM involves derivation of homozygous daughter cells from heterozygous precursors. Placing a receive mutation distal to FRT on the GAL80-minus chromosome arm would render the MARCM-labeled GAL80-minus clones homozygous for the mutation in otherwise heterozygous organisms (Fig. 3a). It allows one to unambiguously determine genes' cell-autonomous functions in the complex nervous system. Analogous strategies can be applied to perform genetic screens using FRT lines that carry various independent mutations.

One can also express various UAS transgenes to manipulate gene functions or alter neuron activities in the MARCM-labeled GAL80-minus clones. If no proper mutation is available, it is possible to silence an endogenous gene by targeted RNAi or via expression of a dominant-negative protein. To ascribe specific phenotypes to a given mutation, one can rescue the homozygous mutant clones using various transposed genomic fragments or a UAS wild-type transgene. One can further modify the rescuing construct to conduct isoform-specific rescue or structure–function analysis of the involved gene. Note in such applications that the added genetic elements need to be placed outside the GAL80-containing chromosome arm, to ensure presence of the transgene(s) in the MARCM-labeled GAL80-minus cells.

Given its genetic design, conventional MARCM is not the correct labeling technique for studying non-cell-autonomous functions of genes. However, if the mutation is placed on the same chromosome arm with GAL80 transgene, the resulting labeled cells will become homozygous wild-type and may settle near the sister cells that are homozygous mutant. This MARCM variant, named reverse MARCM, may allow one to unveil the non-cell-autonomous effects of a mutation. By contrast, twin-spot MARCM permits differential labeling of the paired sister clones in the same mosaic organisms. One can simultaneously study cell-autonomous and non-cell-autonomous gene functions, and better determine the prospective fate of the mutant clone based on the fate of its coupled wild-type sister clone.

5. Typical/ Anticipated Results

5.1. FLP-Out Clones

In the genetic design of promoter > first marker > second marker, the cell population labeled by the first marker reflects the activity pattern of the promoter situated upstream of the FLP-out cassette. Following excision of the FLP-out cassette, the promoter-active cells will then express the second marker and exist as FLP-out clones that can be readily distinguished from cells labeled by the first marker. Patterns of FLP-out clones may vary depending on how FLP is induced. A transient induction of FLP is expected to yield FLP-out clones stochastically (Fig. 2b). By contrast, some patterned continuous expression of FLP can derive reproducible FLP-out clones composed of those promoter-active cells that have lost the FLP-out cassette.

If FLP was induced during organism development, the final patterns of FLP-out clones will further depend on the proliferation pattern of the tissue where clones reside. When FLP-out is occurring in a progenitor cell, all the promoter-active cells derived from that progenitor will be labeled as a FLP-out clone. In the fly CNS, one might therefore obtain multicellular NB clones, two-cell GMC clones, or isolated single-cell clones following induction of FLP in the developing nervous system. The clonally unrelated single-cell clones could interfere with cell lineage analysis by FLP-out strategies. Lastly, various computer algorithms are available for registering the adult *Drosophila* brains from different individuals to each other, allowing direct comparisons of stochastic single-cell clones regardless of differences in brain size and shape (30–32).

5.2. MARCM Clones

Unlike FLP-out, MARCM clones exclusively derive from mitotically active precursors that are no longer present in the adult *Drosophila* nervous system. To label neurons by MARCM thus requires FLP expression in neural precursors of the developing nervous system. One can obtain a multicellular NB clone or a two-cell GMC clone following mitotic recombination in a NB, or derive single-cell clones from GMCs (Fig. 3b, c). For a given neural lineage, NB clones generated during early neurogenesis should contain more cells than those made later. And different two-cell or single-cell clones are anticipated when clones were induced at different times of the lineage development, as distinct neurons are made by a common progenitor in an invariant sequence. Systematic analysis of temporally induced MARCM clones has allowed researchers to identify single neurons based on their lineage origin and timing of birth (Fig. 3c).

When a particular neuron subset is targeted for MARCM labeling using a subtype-specific GAL4 driver, one can first induce NB clones in newly hatched larvae and determine in adult brains the NB clone(s) that may carry neurons of your interest. For those that remain elusive despite collection of many stochastic clones, one should explore if they were born prior to larval hatching by inducting MARCM clones in early embryos. To ultimately label individual neurons in single-cell MARCM clones, one needs to induce mitotic recombination in their intermediate precursors which exist briefly at specific times of development. Initial determination of such cells often requires stage-specific transient induction of FLP through the entire process of neurogenesis. Once the identity and developmental origin are established, one can reproducibly label the same neuron(s) by MARCM via induction of FLP in the precursor(s).

6. Troubleshooting

6.1. FLP-Out/MARCM Parental Flies Are Sick

Solution: Redistribute the genetic elements between two sexes or try to use different insertion lines if possible.

6.2. Little or No FLP-Out/MARCM Progeny

Solution: Arrange the FLP-out/MARCM genetic elements in different combinations. Use young virgin females and males in the cross. To make stock lines healthier, try to use various balancer chromosomes and avoid coexistence of multiple balancers. Finally, scale up the crosses to get least enough samples.

6.3. Few FLP-Out/MARCM Progeny Survive After Heat Shock

Solution: Make sure the temperature of water bath is set at appropriate temperature. Reduce the length of heat shock. Try multiple rounds of tolerable heat shock if clone efficiency becomes an issue.

6.4. No FLP-Out/MARCM Clone Is Seen

Solution: Check the FLP-out/MARCM genetic elements individually and ensure they are properly arranged and function well. For FLP-out experiments, make sure the coexpression of FLP activity and FLP-out cassette in the target cells. As for MARCM experiments, make sure the induction of FLP activity occurs in the progenitors rather than their postmitotic neurons. One can try to induce FLP activity at different developmental times to determine an appropriate condition.

6.5. Low Efficiency of Clone Induction

Solution: Increase the FLP activity by using more potent FLP transgenes (or increase the copy number of FLP transgene) or prolong the heat shock duration if heat shock control is used. If the proliferation duration of target cells is known, try to better synchronize organism development and induce FLP activity at specific developmental time to increase clone efficiency. For MARCM experiments, if genetically feasible, one can try FRT sites on different chromosome arms. If the efficiency remains low, consider rebuild of FLP-out/MARCM lines to ensure intact FRT sites. Last, the efficiency in obtaining different types of clone can differ drastically despite all the conditions have been optimized and checked.

6.6. Too Many FLP-Out/MARMC Background Clones

Solution: Use a more tissue-specific promoter/GAL4 driver to selectively mark the clones of interest. Spatially and/or temporally control the FLP activity to selectively target specific progenitors and/or postmitotic cells. Try a weaker and more stringent FLP transgenes, or reduce the degree of heat shock (for example, FLP-out clones induced by *hs*-FLP can also be generated by 32°C heat shock at low frequency).

7. Conclusion

Single-neuron labeling by genetics offers the unparalleled level of versatility and specificity in the cellular and molecular dissection of the complex brain. It combines binary transgene induction systems with site-specific recombinases to label and/or manipulate single cells within specific neuron subsets in intact organisms. Stochastic single-cell labeling by FLP-out using BrainBow types of reporters allows differential labeling and thus unambiguous tracing of multiple neurons at the same time, greatly facilitating the dense reconstruction of neural circuits. And with MARCM that labels neurons through mitotic recombination, one can determine individual neurons systematically via detailed cell lineage analysis. One can further reproducibly target the same neurons by MARCM for loss-of-function genetic mosaic analysis in intact brains. Additional sophisticated genetic tools can be readily incorporated into these conventional single-cell labeling techniques to (1) refine the neuron pool of interest, (2) improve the specificity of clones, and (3) increase the throughput of analysis. These modular designs endow FLP-out and MARCM with an unlimited potential to realize a comprehensive single-cell analysis in the complex brain.

Acknowledgements

Research in Lee T. Lab is supported by National Institute of Health and Howard Hughes Medical Institute.

References

1. Golgi C (1873) Sulla struttura della sostanza grigia del cervello. Gazetta Medica Lomberda 33:244–6
2. Bossing T, Technau GM (1994) The fate of the CNS midline progenitors in *Drosophila* as revealed by a new method for single cell labelling. Development 120:1895–906
3. Brand A, Perrimon N (1993) Targeted gene expression as a means of altering cell fates and generating dominant phenotypes. Development 118:401–15
4. Spradling AC, Stern D, Beaton A, Rhem EJ, Laverty T, Mozden N, Misra S, Rubin GM (1999) The Berkeley *Drosophila* Genome Project gene disruption project: single P-element insertions mutating 25% of vital *Drosophila* genes. Genetics 153:135–77
5. Hayashi S, Ito K, Sado Y, Taniguchi M, Akimoto A, Takeuchi H, Aigaki T, Matsuzaki F, Nakagoshi H, Tanimura T, Ueda R, Uemura T, Yoshihara M, Goto S (2002) GETDB, a database compiling expression patterns and molecular locations of a collection of Gal4 enhancer traps. Genesis 34:58–61
6. Pfeiffer BD, Jenett A, Hammonds AS, Ngo TTB, Misra S, Murphy C, Scully A, Carlson JW, Wan KH, Laverty TR, Mungall C, Svirskas R, Kadonaga JT, Doe CQ, Eisen MB, Celniker SE, Rubin GM (2008) Tools for neuroanatomy and neurogenetics in *Drosophila*. Proc Natl Acad Sci U S A 105:9715–20
7. Patterson GH, Lippincott-Schwartz J (2002) A photoactivatable GFP for selective photolabeling of proteins and cells. Science 13:1873–77
8. Habuchi S, Ando R, Dedecker P, Verheijen W, Mizuno H, Miyawaki A, Hofkens J (2005) Reversible single-molecule photoswitching in the GFP-like fluorescent protein Dronpa. Proc Natl Acad Sci U S A 102:9511–6
9. Ando R, Hama H, Yamamoto-Hino M, Mizuno H, Miyawaki A (2002) An optical marker based on the UV-induced green-to-red photoconversion of a fluorescent protein. Proc Natl Acad Sci U S A 99:12651–56
10. Lee T, Luo L (1999) Mosaic analysis with a repressible cell marker for studies of gene function in neuronal morphogenesis. Neuron 22: 451–61
11. Yu HH, Chen CH, Lei S, Huang Y, Lee T (2009) Twin-spot MARCM to reveal the developmental origin and identity of neurons. Nat Neurosci 12:947–53
12. Hadjieconomou D, Rotkopf S, Alexandre C, Bell DM, Dickson BJ, Salecker I (2011) Flybow: genetic multicolor cell labeling for neural circuit analysis in *Drosophila melanogaster*. Nat Methods 8:260–6
13. Hampel S, Chung P, McKellar CE, Hall D, Looger LL, Simpson JH (2011) *Drosophila* Brainbow: a recombinase-based fluorescence labeling technique to subdivide neural expression patterns. Nat Methods 8:253–9
14. Struhl G, Basler K (1993) Organizing activity of wingless protein in *Drosophila*. Cell 72: 527–40
15. Basler K, Struhl G (1994) Compartment boundaries and the control of *Drosophila* limb pattern by hedgehog protein. Nature 368: 208–14
16. Ito K, Awano W, Suzuki K, Hiromi Y, Yamamoto D (1997) The *Drosophila* mushroom body is a quadruple structure of clonal units each of which contains a virtually identical set of neurones and glial cells. Development 124:761–71
17. Pignoni F, Zipursky SL (1997) Induction of *Drosophila* eye development by decapentaplegic. Development 124:271–8
18. Wong AM, Wang JW, Axel R (2002) Spatial representation of the glomerular map in the *Drosophila* protocerebrum. Cell 109:229–41
19. Lee T, Luo L (2001) Mosaic analysis with a repressible cell marker (MARCM) for *Drosophila* neural development. Trends Neurosci 24:251–4
20. Lai SL, Lee T (2006) Genetic mosaic with dual binary transcriptional systems in *Drosophila*. Nat Neurosci 9:703–9
21. Griffin R, Sustar A, Bonvin M, Binari R, del Valle Rodriguez A, Hohl AM, Bateman JR, Villalta C, Heffern E, Grunwald D, Bakal C, Desplan C, Schubiger G, Wu C, Perrimon N.

The twin spot generator for differential *Drosophila* lineage analysis. Nat Methods 2009; 6:600–2

22. Potter CJ, Tasic B, Russler EV, Liang L, Luo L (2010) The Q system: a repressible binary system for transgene expression, lineage tracing, and mosaic analysis. Cell 141:536–48
23. Shaner NC, Steinbach PA, Tsien RY (2005) A guide to choosing fluorescent proteins. Nat Methods 2:905–9
24. Wang J, Ma X, Yang JS, Zheng X, Zugates CT, Lee CHJ, Lee T (2004) Transmembrane/juxtamembrane domain-dependent Dscam distribution and function during mushroom body neuronal morphogenesis. Neuron 43:663–72
25. Nicolai LJJ, Ramaekers A, Raemaekers T, Drozdzecki A, Mauss AS, Yan J, Landgraf M, Annaert W, Hassan BA (2010) Genetically encoded dendritic marker sheds light on neuronal connectivity in *Drosophila*. Proc Natl Acad Sci U S A 107:20553–8
26. Han DD, Stein D, Stevens LM (2000) Investigating the function of follicular subpopulations during *Drosophila* oogenesis through hormone-dependent enhancer-targeted cell ablation. Development 127:573–83
27. Osterwalder T, Yoon KS, White BH, Keshishian H (2001) A conditional tissue-specific transgene expression system using inducible GAL4. Proc Natl Acad Sci U S A 98:12596–601
28. Roman G, Endo K, Zong L, Davis RL (2001) P[Switch], a system for spatial and temporal control of gene expression in *Drosophila melanogaster*. Proc Natl Acad Sci U S A 98:12602–7
29. McGuire SE, Mao Z, Davis RL (2004) Spatiotemporal gene expression targeting with the TARGET and gene-switch systems in *Drosophila*. Sci STKE 2004(200):pl6
30. Jefferis GSXE, Potter CJ, Chan AM, Marin EC, Rohlfing T, Maurer CR, Luo L (2007) Comprehensive maps of *Drosophila* higher olfactory centers: spatially segregated fruit and pheromone representation. Cell 128:1187–203
31. Lin HH, Lai JSY, Chin AL, Chen YC, Chiang AS (2007) A map of olfactory representation in the *Drosophila* mushroom body. Cell 128:1205–17
32. Peng H, Ruan Z, Long F, Simpson JH, Myers EW (2010) V3D enables real-time 3D visualization and quantitative analysis of large-scale biological image data sets. Nat Biotechnol 28: 348–53

Chapter 5

Neuronal Morphology in the *Drosophila* Embryo: Visualisation, Digital Reconstruction and Quantification

Matthias Landgraf and Jan Felix Evers

Abstract

Studying the formation of neural networks requires a thorough understanding of their constituent neurons, their development, connectivity and electrical properties. Neuronal morphology is a key element, encompassing parameters important for neuronal function: projection patterns and synaptic termination zones inform on connectivity, while ontogeny is reflected by cell body locations. *Drosophila* is one of the most successful genetic model organisms for studying nervous system development and function. It presents a unique combination of neural networks of intermediate complexity, which are composed of identified neurons that can be genetically manipulated and interrogated throughout their development by imaging and electrophysiology. Here, we present approaches for studying neuronal morphology in the *Drosophila* embryo. The methods are applicable to any cell, though by way of example we focus on motor neurons in the ventral nerve cord. Specifically, we outline and discuss genetic approaches for visualising individual neurons, their detailed neuritic arborisations and putative synaptic sites. We present a method for analysing quantitatively the topology of complex branched neuronal structures through digital reconstruction. This combination of genetic single-cell labelling methods and semi-automated, highly accurate digital 3D reconstructions of complex cell morphologies has already opened up to investigation new areas of nervous system development and it will be central to future progress in our understanding of neural network development.

Key words: *Drosophila*, Genetic model, Nervous system, Neuronal cell morphology, Digital reconstruction, Reconstruction algorithm, Amira, FLP recombinase

1. Introduction

Neurons are the fundamental building blocks of nervous systems. Santiago Ramon y Cajal quite accurately surmised that neuronal structure, used as a key indicator for cell type classification, may be intimately linked to function (1–3). How diverse neuronal projection patterns are generated has been under intense investigation. Already in 1978, pioneering work by Levinthal and Macagno

Bassem A. Hassan (ed.), *The Making and Un-Making of Neuronal Circuits in Drosophila*, Neuromethods, vol. 69,
DOI 10.1007/978-1-61779-830-6_5, © Springer Science+Business Media, LLC 2012

challenged a widely held view that invertebrate nervous systems are largely "hard-wired" (4); see also ref. (5). By reconstructing the morphology of identical neurons on both left and right sides of the nervous system in isogenic *Daphnia*, they demonstrated that the detailed patterns of neuronal arbors are variable and therefore not genetically encoded but epigenetically regulated (e.g. through cell–cell interactions) (4). This has since been underpinned by other reconstruction data of neurons (6, 7). The fact that neurons in the nervous systems of arthropods can be individually identified presents a unique advantage because it allows one to return time and again to the same neurons at different developmental stages and experimental conditions. This is particularly important when studying how epigenetic factors, such as cell–cell interactions including synaptic transmission, act on cell intrinsic programs of differentiation to sculpt neuronal shapes and connectivity suitable for network function (8–13).

Neuronal identity and thus basic cell morphology is progressively specified during development. The developmental origins of peripheral sensory neurons in the *Drosophila* embryo have been well defined (14–19). For central neurons, through painstaking manual and genetic labelling most neuronal lineages have been charted (20–25). However, only for a few lineages have details been established about birth order, transmitter type and projection patterns of individual neurons (26–30).

Methods for visualising and tracing neurons have progressed considerably since the days of Golgi and Cajal, when silver stained neurons were artistically sketched by hand, some with the aid of camera lucida (31). Though the resolution of Golgi's *reazione negra*, improved by Cajal, is magnificent, the experimenter had no control over which cells were labelled. For targeted applications, horseradish peroxidase (HRP) was introduced as a retrograde neuronal tracer in the late 1960s (32, 33). A principal drawback of HRP was that detection required cells to be fixed and subjected to an enzymatic reaction. The need for marking living cells that could also be probed electrophysiologically motivated a search for fluorescent tracer dyes in the 1970s and 1980s, such as Lucifer Yellow (34), Rhodamine (35) and various fluorescent compounds suitable as retrograde tracers (36), including lipophilic carbocyanine dyes (DiI and DiO) (37). With the advent of transgenic methods, genetically encoded reporters of cell morphology were developed. β-Galactosidase and fluorescent protein based reporters are most commonly used. Initially, these were cytoplasmically localised, which is suboptimal for revealing long and thin neurites. To improve the ability to resolve such structures, reporters were targeted to cytoskeletal components (by coupling to microtubule associate proteins Tau or Kinesin (38, 39)) and to the cell membrane, by including transmembrane (e.g. CD2 or CD8) (40, 41) or membrane targeting (farnesylation or myristoilation) motifs (42).

In order to quantify neuronal cell shapes, attempts have been made for well over 45 years to automate the process of tracing neurons, digitise morphologic parameters and evaluate these computationally. Initially, computers interacting with microscopes allowed scientists to record neuronal structures by manually entering cell-shape coordinates point for point (43). With leaps in computational power and digital imaging, scientists now acquire digital image stacks of their specimens and digitally trace or reconstruct cells subsequently. The principal challenge remains: to achieve computational solutions that minimise manual intervention, yet attain maximal accuracy (for a review see ref. (44)).

In this chapter, we outline genetic strategies for visualising single neurons in the *Drosophila* embryonic nervous system and a computational method for digitally reconstructing and quantifying their morphology.

2. Materials and Reagents

Fly stocks: These are specific to the research question.

FLP recombinase sources:

1. *RN2-FLP*-14b—expresses FLP recombinase in the aCC and RP2 motor neurons and the pCC interneuron.
2. *w[1118]; MKRS, P[hsFLP]86E/TM6B, Tb[1]*—hs-FLP source with relatively low background levels (Bloomington stock centre).
3. *P[hsFLP]1, y[1] w[1118]; Dr[Mio]/TM3, ry[*] Sb[1]*—hs-FLP source with relatively low background levels (Bloomington stock centre).

FLP-Conditional driver lines:

1. *tub84b-FRT-CD2-FRT-Gal4*—a conditional generic reporter to be used in combination with a source of FLP recombinase (45–47).
2. *Act5C-FRT-CD2-FRT-Gal4*—can lack expression in some neurons (29, 45, 48).
3. *elav-FRT-CD2-FRT-Gal4*—neuron-specific (M. Gonzales-Gaitan, personal communication).

Reporter stocks:

1. *UAS-myr-mRFP1* (Bloomington stock centre, by Henry Cheng (49))
2. *UAS-pm-Venus* (50)
3. *10xUAS-myr-GFP*—very high expression levels (51)

4. *UAS-DenMark*—mCherry tagged reporter that localises to the somato-dendritic compartment of neurons (52)

Software: Amira is distributed by Visage Imaging (http://www.visageimaging.com/) and up to date information can be obtained online (http://www.amira.com/). In addition to the main licence, we required the Microscopy Pack to import microscopy data from a range of proprietary formats. To reconstruct neuronal morphology, we freely distribute a reconstruction plug-in (53, 54) that can be downloaded from our web site http://www.zoo.cam.ac.uk/zoostaff/ndd/software.

Hardware: Amira software and the associated reconstruction module are available for Windows (XP/Vista/7, 32-bit and 64-bit editions) and Mac OS and Linux operating systems. Current and recently acquired computers are generally suitable for operating the program, but as minimum requirements we recommend 1 GB RAM and an OpenGL certified graphics card with at least 256 MB.

3. Procedures for Neuron Labelling and Analysis of Cell Morphology

3.1. FLPout-Based Strategies for Single Cell Labelling

The ability to selectively visualise neurons of interest is a prerequisite for studying their morphology. Labelling neurons in the *Drosophila* embryo can be achieved with a number of different methods. There is as yet no "one size fits all" solution as far as we are aware, and each approach has specific advantages and disadvantages. The use of lipophilic fluorescent tracer dyes to manually label specific neurons has been described previously (55). Here, we focus on a genetic, recombinase mediated strategy for (stochastic) single cell labelling: the so-called "FLPout" method. And we also discuss the importance of choosing appropriate transgenic reporters. Other genetic single cell labelling strategies are available, as discussed in Chap. 4. There are many advantages associated with labelling neurons genetically: it circumvents requirements for specific manual skills, often the biggest perceived hurdle, and the need for specialised microscopy and cell manipulation equipment. Moreover, at least in principle, any neuron can be labelled, even those deep in the cortex that would be rather inaccessible to microelectrodes. The expression of genetically encoded fluorescent reporters allows live imaging of cells (and their activity) as the nervous system develops.

Ideally, cells are visualised in their entirety and physically separate from others. Because the nervous system of *Drosophila* is bilaterally symmetric and is composed of multiple reiterated, albeit regionally specialised segments, this is difficult to achieve with standard genetic (e.g. Gal4, LexA or QF) expression lines. Even with the most specific expression lines, which target gene expression to a single neuron per half segment, neurites of cells are likely to overlap with those of their contralateral and/or segmental homologues.

Overlap of neurites from different cells makes it difficult, if not impossible, to determine the detailed neuron morphology. One strategy is to fractionate expression patterns so that only one or very few, physically separate, neurons are labelled at a time. Two effective approaches are (1) the MARCM system (see Chap. 4) and (2) recombinase (FLP) based approaches as outlined here.

The use of yeast derived FLP recombinase for generating mosaic expression in *Drosophila* was pioneered by Golic and Lindquist (56) and has been subsequently refined by others (46, 47, 57–59). Briefly, a transgene is constructed in which an "FRT-cassette" (two direct FRT repeats flanking a transcriptional stop) separates the promoter from sequence coding for a transcriptional activator (e.g. Gal4, LexA or QF) or reporter (60–63). Expression of the transcriptional activator or reporter is induced upon FLP-mediated excision of the "FRT-cassette"; expression of FLP recombinase is generally induced by heat shock using a hs-FLP transgene. As long as FLP levels are limited, stochastic mosaic expression can be achieved in individual or small subsets of cells, suitable for determining their morphology (see Note 3.1).

We have had particularly good experience with a FLPout strategy when expressing FLP under control of cell type-specific regulatory regions (e.g. "RN2" of the *even-skipped* gene (64)) and using a generic strong driver line that is conditional on FLP activity, *tub84B-FRT-CD2-FRT-Gal4* (45–47); other variants are *Act5C-FRT-CD2-FRT-Gal4* (29, 45, 48) (lacks expression in some neurons) and *elav-FRT-CD2-FRT-Gal4*, which is neuron-specific (M. Gonzales-Gaitan, personal communication). Advantages of this strategy are that: (1) multiple transgenic reporter genes can be combined in a single, generically useful conditional expression stock; (2) expression is restricted to the cells of interest, as determined by the FLP expression line and is generated automatically without the need of manual heat shock protocols; (3) the frequency of FLPout events can be tuned by exploiting the temperature sensitivity of FLP recombinase, with activity levels being optimal around 29–30°C. Because FLP-mediated induction of the transcriptional activator is stochastic, it is important to be aware that onset of expression and thus duration and levels can vary. It is therefore prudent to determine when expression is induced. This stochastic element to cell labelling also applies to the recently published *Drosophila* equivalents of the "Brainbow" strategy initially developed in the mouse (65–67). A useful resource for FLP based labelling strategies is a collection of enhancer trap-FLP lines generated by Bohm et al., which provides experimental access to many different subsets of neurons (68).

3.2. Genetic Cell Morphology Reporters

As far as reporters for neuron morphology are concerned, our reporters of choice are membrane targeted, myristoylated (myr) reporters, such as *UAS-myr-mRFP1* (from Henry Cheng (49)) and *UAS-pm-Venus* (50). Recently, in an exemplary systematic approach Pfeiffer et al. have generated optimised expression tools for *Drosophila*, including *10xUAS-myr-GFP* (51). For dendritic morphology the recent mCherry tagged *UAS-DenMark* reporter is equally good, albeit lower in fluorescence signal than other *UAS-myr-XFP* transgenes. The DenMark reporter has the additional advantage that it localises to the somato-dendritic compartment of neurons (initially also to the proximal primary neurite, but over time this becomes progressively refined; see Chap. 2) so that in principle it can be used in the same cell in combination with RFP tagged presynaptic reporters (52). In our hands, at least for work in the embryo, where duration and levels of transgene expression are limited, we find that cytoplasmically localised fluorophores fail to satisfactorily label small diameter neurites, particularly of axon (collateral) growth cones and dendritic arbors. Fluorophores targeted to the plasma membrane by farnesylation or fusion to mCD2 or mCD8 are slow to diffuse to distal parts of the cell and lead to rather uneven labelling with excessively high concentrations in somata (J.F.E. and M.L., unpublished).

The question of whether reporters induce structural artefacts is best addressed when the same cell(s) can be visualised by manual labelling. We have found no indication of cell morphology artefacts being induced by *UAS-myr-mRFP1* and *UAS-DenMark* expression. Very high levels of *UAS-mCD8GFP* expression can, at least in some embryonic motor neurons, lead to aberrant dendritic arbor growth. Therefore, caution is generally prudent when using transgenic reporters (J.F.E. and M.L., unpublished).

Localisation of reporter genes can also be influenced by their (fluorophore) tag. Although we have not yet carried out quantifications, we find that for red fluorescent proteins, tdTomato provides one of the brightest and most evenly distributed tags, while mRFP1 and somewhat more so mCherry are retained in puncta in the cell body (presumably in lysosomes) and thus fewer fluorophores diffuse into the remainder of the cell. For presynaptic reporters (see Chaps. 2 and 3), we find that within the embryonic and early larval CNS tags can affect their localisation: Myc, HA, Cerulean, mRFP1 and mCherry permit excellent localisation when tagged to Synaptotagmin (69) or Bruchpilot (49, 70); in contrast, EGFP and Venus often lead to unexpectedly low and diffuse signals in central synaptic terminals, despite having been shown to work well at peripheral neuromuscular synapses. It is conceivable that the underlying reason for such localisation issues might be the residual capacity for dimerisation contained in EGFP and Venus fluorophores (71).

3.3. Neuron Morphology: Digital Reconstructions and Analysis

3.3.1. Providing a Context for Studying Neurons

Ultimately, one would like to be able to view each neuron in relation to all adjoining cells. This is currently not possible. However, there are straightforward markers that will provide a basic framework within which neurons can be analysed effectively. Visualisation of all neurons in the peripheral and central nervous system with fluorescently conjugated anti-HRP allows one to assign approximate segmental identities, cell body coordinates in the periphery or central cortex, principal neurite trajectories and termination zones: choice of nerve and nerve root for afferent and efferent neurons; within the CNS neuropile: ipsilateral vs. contralateral, anterior vs. posterior commissure choice, approximate medio-lateral and dorso-ventral coordinates neuropile. More accurate assignment of trajectories can be achieved by studying neurons in the context of relatively invariant axon tracts, as visualised by anti-Fasciclin2 or anti-Neurotactin staining (72–76). Similarly important is the assignment of which neurites are putatively pre- and/or postsynaptic, which can be achieved with transgenic reporters for neuronal compartments and synaptic specialisations.

Ultimately, one would like to analyse neuron morphology in the context of synaptic connections with other partner neurons. To achieve this level of resolution, one can use a bimolecular fluorescence reconstitution strategy, called "GRASP" (GFP Reconstitution Across Synaptic Partners). This requires the simultaneous use of two independent expression systems (Gal4/UAS and LexA/LexAop), both fairly specific, one for the presynaptic cell(s) and the other for the(ir) postsynaptic partner(s). As currently available for use in *Drosophila*, the "GRASP" system reports merely contacts between cell membranes rather than synaptic sites. However, in our experience sites of synaptic contacts can be reasonably accurately identified by combining the "GRASP" method with a synaptic reporter, e.g. *bruchpilot-RFP* for presynaptic active zones (49, 70).

3.3.2. Quantitative Analysis of Neuron Morphology

Quantification of neuron morphology, particularly of branched dendritic trees, is a matter of balancing the requirement for accuracy with the specific question in mind. For example, directional growth of neurites can be analysed at a rather coarse level, such as scoring growth angles or evaluating the likelihood of distribution in specific sectors (49). In contrast, studying mechanisms that regulate the growth and branching patterns of neurites requires faithful reconstruction and, through this, quantification of their topology.

Digital 3D reconstructions are invaluable for the extraction and statistical evaluation of cell morphological parameters. A number of software packages are available for this task, each with its own advantages and disadvantages. An awkward but important issue is that dendrites of many types of neurons in the CNS, particularly in invertebrates, can grow so densely that branches appear to form loops. This is clearly not the case, at least generally, yet can be very difficult to resolve clearly, even with confocal microscopy. This morphological characteristic is a challenge to algorithms that aim to fully automate the reconstruction process. To date, it has not been solved in a satisfactory manner. As far as we are concerned, careful visual inspection by the scientist is necessary during the reconstruction process, so as to critically determine the actual topology at points of dense intersection. All automated tools we tested on *invertebrate* central dendrites generated incomplete reconstructions. Worse still, errors generating incorrect connectivity between branches are introduced, which require time consuming manual proofreading and editing.

For these reasons, we prefer custom semi-automatic reconstruction techniques developed by Evers and Schmitt and colleagues: HxSkeletonize, a plug-in module for Amira (Visage Imaging, Germany) which we maintain and distribute freely at http://www.zoo.cam.ac.uk/zoostaff/ndd/software (53, 54). This seems to be the best compromise between speed (part automation) and accuracy, fully harnessing the resolution of the confocal image stacks. The software evaluates fluorescence vectors within confocal image stacks, thus homing in on the source of signal and permitting work with specimens that have relatively low signal to noise ratios. The user manually defines the connectivity between dendritic branch points, and the software automatically traces the connecting branches (see web link for detailed instructions and Fig. 1 illustrating the process and Fig. 2 illustrating the resultant visualisation of a reconstructed neuron).

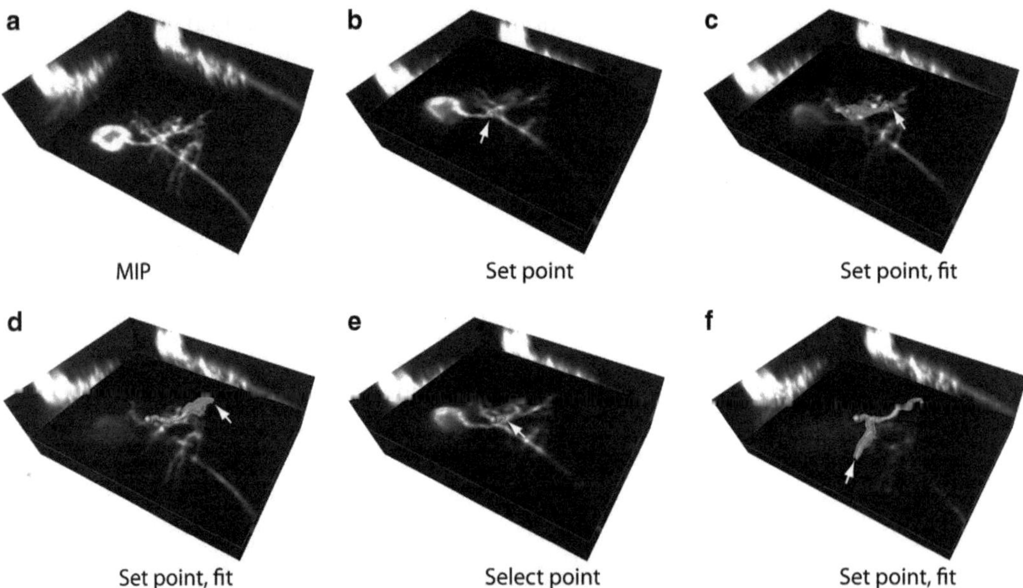

Fig. 1. Process for reconstructing neurons. (**a–f**) High magnification views of parts of the images stack help the user locate neurites in the three-dimensional space and to specify points along neurites by clicking these. (**a**) Maximum intensity projection of the data. (**b**) Selection of the first point (*arrow*) by clicking on the image. (**c**) When the next point along the neurite is similarly selected it is automatically connected with the previously selected point and fitted to the data. (**d**) Repeat of steps (**b**) and (**c**). (**e**) Branch points are selected by the user (*arrow*) and branches are extended by selection of a new endpoint in (**f**).

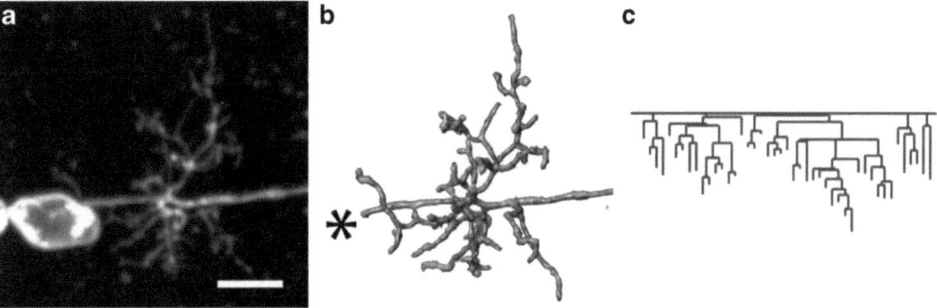

Fig. 2. Visualisation of neurons and reconstructions. (**a**) Maximum intensity projection view of a confocal image stack (RP2 motor neuron in early first instar larva). (**b**) Volume reconstruction of RP2 dendritic tree (**a**) at optical resolution displayed as surface view (algorithmically generated with HxSkeletonize from tubular reconstruction, compare Fig. 1). (**c**) Dendrogram displaying the topology of RP2 dendritic tree as in (**a**). The *topmost line* represents the primary neurite and axon. Distance from the primary neurite is plotted along the *y*-axis. Scale bar: 5 μm.

The process of digitally reconstructing neurons generates almost every imaginable type of quantitative cell morphology data, which can then be mined for analysis. Parameters useful for comparison include total neurite/arbor length; number of neurite/dendritic segments, branch points, branch orders, dendritic/filopodial tips; lengths of neurite/dendritic segments at different branching orders; distribution of neurites within the three-dimensional neuropile using Sholl analysis. For example, we applied this strategy to dendritic arbors of identified motor neurons, so as to address which aspects of a neuron are invariant (indicative of being genetically specified) and which variable (suggestive of epigenetic regulation) (4, 6, 47, 49). Similarly, one can precisely and quantitatively define the morphological features that are affected by different experimental conditions and those which distinguish different neurons. For example, we found that total tree length and number of segments/branch points are distinctive for neuronal types; in contrast, the detailed dendritic tree architecture, including the number of primary branches, can be extremely variable (13, 47, 49).

In addition, these reconstructions can be used further to automatically generate volume reconstructions at optical resolution with the HxSkeletonize plug-in (Fig. 2). This feature has not yet been implemented in other reconstructions frameworks. The advantage is that it allows analysis of neurons in the context of other cells and specialisations: structures visualised in different confocal channels can be mapped with respect to the reconstructed cell. For example, presynaptic sites of putative partner terminals can be mapped onto the surface of dendritic trees and their distribution and density determined (13, 49, 53, 54, 77, 78). Clearly, for these types of application and analysis, resolution is of the essence and everything has to be done to optimise specimen preparation and imaging conditions. The diffraction limited resolution achieved by confocal microscopes is not sufficient to resolve actual synapses; co-localisation data of synaptic markers are indicative rather than definitive and therefore have to be interpreted with due care.

3.4. Naming Neurons: Nomenclature

Almost inextricably linked to the morphology of neurons are descriptor terms. For peripheral sensory neurons, every cell has been charted and catalogued based on position, dendritic morphology, modality, type and sub-class (10, 14, 16, 79, 80). For central neurons, the situation has not yet been resolved. Until now, descriptions have focused on the most basic characteristics: whether a neuron has efferent projections (e.g. motor neurons), and if so, the nerve root of exit from the CNS; whether it has contralateral projections, and if so, through which commissure; whether its axon projects anteriorly or posteriorly (20).

Ultimately, one would like to use names that unequivocally characterise neuronal origin, projections, terminations, connectivity and function in a network context. It will be years before such detailed knowledge will be available for most neurons in the CNS. To deal with this challenge, the most parsimonious solution will be an openly accessible database in which, for every neuron, ontogeny, overall morphology and gene expression data are described together with metadata, putting it in the context of network connections. The benefit of such a database will be that it is adaptable; it can grow as new data become available and existing data are validated. One will be able to refer to neurons by accession numbers (in addition to historical names), which might be better than trying to create a nomenclature that will likely be cumbersome and bound to be superseded by other versions, as our understanding of nervous system development and function increases.

4. Notes

4.1. Notes for 3.1

Most, if not all, current hs-FLP transgenics suffer from varying levels of uninduced basal expression. In our hands, hs-FLP_{122} and hs-FLP_{22} have relatively high levels, while hs-FLP_{1} and hs-FLP_{86E} show relatively low levels of constitutive "leaky" expression. These can be assessed by quantifying excision events generated at different non-heat-shock temperatures, relatively few at 18°C, and increasingly more at progressively higher temperatures (21–25°C), presumably due to the combined temperature sensitive action of both the heat-shock regulatory regions and FLP activity (optimal activity around 30°C (81)). Therefore, when testing a source of hs-FLP, include controls without heat shock kept at 18 and 25°C. Mild heat shocks of 5–10 min at 32°C can be very effective for the purpose of generating mosaic expression. Where strong heat shocks are desired, two consecutive heat shocks separated by a brief period of 15 min are very effective.

4.2. Notes for 3.2

Expression levels: When working with the embryonic nervous system, the duration for transgene expression is often limited and expression levels become critical. Generally, the best way of increasing expression levels is to introduce additional copies of the reporter transgene. Moreover, many of the binary expression systems are temperature sensitive and for Gal4/UAS based systems 29°C provides the highest level of Gal4 activity compatible with normal development. Where expression of the transcriptional activator is weak and/or inconsistent, introduction of a second copy of the transgene often works well; alternatively, introduction of a feedback loop (*UAS-FLP*, *tub84b-FRT-CD2-FRT-Gal4* or equivalent) can boost (and maintain) expression levels.

Genetic backgrounds: Some genetic backgrounds (e.g. certain mutations) do not obviously affect cell morphology per se, but cause subtle delays in the general progression of development so that the timing for staging animals has to be adjusted and normalised for accurate comparisons between control and experimental specimens (J.F.E. and M.L., unpublished).

Fixation, embedding, pH sensitivity: Some red-light-emitting fluorophores, e.g. mRFP, mCherry and tdTomato, preserve relatively high levels of fluorescence when fixing with methanol-free formaldehyde, though less successfully so when fixing with paraformaldehyde. Most commonly used genetically encoded fluorophores are sensitive to changes in pH and fixation and/or mounting of specimens in media with suboptimal pH will also lead to significantly reduced fluorescence (71).

4.3. Notes for 3.3

Segmental landmarks: In the CNS, neuromere boundaries are difficult to delineate but antibody staining against Engrailed marking the posterior part of each neuromere and sets of Hox genes provide indications of (para)segmental boundaries in the cortex.

GRASP system: When using the "GRASP" system in the embryo two factors are important: first, expression levels need to be high and in our hands this can be achieved by using the VP16 version of the LexA activator (62). Second, high resolution confocal imaging is necessary so as to approach as best as possible the actual dimensions of neurites (0.25 μm average diameter) and synapses in the CNS of *Drosophila* larvae (presynaptic sites range from 0.15 to 0.3 μm in diameter) (82).

Reconstructing neurons: The software readily accepts image and associated meta-data from the main confocal imaging companies. For some generic file formats, e.g. tiff stacks, the user will have to enter voxel dimensions.

Fitting parameters: Fitting parameters have to be optimised for a given imaging regime to obtain good quality reconstructions. Internal factors "loc" and "rad" control the degree of smoothness of the reconstruction; higher values increase the coupling force between neighbouring elements. "step" sets the step size at which image data are sampled. "External factors" affect the amount of adjustment made during iterative optimisation of reconstruction accuracy. Larger values will result in bigger changes per fitting cycle. Excessively large values result in failure to iteratively optimise reconstruction accuracy (reconstruction jumps out or oscillates around neuronal branches). In such cases you will have to reduce the values. Excessively small values will result in subtle changes only. As a first approach, apply the same value for all external parameters. Settle with one set of parameters for a specific imaging protocol (objective, scan resolution, embedding media and bit depth).

"Fast fits" are only necessary when initialisation is outside of the structure or snaxels do not assume correct positions when doing an "exact fit". Repeat "exact fit" procedure until only minor changes are visible (iterative process). Resetting the diameter to approximately the right target diameter helps the algorithms to converge on good reconstructions. To optimise a tree reconstruction at the very end, select the whole tree and do the "exact fit" procedure once more.

Check the reconstruction for completeness with reference to the image stack. To check that all branches are actually connected and no unnecessary branch points have been introduced, display the reconstruction as snaxels with very small diameter and branch points being differentially highlighted (also by size); then zoom in and check that all parts are connected, that no circular connections have been generated by accident and that branch points are not clustered next to each other. Manual selection of snaxels allows the user to correct such issues.

Displaying neuronal morphology: Dendritic morphology is not easy to depict as maximum intensity projection views from original image stacks, particularly when diameters of branches are very disparate. 3D reconstructions get around this issue. Reconstructions are often visualised as wire models, but these make it hard to understand the structure of the tree. Tubular reconstructions represent a fair approximation to actual morphology whilst providing clarity. Dendrograms allow to visualise the topology of dendrites: connectivity and branch orders within the tree and make it easy to spot distinct subtrees (possible computational subunits).

5. Future Perspectives

The future will see exciting new developments in areas related to the visualisation and quantification of neuronal cell morphology.

5.1. Genetic Cell Labelling Strategies

The ability to genetically label single, physically isolated cells in the CNS is central to the success of accurately capturing neuron morphology. At the moment, recombinase mediated conditional labelling (as outlined above) is one of the most frequently used approaches. Much needed improvements on current methods include the following: (a) to simultaneously yet differentially express multiple cell and synaptic reporters in different neurons, (b) to implement multiple independent stochastic labelling systems, (c) to overcome the stochastic nature of recombinase based strategies, so as to be able to reliably target specific cells, (d) to improve the ease of use and subcellular targeting of current genetically encoded reporters for electron microscopy (e.g. membrane targeted HRP (83)).

5.2. Reporters for Synaptic Connectivity

The connections that neurons make ultimately constitute an important part of their morphology, and certainly their function. Yet, determining neuronal connectivity in *Drosophila* has remained a real challenge. Serial electron-micrograph reconstructions of considerable parts of the larval CNS are underway, and these have already generated impressive and exciting data on commonly found wiring motifs and other aspects (82). At the same time, this type of analysis is far too labour-intensive to consider for the analysis of how different genetic backgrounds affect connectivity patterns, how these change over developmental time or are affected by mutations and diverse manipulations. Therefore, methods are required that allow investigation of synaptic connections between identified neurons, ideally suitable for live imaging, so that changes can be recorded in real time. A first step in this direction has been the so-called "GRASP" system, first implemented in *Caenorhabditis elegans* (84), then transferred to *Drosophila* (59). At this time, "GRASP" reports cell-cell contacts by reconstitution of GFP fluorescence. Developing this system into a bona fide reporter for synaptic connections will be a major advance. Trans-synaptic tracers, as used in vertebrates, would be a major advance for tracing connectivity in the fly. Ideally, these would be combined with sensitive opto-genetic methods reporting the activity of individual synaptic sites, as pioneered with genetically encoded pH-sensitive fluorescent reporters such as synapto-Phluorin (85).

5.3. Cell Tracing Methods and Integration of Morphology Data into a Coherent Atlas of Connectivity

The third main area of methods development is the implementation of computational strategies for improving the ease and accuracy with which complex neurons can be digitally traced, reconstructed and integrated with other such data into a coherent anatomical atlas. Recent studies successfully combined cell segmentation and/or reconstruction algorithms with warping algorithms so as to fit digitally segmented or reconstructed cell morphologies into common reference brains: http://www.flycircuit.tw (86); http://flybrain.stanford.edu/ (87–90). These methods allow one to generate predictions for likely network wiring diagrams, which can then be tested experimentally. This strategy has thus far been used primarily in the adult nervous system and it has yet to be applied systematically to networks in the *Drosophila* embryo and larva. The relative ease with which identified neurons can be reliably labelled and analysed using light microscopy would ideally be integrated with the resolution that electron microscopy affords. To this end, Cardona et al. have developed algorithms embedded in the EM-track2 software as an interface between these two levels the of resolution and analysis (82). This interface between imaging and analysis methods will undoubtedly remain a hotbed for future developments. Ultimately, studies on cell morphology will converge to models of neuronal networks. As research activity in this area increases, new methods and approaches will be developed and new data sets produced. One challenge will be to harmonise data sets generated by different means so that these can be integrated into a common, standardised reference framework to facilitate comparisons and achieve complete coverage of the nervous system, as has been attempted for neural network models (91). At the time of going to press major advances had been published, detailing the characterisation of embryonic neurons (92) and new genetic tools to visualise these cells (93–94).

References

1. Cajal SR (1995) Histology of the nervous system, vol 1. Oxford University Press, New York
2. Kim IJ, Zhang Y, Yamagata M, Meister M, Sanes JR (2008) Molecular identification of a retinal cell type that responds to upward motion. Nature 452:478–482
3. Branco T, Häusser M (2010) The single dendritic branch as a fundamental functional unit in the nervous system. Curr Opin Neurobiol 20:494–502
4. Levinthal F, Macagno E (1976) Anatomy and development of identified cells in isogenic organisms. Cold Spring Harb Symp Quant Biol 40:321–331
5. Murphey RK (1986) The myth of the inflexible invertebrate: competition and synaptic remodelling in the development of invertebrate nervous systems. J Neurobiol 17:585–591
6. Goodman CS (1978) Isogenic grasshoppers: genetic variability in the morphology of identified neurons. J Comp Neurol 182(4):681–705
7. Vonhoff F, Duch C (2010) Tiling among stereotyped dendritic branches in an identified *Drosophila* motoneuron. J Comp Neurol 518:2169–2185
8. Voyvodic JT (1987) Development and regulation of dendrites in the rat superior cervical ganglion. J Neurosci 7:904–912

9. Gao F-B, Brenman JE, Jan LY, Jan YN (1999) Genes regulating dendritic outgrowth, branching, and routing in *Drosophila*. Genes Dev 13:2549–2561
10. Grueber WB, Jan LY, Jan YN (2003) Different levels of the homeodomain protein cut regulate distinct dendrite branching patterns of *Drosophila* multidendritic neurons. Cell 112: 805–818
11. Wu GY, Cline HT (2003) Time-lapse in vivo imaging of the morphological development of *Xenopus* optic tectal interneurons. J Comp Neurol 459:392–406
12. Parrish JZ, Kim MD, Jan LY, Jan YN (2006) Genome-wide analyses identify transcription factors required for proper morphogenesis of *Drosophila* sensory neuron dendrites. Genes Dev 20:820–835
13. Tripodi M, Evers JF, Mauss A, Bate M, Landgraf M (2008) Structural homeostasis: compensatory adjustments of dendritic arbor geometry in response to variations of synaptic input. PLoS Biol 6:e260
14. Bodmer R, Jan Y (1987) Morphological differentiation of the embryonic peripheral neurons in *Drosophila*. Development genes and evolution 196:69–77
15. Dambly-Chaudière C, Ghysen A (1986) The sense organs in the *Drosophila* larva and their relation to the embryonic pattern of sensory neurons. Dev Genes Evol 195:222–228
16. Campos-Ortega JA, Hartenstein V (1985) The embryonic development of *Drosophila melanogaster*. Springer, Berlin
17. Bodmer R, Carretto R, Jan YN (1989) Neurogenesis of the peripheral nervous system in *Drosophila* embryos: DNA replication patterns and cell lineages. Neuron 3:21–32
18. Orgogozo V, Schweisguth F, Bellaïche Y (2001) Lineage, cell polarity and inscuteable function in the peripheral nervous system of the *Drosophila* embryo. Development 128:631–643
19. Grueber WB, Jan LY, Jan YN (2002) Tiling of the *Drosophila* epidermis by multidendritic sensory neurons. Development 129:2867–2878
20. Schmidt H, Rickert C, Bossing T, Vef O, Urban J, Technau GM (1997) The embryonic central nervous system lineages of *Drosophila melanogaster*. II. Neuroblast lineages derived from the dorsal part of the neuroectoderm. Dev Biol 189:186–204
21. Bossing T, Udolph G, Doe CQ, Technau GM (1996) The embryonic central nervous system lineages of *Drosophila melanogaster*. I. Neuroblast lineages derived from the ventral half of the neuroectoderm. Dev Biol 179:41–64
22. Bossing T, Technau GM (1994) The fate of the CNS midline progenitors in *Drosophila* as revealed by a new method for single cell labelling. Development 120:1895–1906
23. Doe CQ, Technau GM (1993) Identification and cell lineage of individual neural precursors in the *Drosophila* CNS. Trends Neurosci 16: 510–514
24. Schmid A, Chiba A, Doe CQ (1999) Clonal analysis of *Drosophila* embryonic neuroblasts: neural cell types, axon projections and muscle targets. Development 126:4653–4689
25. Larsen C, Shy D, Spindler SR, Fung S, Pereanu W, Younossi-Hartenstein A, Hartenstein V (2009) Patterns of growth, axonal extension and axonal arborization of neuronal lineages in the developing *Drosophila* brain. Dev Biol 335:289–304
26. Isshiki T, Pearson B, Holbrook S, Doe CQ (2001) *Drosophila* neuroblasts sequentially express transcription factors which specify the temporal identity of their neuronal progeny. Cell 106:511–521
27. Novotny T, Eiselt R, Urban J (2002) Hunchback is required for the specification of the early sublineage of neuroblast 7-3 in the *Drosophila* central nervous system. Development 129: 1027–1036
28. Karcavich R, Doe CQ (2005) *Drosophila* neuroblast 7-3 cell lineage: a model system for studying programmed cell death, Notch/ Numb signaling, and sequential specification of ganglion mother cell identity. J Comp Neurol 481:240–251
29. Lacin H, Zhu Y, Wilson BA, Skeath JB (2009) dbx mediates neuronal specification and differentiation through cross-repressive, lineage-specific interactions with eve and hb9. Development 136(19):3257–3266
30. Karlsson D, Baumgardt M, Thor S (2010) Segment-specific neuronal subtype specification by the integration of anteroposterior and temporal cues. PLoS Biol 8:e1000368
31. DeFelipe J, Jones EG (1992) Santiago Ramón y Cajal and methods in neurohistology. Trends Neurosci 15(7):237–245
32. Zacks SI, Atushi S (1969) Uptake of exogenous horseradish peroxidase by coated vesicles in mouse neuromuscular junctions. J Histochem Cytochem 17:161–179
33. Kristensson K, Olsson Y, Sjöstrand J (1971) Axonal uptake and retrograde transport of exogenous proteins in the hypoglossal nerve. Brain Res 32(2):399–406
34. Takato M, Goldring S (1979) Intracellular marking with Lucifer Yellow CH and horseradish

peroxidase of cells electrophysiologically characterized as glia in the cerebral cortex of the cat. J Comp Neurol 186:173–188

35. Bonhoeffer F, Huf J (1980) Recognition of cell types by axonal growth cones in vitro. Nature 288:162–164
36. Kuypers H, Catsman-Berrevoets CE, Padt RE (1977) Retrograde anoxal transport of fluorescent substances in the rat's forebrain. Neurosci Lett 6:127–133
37. Honig MG, Hume RI (1986) Fluorescent carbocyanine dyes allow living neurons of identified origin to be studied in long-term cultures. J Cell Biol 103:171–187
38. Callahan CA, Thomas JB (1994) Tau-beta-galactosidase, an axon-targeted fusion protein. Proc Natl Acad Sci U S A 91:5972–5976
39. Wells W, Jan L, Jan Y (1993) Tracing neurons with a kinesin-β-galactosidase fusion protein. Development genes and evolution 222:112–122
40. Dunin-Borkowski OM, Brown NH (1995) Mammalian CD2 is an effective heterologous marker of the cell surface in *Drosophila*. Dev Biol 168:689–693
41. Lee T, Luo L (1999) Mosaic analysis with a repressible cell marker for studies of gene function in neuronal morphogenesis. Neuron 22:451–461
42. Kaltschmidt JA, Davidson CM, Brown NH, Brand AH (2000) Rotation and asymmetry of the mitotic spindle direct asymmetric cell division in the developing central nervous system. Nat Cell Biol 2:7–12
43. Glaser EM, Van Der Loos H (1965) A semi-automatic computer-microscope for the analysis of neuronal morphology. Biomed Eng 12:22–31
44. Meijering E (2010) Neuron tracing in perspective. Cytometry A 77:693–704
45. Pignoni F, Zipursky SL (1997) Induction of *Drosophila* eye development by decapentaplegic. Development 124:271–278
46. Roy B, Singh AP, Shetty C, Chaudhary V, North A, Landgraf M, Vijayraghavan K, Rodrigues V (2007) Metamorphosis of an identified serotonergic neuron in the *Drosophila* olfactory system. Neural Dev 2:20
47. Ou Y, Chwalla B, Landgraf M, van Meyel DJ (2008) Identification of genes influencing dendrite morphogenesis in developing peripheral sensory and central motor neurons. Neural Dev 3:16
48. Pearson BJ, Doe CQ (2003) Regulation of neuroblast competence in *Drosophila*. Nature 425:624–628
49. Mauss A, Tripodi M, Evers JF, Landgraf M (2009) Midline signalling systems direct the formation of a neural map by dendritic targeting in the *Drosophila* motor system. PLoS Biol 7:e1000200
50. Sugimura K, Yamamoto M, Niwa R, Satoh D, Goto S, Taniguchi M, Hayashi S, Uemura T (2003) Distinct developmental modes and lesion-induced reactions of dendrites of two classes of *Drosophila* sensory neurons. J Neurosci 23:3752–3760
51. Pfeiffer BD, Ngo TT, Hibbard KL, Murphy C, Jenett A, Truman JW, Rubin GM (2010) Refinement of tools for targeted gene expression in *Drosophila*. Genetics 186:735–755
52. Nicolaï LJ, Ramaekers A, Raemaekers T, Drozdzecki A, Mauss AS, Yan J, Landgraf M, Annaert W, Hassan BA (2010) Genetically encoded dendritic marker sheds light on neuronal connectivity in *Drosophila*. Proc Natl Acad Sci U S A 107:20553–20558
53. Evers JF, Schmitt S, Sibila M, Duch C (2005) Progress in functional neuroanatomy: precise automatic geometric reconstruction of neuronal morphology from confocal image stacks. J Neurophysiol 93:2331–2342
54. Schmitt S, Evers JF, Duch C, Scholz M, Obermayer K (2004) New methods for the computer-assisted 3-D reconstruction of neurons from confocal image stacks. Neuroimage 23:1283–1298
55. Budnik V, Gorczyca M, Prokop A (2006) Selected methods for the anatomical study of *Drosophila* embryonic and larval neuromuscular junctions. Int Rev Neurobiol 75:323–365
56. Golic KG, Lindquist S (1989) The FLP recombinase of yeast catalyzes site-specific recombination in the *Drosophila* genome. Cell 59: 499–509
57. Struhl G, Basler K (1993) Organizing activity of wingless protein in *Drosophila*. Cell 72:527–540
58. Wong AM, Wang JW, Axel R (2002) Spatial representation of the glomerular map in the *Drosophila* protocerebrum. Cell 109:229–241
59. Gordon MD, Scott K (2009) Motor control in a *Drosophila* taste circuit. Neuron 61:373–384
60. Fischer JA, Giniger E, Maniatis T, Ptashne M (1988) GAL4 activates transcription in *Drosophila*. Nature 332:853–856
61. Brand AH, Perrimon N (1993) Targeted gene expression as a means of altering cell fates and generating dominant phenotypes. Development 118:401–415
62. Lai SL, Lee T (2006) Genetic mosaic with dual binary transcriptional systems in *Drosophila*. Nat Neurosci 9:703–709
63. Potter CJ, Tasic B, Russler EV, Liang L, Luo L (2010) The Q system: a repressible binary

system for transgene expression, lineage tracing, and mosaic analysis. Cell 141:536–548

64. Fujioka M, Emi-Sarker Y, Yusibova GL, Goto T, Jaynes JB (1999) Analysis of an even-skipped rescue transgene reveals both composite and discrete neuronal and early blastoderm enhancers, and multi-stripe positioning by gap gene repressor gradients. Development 126: 2527–2538
65. Livet J, Weissman TA, Kang H, Draft RW, Lu J, Bennis RA, Sanes JR, Lichtman JW (2007) Transgenic strategies for combinatorial expression of fluorescent proteins in the nervous system. Nature 450:56–62
66. Hampel S, Chung P, McKellar CE, Hall D, Looger LL, Simpson JH (2011) *Drosophila* Brainbow: a recombinase-based fluorescence labeling technique to subdivide neural expression patterns. Nat Methods 8:253–259
67. Hadjieconomou D, Rotkopf S, Alexandre C, Bell DM, Dickson BJ, Salecker I (2011) Flybow: genetic multicolor cell labeling for neural circuit analysis in *Drosophila melanogaster*. Nat Methods 8:260–266
68. Bohm RA, Welch WP, Goodnight LK, Cox LW, Henry LG, Gunter TC, Bao H, Zhang B (2010) A genetic mosaic approach for neural circuit mapping in *Drosophila*. Proc Natl Acad Sci U S A 107:16378–16383
69. Robinson IM, Ranjan R, Schwarz TL (2002) Synaptotagmins I and IV promote transmitter release independently of Ca(2+) binding in the C(2)A domain. Nature 418:336–340
70. Wagh DA, Rasse TM, Asan E, Hofbauer A, Schwenkert I, Dürrbeck H, Buchner S, Dabauvalle MC, Schmidt M, Qin G, Wichmann C, Kittel R, Sigrist SJ, Buchner E (2006) Bruchpilot, a protein with homology to ELKS/CAST, is required for structural integrity and function of synaptic active zones in *Drosophila*. Neuron 49:833–844
71. Shaner NC, Steinbach PA, Tsien RY (2005) A guide to choosing fluorescent proteins. Nat Methods 2:905–909
72. Nassif C, Noveen A, Hartenstein V (1998) Embryonic development of the *Drosophila* brain. I. Pattern of pioneer tracts. J Comp Neurol 402:10–31
73. Nassif C, Noveen A, Hartenstein V (2003) Early development of the *Drosophila* brain: III. The pattern of neuropile founder tracts during the larval period. J Comp Neurol 455: 417–434
74. Landgraf M, Sánchez-Soriano N, Technau G, Urban J, Prokop A (2003) Charting the *Drosophila* neuropile: a strategy for the standardised characterisation of genetically amenable neurites. Dev Biol 260:207–225
75. Cardona A, Larsen C, Hartenstein V (2009) Neuronal fiber tracts connecting the brain and ventral nerve cord of the early *Drosophila* larva. J Comp Neurol 515:427–440
76. Truman JW, Schuppe H, Shepherd D, Williams DW (2004) Developmental architecture of adult-specific lineages in the ventral CNS of *Drosophila*. Development 131:5167–5184
77. Evers JF, Muench D, Duch C (2006) Developmental relocation of presynaptic terminals along distinct types of dendritic filopodia. Dev Biol 297:214–227
78. Meseke M, Evers JF, Duch C (2009) Developmental changes in dendritic shape and synapse location tune single-neuron computations to changing behavioral functions. J Neurophysiol 102:41–58
79. Ghysen A, Lewis EB (1986) The function of *bithorax* genes in the abdominal central nervous system of *Drosophila*. Roux's Arch Dev Biol 195:203–209
80. Stocker RF (1994) The organization of the chemosensory system in *Drosophila melanogaster*: a review. Cell Tissue Res 275:3–26
81. Buchholz F, Angrand PO, Stewart AF (1998) Improved properties of FLP recombinase evolved by cycling mutagenesis. Nat Biotechnol 16:657–662
82. Cardona A, Saalfeld S, Preibisch S, Schmid B, Cheng A, Pulokas J, Tomancak P, Hartenstein V (2010) An integrated micro- and macroarchitectural analysis of the *Drosophila* brain by computer-assisted serial section electron microscopy. PLoS Biol 8(10):e1000502
83. Clements J, Lu Z, Gehring WJ, Meinertzhagen IA, Callaerts P (2008) Central projections of photoreceptor axons originating from ectopic eyes in *Drosophila*. Proc Natl Acad Sci U S A 105:8968–8973
84. Feinberg EH, Vanhoven MK, Bendesky A, Wang G, Fetter RD, Shen K, Bargmann CI (2008) GFP Reconstitution Across Synaptic Partners (GRASP) defines cell contacts and synapses in living nervous systems. Neuron 57:353–363
85. Yuste R, Miller RB, Holthoff K, Zhang S, Miesenböck G (2000) Synapto-pHluorins: chimeras between pH-sensitive mutants of green fluorescent protein and synaptic vesicle membrane proteins as reporters of neurotransmitter release. Methods Enzymol 327:522–546
86. Chiang AS, Lin CY, Chuang CC, Chang HM, Hsieh CH, Yeh CW, Shih CT, Wu JJ, Wang GT, Chen YC, Wu CC, Chen GY, Ching YT, Lee PC, Lin CY, Lin HH, Wu CC, Hsu HW,

Huang YA, Chen JY, Chiang HJ, Lu CF, Ni RF, Yeh CY, Hwang JK (2011) Three-dimensional reconstruction of brain-wide wiring networks in *Drosophila* at single-cell resolution. Curr Biol 21:1–11

87. Chou YH, Spletter ML, Yaksi E, Leong JC, Wilson RI, Luo L (2010) Diversity and wiring variability of olfactory local interneurons in the *Drosophila* antennal lobe. Nat Neurosci 13:439–449

88. Jefferis GS, Potter CJ, Chan AM, Marin EC, Rohlfing T, Maurer CR, Luo L (2007) Comprehensive maps of *Drosophila* higher olfactory centers: spatially segregated fruit and pheromone representation. Cell 128:1187–1203

89. Cachero S, Ostrovsky AD, Yu JY, Dickson BJ, Jefferis GS (2010) Sexual dimorphism in the fly brain. Curr Biol 20:1589–1601

90. Yu JY, Kanai MI, Demir E, Jefferis GS, Dickson BJ (2010) Cellular organization of the neural circuit that drives *Drosophila* courtship behavior. Curr Biol 20:1602–1614

91. Gleeson P, Crook S, Cannon RC, Hines ML, Billings GO, Farinella M, Morse TM, Davison AP, Ray S, Bhalla US, Barnes SR, Dimitrova YD, Silver RA (2010) NeuroML: a language for describing data driven models of neurons and networks with a high degree of biological detail. PLoS Comput Biol 6:e1000815

92. Rickert C, Kunz T, Harris K-L et al. (2011) Morphological characterization of the entire interneuron population reveals principles of neuromere organization in the ventral nerve cord of Drosophila. The Journal of neuroscience : the official journal of the Society for Neuroscience 31:15870–15883

93. Nern A, Pfeiffer BD, Svoboda K et al. (2011) Multiple new site-specific recombinases for use in manipulating animal genomes. Proc Natl Acad Sci USA 108: 14198–14203

94. Pfeiffer BD, Truman JW, Rubin GM (2012) Using translational enhancers to increase transgene expression in Drosophila. Proc Natl Acad Sci U S A 109:6626–6631

Part II

Physiology

Chapter 6

Studying Synaptic Transmission at the *Drosophila* Neuromuscular Junction Using Advanced FM 1-43 Technology

Ana Clara Fernandes*, Valerie Uytterhoeven*, and Patrik Verstreken

Abstract

Neurons communicate at synapses by releasing neurotransmitters from synaptic vesicles, and this communication underlies information transfer in neuronal circuits. While classic methodologies including electrophysiology and electron microscopy are still extensively used in molecular synaptic transmission research, advanced imaging tools have taken a center stage. In this chapter, we review techniques to study presynaptic function and describe two advanced protocols that make use of the membrane binding dye FM 1-43, enabling to probe deep into the mechanisms of synaptic vesicle function and dynamics at the *Drosophila* third instar larval neuromuscular junction.

Key words: Neurotransmitter release, Endocytosis, Third instar neuromuscular junction, FM 1-43, Imaging, Photoconversion, Synaptic vesicle

1. Introduction

The neurons in our brain communicate using small transmitter-filled synaptic vesicles. During intense neuronal activity, some neurons fire up to 800 times per second, releasing transmitters each time, and it is this communication that guarantees accurate transfer of information travelling through neuronal circuits, leading to higher brain functions such as behavior, memory and thought. Proper control of neurotransmission is also critical for normal brain function, as is evident from the altered behavior and mood-states induced by narcotics and psychiatric disorders that affect neurotransmission (1–5). Hence, understanding the mechanisms

*These authors Ana Clara Fernandes and Valerie Uytterhoeven contributed equally to this work.

Bassem A. Hassan (ed.), *The Making and Un-Making of Neuronal Circuits in Drosophila*, Neuromethods, vol. 69,
DOI 10.1007/978-1-61779-830-6_6, © Springer Science+Business Media, LLC 2012

underlying neurotransmission will help elucidate normal and pathological brain function.

Looking beyond a single synapse, neurons assemble into circuits that underlie brain function. While neuroscientists have begun visualizing behavioral circuits, as outlined elsewhere in this volume, the function of these circuits remains largely elusive. Using knowledge gathered from single synapses, we may be able to manipulate single neurons within particular circuits and analyze information flow and behavior. Ultimately, the results of such work will help elucidate higher brain function as well.

Maintenance of neurotransmission depends not only on efficient vesicle fusion machinery but also on the continuous availability of vesicles at the synapse during stimulation (6, 7). Given that neuronal cell bodies are often located far from nerve endings, synapses operate in part autonomously, and vesicles that are depleted during stimulation are rapidly and locally replenished for continued neurotransmission. Numerous proteins and lipids implicated in the regulation of presynaptic function have been characterized (8–10), and studying the effect of these molecules on the behavior of synaptic vesicles at presynaptic nerve terminals is thus an important aspect in our understanding of the mechanisms of synaptic activity.

1.1. The Amazing Power of Genetics

Studying the molecular mechanisms of presynaptic function most often requires loss and gain of protein function. Pharmacological inhibition or inhibition of protein function using antibodies and peptides in giant synapse preparations are fast and acutely alter synaptic function, but the observed in vivo effect is not always specific and sometimes hard to reconcile with in vitro observations. Conversely, classical genetic manipulation in model organisms is specific for the protein studied, but the chronic loss of gene function may in some cases result in compensatory cellular changes that mask the full extent of the protein's function, particularly in the regulation of synaptic transmission (11, 12). In the last decade, technology for genetically encoded acute protein inactivation has been introduced to synaptic research (13, 14). Here, a protein of interest is expressed as a fusion protein in a null mutant background and an external trigger activates the tag that incapacitates the protein of interest effectively creating a (local) loss of protein function. The most commonly used methodology is FlAsH-mediated Fluorescein Assisted Light Inactivation (FALI) (13–15), but other methodologies have been developed as well (Shield 1 (16), FK-FALI (17)) and may be applicable to create situations where two or more proteins are to be sequentially inactivated. In FlAsH-FALI, a protein of interest is tagged using a tetracysteine tag (13) that specifically binds FlAsH, a membrane permeable fluorescein derivative (15). When excited by green light FlAsH creates reactive oxygen species within an action radius of only a few angstroms effectively inactivating the nearby protein of interest (18). This technology has been successfully used to study the role of several endocytic proteins in fruit flies (13, 14), and we would like to refer to Habets and Verstreken for the use of FlAsH-FALI at the *Drosophila* larval NMJ (19). Taken together, genetic analyses in combination with acute protein inactivation are exciting developments to further tackle the mechanisms of synaptic transmission.

1.2. Investigating the Molecular Mechanisms of Synaptic Transmission

Classically, studies of synaptic transmission include electrophysiological analyses, where the electrical signal elicited at the postsynaptic terminal in response to presynaptic release reveals aspects of presynaptic function. Quantal parameters such as vesicle size, synaptic vesicle pool size, and release probabilities can be calculated from such datasets and various modes of presynaptic plasticity can be assessed (8, 10, 20, 21). While optogenetics tools that also monitor presynaptic activity are being hastily introduced (see Chap. 3), they do not hold the temporal resolution of electrophysiological recordings. Thus, an ideal model synapse for studying the mechanisms of synaptic transmission should be amenable to extensive electrophysiological manipulation.

Ultrastructural studies to assess synaptic organellar composition are also integral to the study of synaptic function. Defects in neurotransmitter release often result in an increase in synaptic vesicle number (22–24), while defects in vesicle endocytosis or vesicle recycling frequently result in an accumulation of endocytic intermediates and larger cisternae that have formed as a result of inefficient vesicle reformation (12, 14, 25). While electron microscopy and three-dimensional electron tomography are invaluable assets when assessing presynaptic function, tissue is fixed for imaging, thus only allowing researchers to investigate an end-point. In contrast, superresolution (STED) live imaging (26) of synaptic vesicle dynamics is becoming feasible and technology that enables to combine live imaging with ultrastructural electron microscopy analyses, such as Correlative Light Electron Microscopy (CLEM) (27), or photoconversion (described below) is being successfully employed at synapses to follow organelle dynamics.

As described elsewhere in this volume, several activity monitoring fluorescent probes exist and while their temporal dynamics may not surpass electrophysiological recordings, they enable to follow specific presynaptic features that are not always detectable using recording electrodes. Synaptic transmission is critically dependent on membrane voltage changes (action potentials and graded potentials) (28, 29), calcium influx through voltage gated calcium channels (10, 30), and at chemical synapses also on synaptic vesicle fusion and neurotransmitter release (30). While membrane permeable fluorescent molecules that monitor specific synaptic features (e.g., calcium influx) exist and are extensively used, genetically encoded probes hold the advantage that they can be targeted to specific neuronal cell populations in the brain, enabling in situ monitoring of synaptic activity (31). For the study of synaptic transmission, genetically encoded synapto-pHluorin and the membrane binding FM 1-43 dye are invaluable tools that specifically report on synaptic vesicle dynamics (32–34).

Synapto-pHluorin constitutes a fusion protein between a synaptic vesicle-associated protein (the original version of this probe harbors *n*-Synaptobrevin) and a pH sensitive GFP (32). The GFP moiety of the protein is present at the luminal side of synaptic vesicles and is quenched in the acid environment inside the synaptic vesicle. Upon fusion the pH rises and GFP fluorescence is detected. The rise of GFP fluorescence is thus a measure for vesicle fusion and the fall in GFP fluorescence is a measure of synaptic membrane uptake and reacidification when new synaptic vesicles re-form. This probe is widely used, but in some systems (including *Drosophila*) the probe suffers from low signal to noise ratios likely because synaptic vesicles in *Drosophila* are less acid than in vertebrate systems (35). Thus, new developments to improve this tool's dynamic range will be welcomed.

FM 1-43 is a lipophilic dye that when in aqueous environment is nonfluorescent but dramatically increases quantum yield when bound to membranes (36). When added to a synaptic preparation, the probe binds the presynaptic membrane, and upon vesicle formation, is internalized into the membrane of newly formed synaptic vesicles, thus labeling synaptic terminals fluorescently and yielding a measure of vesicle endocytosis (37, 38). Conversely, when FM 1-43 labeled synapses are stimulated in the absence of externally added dye, FM 1-43 is released from fusing vesicles, and thus the decrease in fluorescence is a measure of exocytosis (39, 40). Given that FM 1-43 is readily internalized into synaptic vesicles, various stimulation paradigms have been employed to label specific vesicle pools (reserve pool, readily releasable pool, etc.) (39–41) and to follow their dynamic behavior under various conditions, allowing researchers to study mechanisms of synaptic plasticity. Furthermore, FM 1-43 labels the endocytic route that synaptic vesicles follow after formation at the presynaptic membrane and this feature can be exploited to assess if vesicles travel via intermediate compartments and/or exchange content within nerve terminals at—for example—endosomal stations (42–45). Given the versatility of FM 1-43, in this chapter we will describe how to combine FM 1-43 imaging and electron microscopy and how to perform double labeling of FM 1-43 with other organellar markers, thus significantly expanding the broad applicability of this tool in the study of synaptic transmission. We would also like to refer to Verstreken et al. (37) for "basic" FM 1-43 labeling protocols.

1.3. The Third Instar Larval Drosophila Neuromuscular Junction as an Ideal System to Study Synaptic Transmission

No ideal model synapse to study synaptic transmission exists, but the *Drosophila melanogaster* third instar larval neuromuscular junction (NMJ) harbors a number of unique features that make it very suitable to study synaptic transmission:

(1) The mechanisms of synaptic transmission are very well conserved across species, and almost all vertebrate proteins

implicated in synaptic communication exist in fruit flies (46, 47). The synaptic proteins that are not found in the fruit fly genome are also not found in the *Caenorhabditis elegans* genome and those not found in the fly genome and that have been studied at mammalian synapses do not hold essential functions in the process but rather play a modulatory role (48–50). Thus, essential processes involved in synaptic function are conserved between flies and man.

(2) The morphology of larval NMJs is very well documented, and given that the innervation pattern of motor neurons on muscles is very stereotyped, single cell studies can be compared from animal to animal. Thus, the fruit fly NMJ in third instar larvae allows for defined single cell quantitative analyses.

(3) Given the easy access to this synapse and the superficial location of its boutons that can measure up to 5 μm in diameter, electrophysiology, electron microscopy, and imaging are very feasible. Thus, at this synapse numerous state-of-the-art techniques can be combined, usually allowing us to propose specific functions for the proteins studied.

(4) The molecular genetics tool-box in *Drosophila* is vast (for an overview see (51, 52)) and most genetic tools that have been developed for fruit fly research are applicable at the NMJ. Notably, because motor neurons contact muscle cells, specific promoters to drive transgene expression in neurons (presynaptically at the NMJ) or in muscles (postsynaptically at the NMJ) exist (53). Furthermore, tools to circumvent early lethality that may be associated with more severe mutants have been successfully employed, including RNAi-mediated knock down and FlAsH-mediated FALI (see above) (13, 14, 54, 55).

(5) Numerous "optogenetic" probes and other new technologies perform outstandingly at this synapse, including super resolution imaging, calcium imaging, etc. (32, 56–58).

Given these advantages the larval fruit fly NMJ constitutes an ideal model system to study synaptic transmission, and in this chapter we outline two specific advanced FM 1-43 applications that allow to probe deep into presynaptic function at this synapse.

2. Materials

2.1. Larval Preparation and Dissection

- Modified HL-3 solution: 110 mM NaCl, 5 mM KCl, 10 mM $NaHCO_3$, 5 mM HEPES, 30 mM Sucrose, 5 mM Trehalose, 10 mM $MgCl_2$, pH 7.2
- Sylgard dissection plate

- Insect dissection pins
- Dissection forceps and dissection scissors
- Dissection stereomicroscope

2.2. FM 1-43 Loading

- Stimulation solution (modified HL-3+90 mM KCl+1.5 mM $CaCl_2$, pH 7.2): 25 mM NaCl, 90 mM KCl, 10 mM $NaHCO_3$, 5 mM HEPES, 30 mM sucrose, 5 mM trehalose, 10 mM MgCl.
- FM 1-43FX (photoconversion) or FM 1-43 (colocalization) (4 μM, final concentration) (Invitrogen).

2.3. Photoconversion of 3,3 Diamino-benzidine

- 0.1 M phosphate buffered saline (PBS) pH 7.4
- 2 mg/mL 3,3 diaminobenzidine (DAB, Sigma) dissolved in PBS (See Note 1)
- Fixative: 1% glutaraldehyde and 4% paraformaldehyde in 0.1 M Na-Cacodylate buffer
- Water immersion lens ×40 0.8 NA (or higher)
- Nikon FN1 epifluorescent microscope with an Intensilight (C-HGFL, Nikon) light source filtered through a standard FITC excitation filter (470/30 nm)

2.4. Image Acquisition on a Confocal Microscope and Quantification of Colocalization

- Water immersion lens ×63 1.0 NA
- For GFP/FM 1-43 double labeling: confocal microscope with 488 nm laser excitation and 510/20 nm band pass emission filter (GFP) and 570–620 nm spectral emission using a 510Meta detector (FM 1-43)
- Photoshop and ImageJ for data quantification

3. Methods

FM 1-43 and related FM dyes have been widely used to follow endocytic processes also at synapses, and further developments that make use of FM 1-43 renew the interest in this tool. In the following sections, we describe how FM 1-43 is used to bridge the gap between live fluorescence imaging and electron microscopy by using FM 1-43 fluorescence in synaptic vesicles to photoconvert DAB into an electron dense precipitate visible at EM level, thus allowing us to assess endocytic traffic at single vesicle resolution (43, 59, 60). Furthermore, we also discuss live double-labeling experiments of fluorescent (protein) markers and FM 1-43, enabling subcellular analyses of synaptic vesicle transport (43). These developments significantly expand the usefulness of FM dyes and will help to further probe into the mechanisms of synaptic vesicle function and neurotransmission.

3.1. Loading and Unloading of Synaptic Vesicles with FM 1-43

Dissect the larvae, cut the motor neurons—see Note 2—(also please refer to Verstreken et al. (37) for detailed guidelines on this procedure) and incubate dissected preparations in stimulation solution containing FM 1-43FX (photoconversion) or FM 1-43 (colocalization) (4 μM) for 1 min. Wash the larvae for 5 min with normal modified HL-3 in order to remove noninternalized dye. The samples are now ready for imaging (Fig. 1a). For details on solutions and imaging see Notes 3 and 4, respectively. If unloading of FM 1-43 is desired repeat the same stimulation protocol but using stimulation solution without FM 1-43. For a complete description of the procedures and different loading and unloading paradigms, please see Table 1 in (37).

3.2. DAB Photoconversion into Electron Dense Precipitates

Load the synaptic boutons with FM 1-43 as described in Sect. 3.1 (See Note 5). Pre-fix the labeled samples by incubating them in fixative solution during 15 min at room temperature and wash the larvae in 0.1 M PBS, pH 7.4. Following the pre-fixation of the samples, incubate them 2×5 min in DAB (2 mg/mL)—Note 6. For photoconversion of DAB into electron dense precipitates, illuminate the regions for photoconversion for 20 min through a ×40 0.8 NA water immersion lens using a fluorescent light and a FITC excitation filter (470/30 nm)—Notes 1 and 7. Finally wash samples 3×10 min in PBS pH 7.4. The presence of brown precipitates as a sign of efficient photoconversion can be verified using standard light microscopy (Fig. 1b). Process samples for transmission electron microscopy (TEM) (38)—see Note 8. Figure 1c, d compares standard TEM with TEM on photoconverted samples, respectively. In Fig. 1d several small synaptic vesicles containing DAB precipitates (arrows), indicate FM 1-43 photoconversion. For details on quantification see Note 9.

3.3. Colocalization of FM 1-43 with GFP Tagged Proteins

Collect larvae expressing the GFP-tagged protein of interest and label the NMJs with FM 1-43 as described in Sect. 3.1—see Notes 10 and 11. Image the larva using an upright confocal microscope with a ×63 NA 1.0 water immersion objective (×40 NA 0.8 works fine as well but yields less signal), using an excitation wavelength of 488 nm (excites both FM 1-43 and GFP—see Fig. 2) and detect both signals separately by using a 510/20 nm emission filter for GFP and 570–620 nm emission settings on a 510Meta detector for FM 1-43. Although these settings ensure minimal cross talk, controls with NMJ samples labeled with only GFP or samples labeled with only FM 1-43, as outlined in Note 11 are essential. See also Note 12 for imaging with other fluorophores.

To quantify the percentage of colocalization between GFP and FM 1-43 in the images acquired, ImageJ can be used. Apply a suitable threshold using the standard threshold module in ImageJ such that the individually thresholded images include the labeled areas for GFP and for FM 1-43 and multiply the thresholded images in

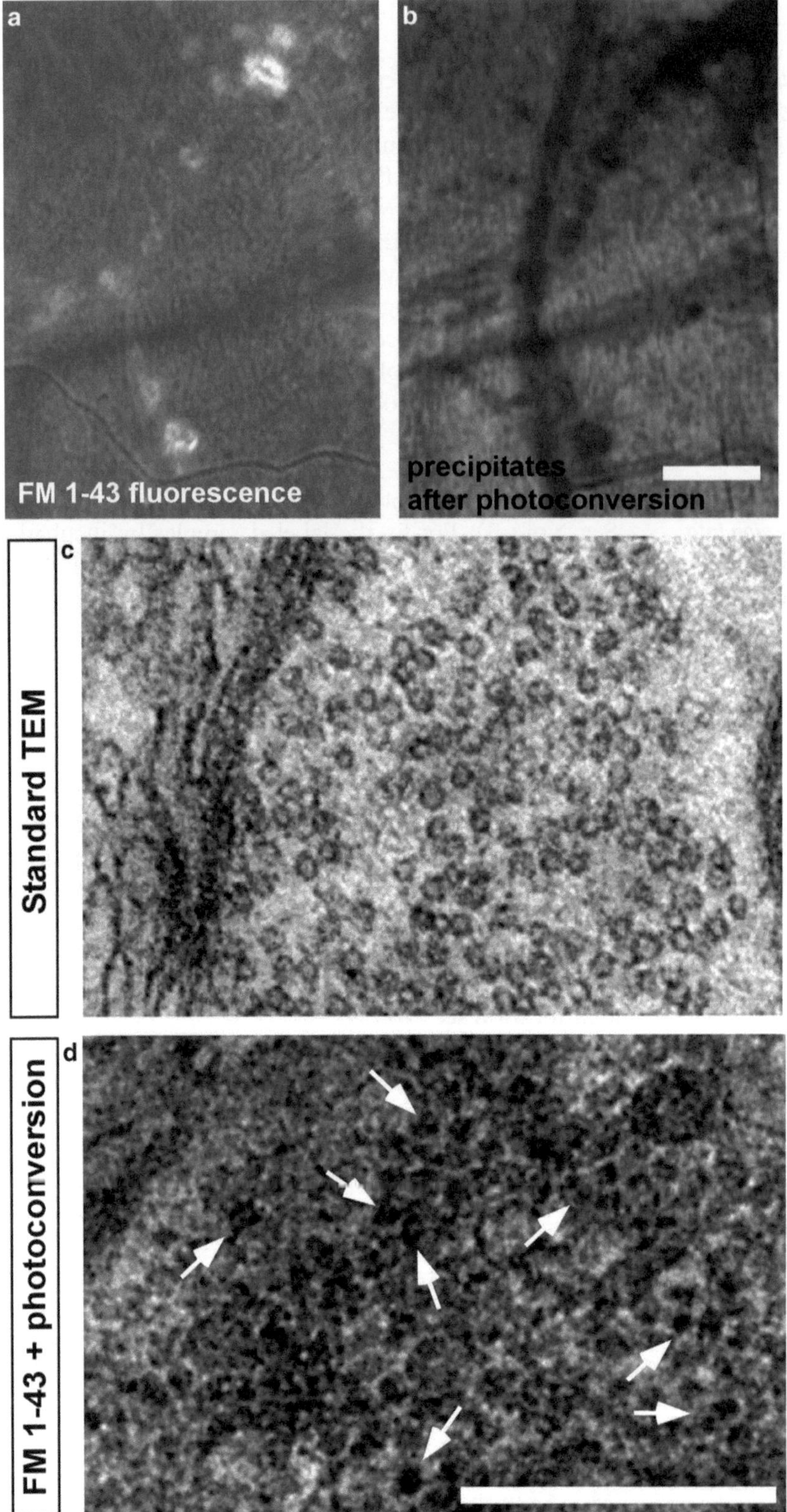

Fig. 1. Photoconversion of DAB in FM 1-43 loaded synapses. (**a**) FM 1-43 dye uptake in a control sample. For FM 1-43 loading, dissected larvae were incubated in modified HL-3 and incubated for 1 min in stimulation solution with 4 μM FM 1-43. (**b**) After photoconversion DAB precipitates in boutons are visible as dark brown spots on the muscle. (**c**) Standard TEM of control boutonic profiles and (**d**) TEM following the photoconversion of DAB using internalized FM 1-43. Numerous vesicles (*arrows*) contain photoconverted DAB. Scale bare in (**b**) for (**a**) and (**b**): 20 μm. Scale bar in (**d**) for (**c**) and (**d**): 0.5 μm.

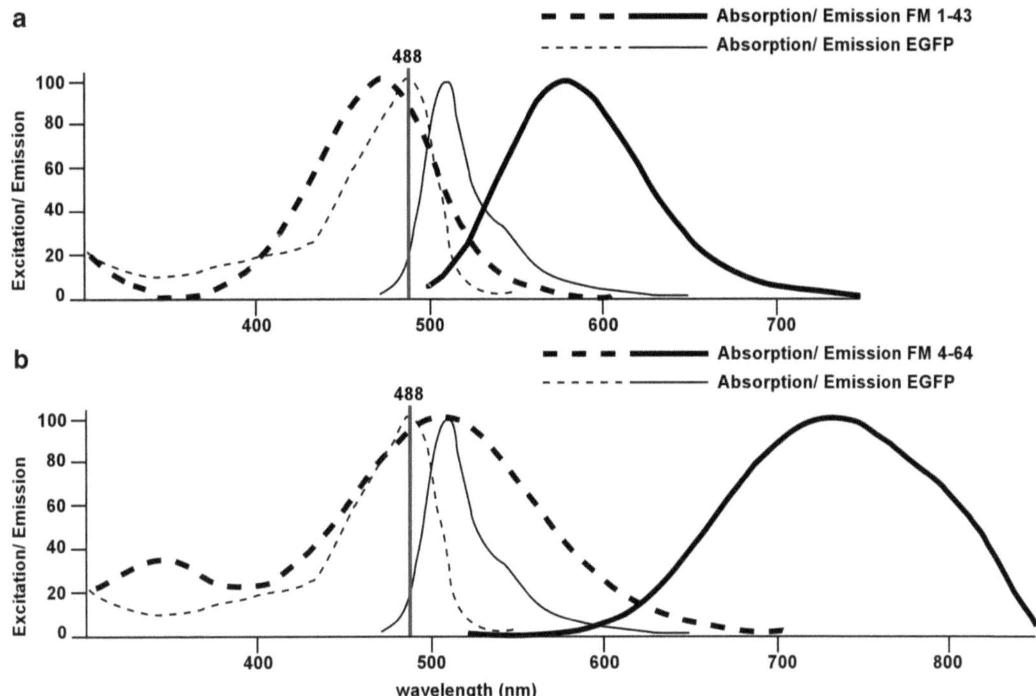

Fig. 2. Spectral properties of FM 1-43, FM 4-64 and EGFP. (**a**, **b**) Spectral properties of EGFP and FM 1-43 (**a**) or the red shifted FM 4-64 (**b**) when bound to lipids as well as the 488-nm laser excitation line.

Photoshop (calculations) to define the overlapping region and save the image as a tiff file. To determine the total number of overlapping pixels, determine the "integrated density" (the sum of pixels that were set to "1" upon thresholding and thus define the overlap) in the ImageJ "measure" module. Also determine the total number of FM 1-43 labeled pixels in the thresholded FM 1-43 image using "integrated density." To calculate the percentage of FM 1-43 labeled area that overlaps with GFP, divide the "integrated densities" of the overlapping area and the FM 1-43 labeled area—Note 13.

4. Notes

1. DAB can be dissolved in PBS beforehand and aliquots can be frozen for later use. The dissolved DAB should be filtered through a syringe filter (Acrodisc, 0.2 μm, Life Sciences) to clear it from precipitates and filtering should be performed each time right before using the solution.
2. FM 1-43 staining is performed by stimulating the motor neurons to undergo exo/endocytic cycles allowing to label synaptic vesicle membrane that was exposed to the synaptic cleft.

For the experimenter to define the stimulation conditions independently from endogenous activity it is important that the nerves emerging from the ventral nerve cord to the different segments are cut. Furthermore, it is also advisable to dissect controls and mutants on one sylgard plate and label them together under the same conditions. If electrical nerve stimulation is used, it is advisable to alternate a control and a mutant preparation to control for time dependent changes in setup, solution, etc.

3. All the solution used for FM 1-43 staining should be prepared fresh daily. Older solutions often show poor results. While freezing the modified HL-3 may be an option, it is usually faster to prepare a fresh batch than thawing one. The washing steps should be performed gently; never poor solution directly on top of the dissected larva, and depending on the conditions, washing times may vary and should be carried out under dim light conditions. Usually, longer washing times reduce background without significantly affecting the fluorescent FM 1-43 signal.
4. The NMJs enervating muscles 4, 6, 7, 12, and 13 are well characterized, easy to access and often imaged. Across animals, it is advisable to consistently image the same NMJ in the same hemisegment.
5. When performing FM 1-43 staining prior to photoconversion of DAB, it is advisable to assess proper fluorescent labeling of boutons using an epifluorescent microscope prior to photoconversion.
6. It is important to replace the DAB solution once (after 5 min) during the preincubation period and again prior to illumination to limit unwanted DAB precipitation on the preparation.
7. In order to photoconvert DAB, select a region with many FM 1-43FX labeled boutons and do not move the sample anymore after you started illumination. Photoconversion is achieved using a ×40 water immersion objective; ×60 is suitable too although the photoconverted area becomes smaller and thus less tissue can be processed for TEM. Light delivered through lower magnification objectives is not strong enough for efficient photoconversion of DAB.
8. The photoconverted area which can be seen as a round brown spot on the larvae can be easily located (e.g., with a stereomicroscope). After processing of the larvae for TEM, this area can be trimmed before embedding facilitating the identification of relevant boutons in EM.

9. You can evaluate the electron density of the photoconverted DAB by quantifying the mean gray values of the vesicular area corrected for background after photoconversion and in standard TEM.
10. The use of the UAS-GAL4 system (or other binary systems) allows the expression of proteins under control of specific promoters. At the NMJ, choosing for muscle or neuronal specific drivers, depending on the experiment, increases the signal to noise ratio.
11. When performing FM 1-43 and GFP colocalization experiments it is important to control for false positive colocalization. This can be achieved in two ways:
    (a) Perform experiments where samples are labeled only with GFP or only with FM 1-43 and image both channels as indicated in Sect. 3.3. When excited with a 488-nm laser, no signal should be detected with the 510/20 nm band pass emission filter in samples labeled with only FM 1-43 while boutonic FM 1-43 labeling should be visible with the 570–620 nm emission settings. Conversely, in GFP expressing animals that are not labeled by FM 1-43, GFP labeling should be detectable with the 510/20 nm band pass emission filter, but no signal should be visible with the 570–620 nm emission filter. If some bleed-through would be detectable, adjust the acquisition settings and use these for further imaging of double labeled samples as well.
    (b) Include a negative control where you do not expect much colocalization. For example, in *skywalker* mutants much of the internalized FM 1-43 colocalizes with endosomal markers (Rab5-GFP or 2xFYVE-GFP) at synaptic boutons (Fig. 3a), while FM 1-43 internalized in synapses treated with chlorpromazine does not (Fig. 3b) (43). This negative control indicates minimal bleed-through FM 1-43 in the GFP channel.
12. While this protocol describes the procedure for imaging and calculations of FM 1-43 with GFP, given adjusted excitation and emission settings, the use of other XFPs is possible as well (e.g., CFP); conversely, the red shifted FM 4-64 or FM 5-95 may be used as well in combination with RFP/dsRed variants.
13. ImageJ offers a few plug-ins for colocalization analysis but given the nonspecific and often complex labeling patterns of FM 1-43, semiquantitative analyses yield more consistent results.

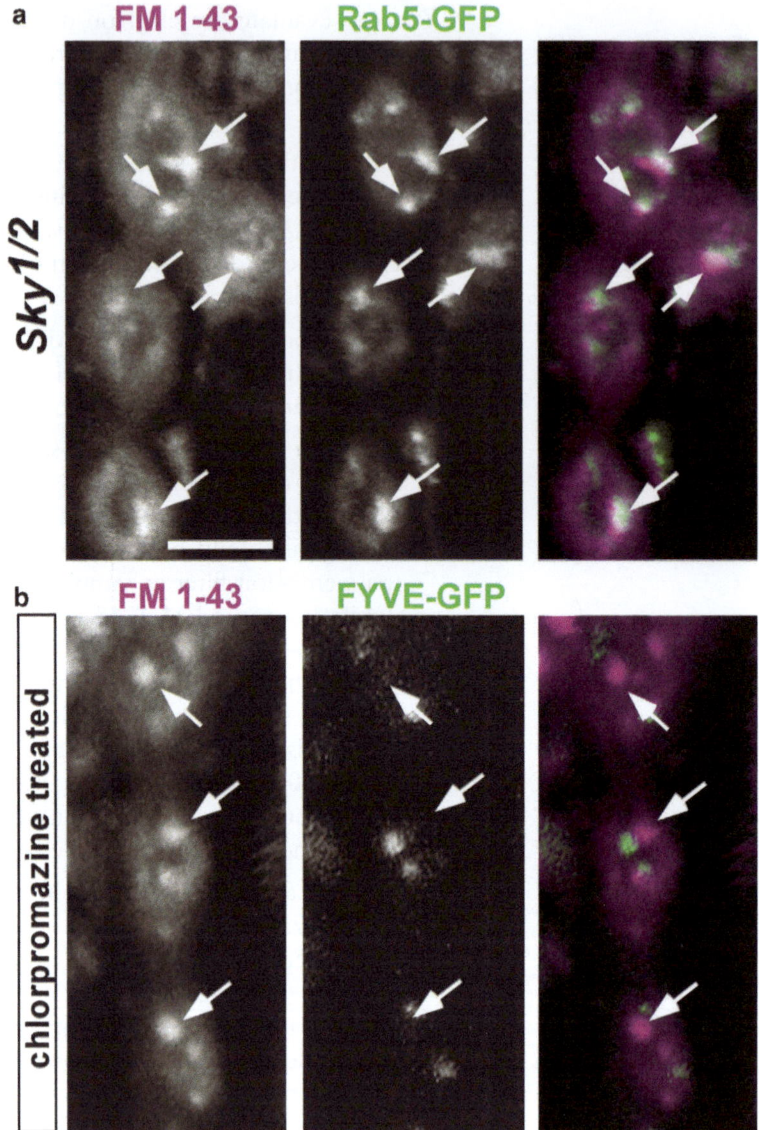

Fig. 3. Colocalization of FM 1-43 with endosomal markers. (**a**, **b**) FM 1-43 loading (magenta) using 1 min of 90 mM KCl stimulation in *skywalker* (*sky*$^{1/2}$) mutants (**a**) and in wild type boutons pretreated with chlorpromazine (**b**) both expressing endosomal markers Rab5-GFP (**a**) or FYVE-GFP (**b**), respectively. (**a**) The FM 1-43 labeling observed in *sky* mutants visualize synaptic vesicles that are forced to cycle via endosomes. (**b**) Animals were pretreated with chlorpromazine to induce the uptake of large membrane sheets (14) and were then loaded with FM 1-43. *Arrows* in (**a**) and (**b**) indicate the location of the FM 1-43 accumulations. Scale bar in (**a**) for (**a**, **b**): 2.5 μm.

Acknowledgements

We thank members of the Verstreken lab for constructive comments and Katarzyna Miskiewicz for providing the TEM data presented in Fig. 1. The work in the Verstreken lab is supported by a Marie Curie Excellence grant (MEXT-CT-2006-042267); and ERC Starting Grant (260678), FWO grants (G074709, G094011 and G095511), the Research Fund KU Leuven, a Methusalem grant of the Flemish Government and KULeuven, the Francqui Foundation, VIB, an Institute for the Promotion of Innovation through Science and Technology in Flanders (IWT-Vlaanderen) O&O grant, and an IWT fellowship to VU.

References

1. Engel AG (1991) Review of evidence for loss of motor nerve terminal calcium channels in Lambert-Eaton myasthenic syndrome. Ann N Y Acad Sci 635:246–258
2. Yao PJ (2004) Synaptic frailty and clathrin-mediated synaptic vesicle trafficking in Alzheimer's disease. Trends Neurosci 27(1): 24–29
3. Zhang C et al (2009) Presenilins are essential for regulating neurotransmitter release. Nature 460(7255):632–636
4. Milnerwood AJ, Raymond LA (2010) Early synaptic pathophysiology in neurodegeneration: insights from Huntington's disease. Trends Neurosci 33(11):513–523
5. Esposito G, Fernandes AC, Verstreken P (2012) Synaptic vesicle trafficking and Parkinson's disease. Dev Neurobiol 72(1):134–144
6. Schweizer FE, Ryan TA (2006) The synaptic vesicle: cycle of exocytosis and endocytosis. Curr Opin Neurobiol 16(3):298–304
7. Haucke V, Neher E, Sigrist SJ (2011) Protein scaffolds in the coupling of synaptic exocytosis and endocytosis. Nat Rev Neurosci 12(3): 127–138
8. Jahn R, Lang T, Sudhof TC (2003) Membrane fusion. Cell 112(4):519–533
9. Sudhof TC (2008) Neurotransmitter release. Handb Exp Pharmacol 184:1–21
10. Sudhof TC (2004) The synaptic vesicle cycle. Annu Rev Neurosci 27:509–547
11. Schluter OM et al (2004) A complete genetic analysis of neuronal Rab3 function. J Neurosci 24(29):6629–6637
12. Hayashi M et al (2008) Cell- and stimulus-dependent heterogeneity of synaptic vesicle endocytic recycling mechanisms revealed by studies of dynamin 1-null neurons. Proc Natl Acad Sci U S A 105(6):2175–2180
13. Marek KW, Davis GW (2002) Transgenically encoded protein photoinactivation (FlAsH-FALI): acute inactivation of synaptotagmin I. Neuron 36(5):805–813
14. Kasprowicz J et al (2008) Inactivation of clathrin heavy chain inhibits synaptic recycling but allows bulk membrane uptake. J Cell Biol 182(5):1007–1016
15. Griffin BA, Adams SR, Tsien RY (1998) Specific covalent labeling of recombinant protein molecules inside live cells. Science 281(5374): 269–272
16. Banaszynski LA et al (2006) A rapid, reversible, and tunable method to regulate protein function in living cells using synthetic small molecules. Cell 126(5):995–1004
17. Marks KM, Braun PD, Nolan GP (2004) A general approach for chemical labeling and rapid, spatially controlled protein inactivation. Proc Natl Acad Sci U S A 101(27): 9982–9987
18. Beck S et al (2002) Fluorophore-assisted light inactivation: a high-throughput tool for direct target validation of proteins. Proteomics 2(3):247–255
19. Habets RL, Verstreken P (2010) Acute inactivation of proteins using FlAsH-FALI. In: Zhang B, Freeman MR, Waddell S (eds) *Drosophila* neurobiology—a laboratory manual. Cold Spring Harbour Press, New York
20. Dittman J, Ryan TA (2009) Molecular circuitry of endocytosis at nerve terminals. Annu Rev Cell Dev Biol 25:133–160
21. Hallermann S et al (2010) Mechanisms of short-term plasticity at neuromuscular active zones of drosophila. HFSP J 4(2):72–84

22. Richmond JE, Davis WS, Jorgensen EM (1999) UNC-13 is required for synaptic vesicle fusion in *C. elegans*. Nat Neurosci 2(11):959–964
23. Long AA et al (2008) Presynaptic calcium channel localization and calcium-dependent synaptic vesicle exocytosis regulated by the Fuseless protein. J Neurosci 28(14):3668–3682
24. Broadie K et al (1995) Syntaxin and synaptobrevin function downstream of vesicle docking in *Drosophila*. Neuron 15(3):663–673
25. Verstreken P et al (2009) Tweek, an evolutionarily conserved protein, is required for synaptic vesicle recycling. Neuron 63(2):203–215
26. Willig KI et al (2006) STED microscopy reveals that synaptotagmin remains clustered after synaptic vesicle exocytosis. Nature 440(7086): 935–939
27. van Rijnsoever C, Oorschot V, Klumperman J (2008) Correlative light-electron microscopy (CLEM) combining live-cell imaging and immunolabeling of ultrathin cryosections. Nat Methods 5(11):973–980
28. Rosen SC et al (2000) Outputs of radula mechanoafferent neurons in Aplysia are modulated by motor neurons, interneurons, and sensory neurons. J Neurophysiol 83(3):1621–1636
29. Liu Q, Hollopeter G, Jorgensen EM (2009) Graded synaptic transmission at the *Caenorhabditis elegans* neuromuscular junction. Proc Natl Acad Sci U S A 106(26):10823–10828
30. Katz B (1966) Nerve, muscle and synapse. McGraw-Hill series in the new biology, McGraw-Hill, New York, p 193
31. Seelig JD et al (2010) Two-photon calcium imaging from head-fixed *Drosophila* during optomotor walking behavior. Nat Methods 7(7):535–540
32. Miesenbock G, De Angelis DA, Rothman JE (1998) Visualizing secretion and synaptic transmission with pH-sensitive green fluorescent proteins. Nature 394:192–195
33. Betz W, Mao F, Bewick G (1992) Activity-dependent fluorescent staining and destaining of living vertebrate motor nerve terminals. J Neurosci 12(2):363–375
34. Ramaswami M, Krishnan KS, Kelly RB (1994) Intermediates in synaptic vesicle recycling revealed by optical imaging of *Drosophila* neuromuscular junctions. Neuron 13(2):363–375
35. Sturman DA et al (2006) Nearly neutral secretory vesicles in *Drosophila* nerve terminals. Biophys J 90(6):L45–L47
36. Richards DA, Bai J, Chapman ER (2005) Two modes of exocytosis at hippocampal synapses revealed by rate of FM1-43 efflux from individual vesicles. J Cell Biol 168(6):929–939
37. Verstreken P, Ohyama T, Bellen HJ (2008) FM 1-43 labeling of synaptic vesicle pools at the *Drosophila* neuromuscular junction. Methods Mol Biol 440:349–369
38. Ramachandran P, Budnik V (2010) Fm1-43 labeling of *Drosophila* larval neuromuscular junctions. Cold Spring Harb Protoc 2010(8):pdb.prot5471
39. Delgado R et al (2000) Size of vesicle pools, rates of mobilization, and recycling at neuromuscular synapses of a *Drosophila* mutant, shibire. Neuron 28(3):941–953
40. Kuromi H, Kidokoro Y (2005) Exocytosis and endocytosis of synaptic vesicles and functional roles of vesicle pools: lessons from the *Drosophila* neuromuscular junction. Neuroscientist 11(2):138–147
41. Kidokoro Y et al (2004) Synaptic vesicle pools and plasticity of synaptic transmission at the *Drosophila* synapse. Brain Res Rev 47(1–3): 18–32
42. Wucherpfennig T, Wilsch-Brauninger M, Gonzalez-Gaitan M (2003) Role of *Drosophila* Rab5 during endosomal trafficking at the synapse and evoked neurotransmitter release. J Cell Biol 161(3):609–624
43. Uytterhoeven V et al (2011) Loss of skywalker reveals synaptic endosomes as sorting stations for synaptic vesicle proteins. Cell 145(1):117–132
44. Murthy VN, Stevens CF (1998) Synaptic vesicles retain their identity through the endocytic cycle. Nature 392(6675):497–501
45. Zenisek D, Steyer JA, Almers W (2000) Transport, capture and exocytosis of single synaptic vesicles at active zones. Nature 406(6798):849–854
46. Lloyd TE et al (2000) A genome-wide search for synaptic vesicle cycle proteins in *Drosophila*. Neuron 26(1):45–50
47. Littleton JT (2000) A genomic analysis of membrane trafficking and neurotransmitter release in *Drosophila*. J Cell Biol 150(2): F77–F82
48. McMahon HT et al (1996) Synaptophysin, a major synaptic vesicle protein, is not essential for neurotransmitter release. Proc Natl Acad Sci U S A 93(10):4760–4764
49. Lao G et al (2000) Syntaphilin: a syntaxin-1 clamp that controls SNARE assembly. Neuron 25(1):191–201
50. Groffen AJ et al (2010) Doc2b is a high-affinity Ca2+ sensor for spontaneous neurotransmitter release. Science 327(5973):1614–1618
51. Venken KJT, Bellen HJ (2005) Emerging technologies for gene manipulation in *Drosophila melanogaster*. Nat Rev Genet 6(3):167–178

52. Bellen HJ, Tong C, Tsuda H (2010) 100 years of *Drosophila* research and its impact on vertebrate neuroscience: a history lesson for the future. Nat Rev Neurosci 11(7):514–522
53. Elliott DA, Brand AH (2008) The GAL4 system: a versatile system for the expression of genes. Methods Mol Biol 420:79–95
54. Dietzl G et al (2007) A genome-wide transgenic RNAi library for conditional gene inactivation in *Drosophila*. Nature 448(7150):151–156
55. Ni JQ et al (2009) A *Drosophila* resource of transgenic RNAi lines for neurogenetics. Genetics 182(4):1089–1100
56. Reiff DF et al (2005) In vivo performance of genetically encoded indicators of neural activity in flies. J Neurosci 25(19):4766–4778
57. Tian L et al (2009) Imaging neural activity in worms, flies and mice with improved GCaMP calcium indicators. Nat Methods 6(12): 875–881
58. Kittel RJ et al (2006) Bruchpilot promotes active zone assembly, Ca2+ channel clustering, and vesicle release. Science 312(5776): 1051–1054
59. Akbergenova Y, Bykhovskaia M (2009) Stimulation-induced formation of the reserve pool of vesicles in *Drosophila* motor boutons. J Neurophysiol 101(5):2423–2433
60. Vijayakrishnan N, Woodruff EA 3rd, Broadie K (2009) Rolling blackout is required for bulk endocytosis in non-neuronal cells and neuronal synapses. J Cell Sci 122(Pt 1):114–125

Chapter 7

Optical Recording of Visually Evoked Activity in the *Drosophila* Central Nervous System

Dierk F. Reiff

Abstract

Drosophila has become a powerful experimental animal for the analysis of neuronal circuits and computations underlying innate behavior. In *Drosophila*, perturbational genetics is currently combined with the direct recording of neural activity in the CNS and eventually the quantitative analysis of behavior. Any deviation of the recorded response from the normal response is indicative of the functional role of the manipulated neurons and mechanisms in a specific computation or behavior. In these experiments, strong correlation is established by directly recording the membrane potential with electrodes (whole cell recording from the soma) or by optical recording changes in the concentration of intracellular calcium. In addition to recordings from the soma, optical measurements provide access to subcellular compartments as well as large ensembles of visual interneurons. Furthermore, optical recordings are not limited by the small size of the cell body, and neurons located deep inside the brain can be analyzed by using two-photon laser scanning microscopy (2PLSM). The latter aspect is of particular importance, as studying vision in flies requires that the large compound eyes covering almost the entire head of the fly remain fully intact. However, optical imaging of sensory processing in the fly visual system comes along with inherent difficulties of the approach: Fluorescence excitation causes blinding of the fly and photons from the visual stimulus enter the detection pathway and corrupt the recorded signals. In this chapter, I describe a method and guidelines suitable to bypass these problems. Genetic targeting of a population of visual interneurons is used to express a genetically encoded fluorescent indicator for intracellular calcium (GECI). The GECI molecules are expressed in the soma as well as all subcellular compartments. Thus, the requirement of dye application is overcome, and ultimately, a functionally homogeneous population of neurons can be analyzed with high spatial resolution. Fluorescence of the GECI is excited and recorded using in vivo 2PLSM that helps to prevent direct excitation of photoreceptors by laser light. Optical recordings are performed during visual stimulation and sensory processing of the fly. By separating fluorescence recording and visual stimulus presentation in time, even most subtle changes in GECI fluorescence are captured, while the visual stimulus is excluded from the recorded fluorescence signal.

Key words: Vision, Visual processing, Sensory coding, Optical imaging, 2-Photon, Calcium, Neurogenetics, Physiology, Motion detection, Genetic probes, Interlaced stimulus presentation

Bassem A. Hassan (ed.), *The Making and Un-Making of Neuronal Circuits in Drosophila*, Neuromethods, vol. 69, DOI 10.1007/978-1-61779-830-6_7, © Springer Science+Business Media, LLC 2012

1. Introduction

The neuronal mechanisms and computations underlying vision in flies have long attracted attention in neuroscience (reviewed in (1, 2)). Extracellular population recordings like the electroretinogram (3) and whole cell patch clamp recordings on isolated photoreceptors of *Drosophila* (4, 5) have been combined with molecular genetics and the study of mutant flies. In these elegant experiments, the molecular mechanisms of invertebrate phototransduction have been revealed. In contrast, the subsequent neuronal processing steps in the optic lobe have so far mostly escaped rigorous analysis. Thus, the basic computations underlying the detection of the polarity of brightness changes, color, form, and motion remain to be discovered. This is more surprising, as detailed information exists on the anatomy of the peripheral fly visual system. The fly visual system comprises about 750 ommatidia that constitute each compound eye. Each individual ommatidium has its own little lens and hosts 6 outer and 2 inner photoreceptors and 11 accessory cells. The outer and inner photoreceptors accomplish the first neuronal processing step in vision: they convert light energy into the flow of mostly cations through the Transient Receptor Potential (TRP and TRPL) channels which generates the receptor potential and triggers the release of Histamine (6) from photoreceptor terminals. The photoreceptors are highly optimized for this task, as they are able to detect single photons. At the same time, photoreceptors adapt their responsiveness to intensity differences of about ten orders of magnitude. Thus, in addition to sensory input from the visual surround, photoreceptors are highly susceptible for the light that is used to excite GECIs or synthetic probes for calcium in optical imaging experiments.

Downstream of the photoreceptors, the fly visual system comprises four successive optic neuropiles: the lamina, medulla, lobula, and the lobula plate. The visual interneurons within these structures are arranged in a highly repetitive and retinotopically organized manner reflecting the special layout of the compound eyes and of the visual scene. The axons of the outer photoreceptors (R1–6) project with their axons to the lamina according to the principles of neuronal superposition (7). In the lamina, T-shaped ribbon synapses transmit luminance signals on the dendrites of the large lamina monopolar cells L1, L2, and the smaller L3 cell that all express Histamine gated chloride channels encoded by the gene *ort* (8–10). Outer photoreceptors express the spectrally broadly sensitive rhodopsin Rh1. Together with their target cells L1–3, they are thought to establish the peripheral side of the achromatic motion detection pathway (11, 12). Inner photoreceptors express different and spectrally more narrowly tuned rhodopsins. R7 expresses Rh3 or Rh4 and R8 expresses Rh5 or Rh6. R7/8 is considered the color detection system (11, 13, 14). However, recent electron microscopic studies suggest that the separation of color and motion pathways is not complete at the level of postsynaptic neurons in the medulla (15).

Based on these anatomical and developmental findings, a series of behavioral and genetic studies has been performed, suggesting functional roles for particular types of neurons in the detection of color and motion as well as the control of particular visually driven behaviors (12–14, 16, 17). In these studies, causality between neurons and the execution of particular behaviors was established by genetically manipulating activity in identified neurons and monitoring the behavioral performance of animals under precisely defined stimulus conditions. Yet, detailed information on the encoding of information in identified neurons and the nature of the transmitted signal was still missing. This gap is now being closed by several recent studies on fly motion vision. Combining defined visual stimuli and perturbational genetics with direct electrical or optical recording of activity, these studies (18–24) show how *Drosophila* sensory neurons change their responsiveness and tuning during active locomotion and provide new insights into the biological implementation of a classical model in neuroscience, the Hassenstein–Reichardt model (HRM) of visual motion detection (25, 26). This long-standing model provides the algorithmic framework for a quantitative description of visual motion detection in flies and accurately reproduces cellular and behavioral responses to motion stimuli in surprising detail (27). Combining genetics with direct recording, it has now been shown that motion detection in *Drosophila* starts with splitting the visual input into two parallel channels for the detection of brightness increments (ON) or decrements (OFF). L1 transmits mostly ON information and mediates the detection of moving ON-edges; L2 transmits mostly information on light OFF and mediates the detection of moving OFF edges. Thereby, L1 and L2 perform an incomplete half-wave rectification of the input signal (19, 20) and feed their signals into two parallel, none interacting motion detectors subunits. This breakthrough currently serves as a base for many future studies and allowed Eichner et al. to deduce a new internal structure of the HRM that is likely closer to the biological hardware (28). These experiments in behaving as well as nonbehaving fruit flies demonstrate the great power of combining genetics with direct recording of activity for deciphering encoding and processing of information in the small neuronal circuits of the *Drosophila* visual system.

To facilitate such studies, this chapter provides guidelines for a combined genetic and optical approach to record activity from visual interneurons while the fly is processing sensory information. Tricks and techniques are provided to overcome most of the problems associated with optical recording of activity in the visual system of small experimental animals during the presentation of bright and time-varying visual stimuli. Information on building and operating a 2PLSM, recent advances in the design of GECIs as well as on genetic studies in flies have extensively been provided elsewhere. Excellent information is provided in the laboratory manuals "*Drosophila* Neurobiology" and "Imaging," both from Cold Spring Harbor Laboratory Press (29, 30).

2. Materials

2.1. Materials for the Preparation of the Fly

Metal holder (Fig. 1, custom-built, aluminum carrying a metal insert)

Plexiglas fly holder with magnet (Fig. 1)

Aluminum foil

Medical grade silicone (Unimed, Lausanne/Switzerland)

Bee wax

Wax melter, custom-built with thin wire (alternatively purchased from Almore International).

Dissecting microscope (×160 total magnification)

Forceps Dumont # 3

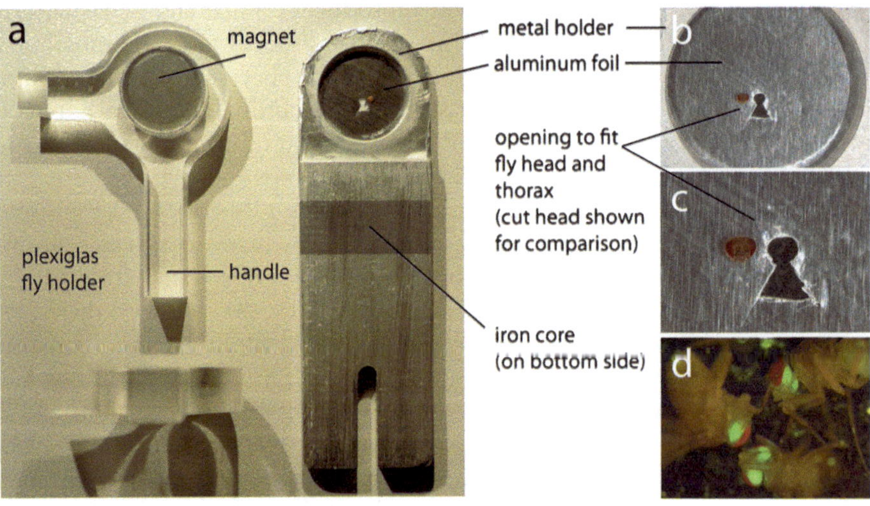

Fig. 1. Recording platform. (**a**) Metal and Plexiglas holder used to mount the fly. The metal holder has an iron core allowing for flexible and reversible attachment of the Plexiglas holder via its small magnet. This simple design allows positioning of the fly such that the head of the fly precisely fits into the whole in the aluminum foil. (**b**) View on the top of the recording chamber. Aluminum foil is glued to the bottom surface of the metal holder. The outer ring of the metal forms a well on top of the aluminum foil that will hold the saline in place. At the center of the recording chamber aluminum foil has been cut out. (**c**) Close-up of the center and cut hole. The round hole in the aluminum foil precisely fits the size and shape of the fly head (fly head shown next to the whole for comparison). There is an additional delta shaped cut-out that allows insertion of the thorax of the fly. (**d**) Experimental flies have been placed under a fluorescence microscope to check GECI expression. Green fluorescence of the genetic indicator TN-XXL highlights the optic lobe of the flies.

Forceps Dumont # 5 (to be sharpened)
Sharpening stone
Fine needle (27G x¾ in.; 0.4 × 20 mm)
Bucket with ice
Small vial with plug
GECI expressing flies (see fly rearing and crossing)

2.2. Extracellular Saline

The composition of the saline used for optical recording experiments has been described by Wilson et al. (31) (see supplement, ~280 mOsm, 1.5 mM calcium). To increase the magnitude of fluorescence changes exhibited by GECIs during neuronal activity the calcium concentration can be increased to 3 mM. We keep 24 mL aliquots at 104% of the final concentration without $CaCl_2$ and $NaHCO_3$. These aliquots can be kept in a –20 C freezer for several months. Thaw an aliquot before starting the experiment and add 500 μL of a 50 × $NaHCO_3$ stock solution and 500 μL of a 50 × $CaCl_2$ stock solution. Alternatively, 500 μL of a 100 × $CaCl_2$ stock solution can be added to increase the final calcium concentration to 3 mM (we do not compensate for the slight increase in osmolarity). The pH is adjusted to ~7.3. We observed that permanent perfusion of the preparation with bubbled saline is not necessary. In our hands, it is sufficient to replace a fraction of the bath solution every 20 min. This way, stable recordings can be obtained over several hours. Table 1 presents a saline recipe.

Table 1
Saline recipe

	Final (mmol/L)	104% Stock (24 mL)	Add before experiment
NaCl	103.00	107.12	
KCl	3.00	3.12	
TES	5.00	5.20	
Trehalose	10.00	10.40	
Glucose	10.00	10.40	
Sucrose	7.00	7.28	
NaH_2PO_4	1.00	1.04	
$MgCl_2$	4.00	4.16	
50 × $NaHCO_3$	26.00		add 500 μL
(100 × $CaCl_2$	3.00		add 500 μL)
50 × $CaCl_2$	1.50		add 500 μL

2.3. Imaging Equipment

Two-Photon Laser Scanning Microscope (32): We use a previously described custom-built TPLSM ((20, 33) design kindly provided by Denk, MPI Heidelberg) which allows for wide-field or Two-Photon imaging through the same objective (typically ×63/0.90 n.a., water immersion, IR Achroplan; Zeiss, Jena, Germany). Two-Photon fluorescence is excited by a mode-locked Ti–Sapphire laser (<100 fs, 80 MHz, 700–1,000 nm; pumped by a 10 W Millenia laser; both Tsunami; Spectraphysics, Darmstadt, Germany). During experiments, laser intensity is held constant at typically 5 and maximally 10 mW at the specimen. CFP is selectively excited at 825 nm if TN-XXL (34) is expressed as a genetic calcium indictor protein. CFP and YFP emission are recorded simultaneously using two separate band-pass filters (CFP: HQ 535/30, YFP: 485/30; both oriented at 18°) and subsequent photomultipliers. Alternatively, GCaMP 3 (35) can be expressed, which typically provides higher fractional fluorescence changes.

2.4. Visual Stimulation and Light-Emitting Diode Arena

Visual stimuli are presented to the eyes of the fly using a light-emitting diode (LED) arena. The basic design of the LED arena has been developed by Reiser and Dickinson (36). The Dickinson Laboratory provides detailed building-plans in the open-source information (http://www.dickinson.caltech.edu/panelspage). Modifications of this design have been described in two previous reports (18, 20). In summary, our custom-built LED arena is composed of premanufactured 8×8 LED panels (TA08-81GWA dot matrix displays; Kingbright, Walnut, CA), each harboring 64 individual green 568-nm LEDs. The arena covers about 170° in azimuth and 80° in elevation, with an angular resolution of about 1.4° between the centers of adjacent LEDs. In our configuration, each dot matrix display is controlled by an ATmega168 microcontroller (Atmel, San Jose, CA, USA) combined with a ULN2804 line driver (Toshiba America Inc, NY, USA) acting as a current sink. All dot matrix displays are in turn controlled and synchronized via an I2C interface by an ATmega128 (Atmel, San Jose, CA, USA) based main controller board which reads in pattern information from a compact flash (CF) memory card. We use Matlab (Mathworks Inc., USA) for programming and generation of the patterns as well as for sending the serial command sequences via RS-232 to the main controller board. The arena is capable of a frame rate of 660 frames/s and is operated with four intensity levels only. The actual frame rate is 500 frames/s determined by the scanning movement of the x-mirror of the laser scanning microscope. A TTL signal generated every 2 ms at the beginning of each line-scan of the horizontal scanning mirror is used to trigger the individual panels of the LED arena. This design enables the precise presentation of the visual stimulus during 330 μs of the 400 μs fly-back period of the horizontal scanning mirror. The fly-back occurs after the x-mirror having reached its final position after 1.6 ms scanning. The resulting temporal switching at 500 Hz between fluorescence recording and stimulus presentation is well above the flicker-fusion frequency of the *Drosophila* eye. Using the above-mentioned dot matrix displays, image sequences of 10 cd/m^2 mean brightness (corresponding to 60 cd/m^2 during light-ON) can be generated. It has to be mentioned that the used LEDs no longer conform to the current state of LED design. Recent improvements in the doping of semiconductors resulted in much brighter LEDs that allow for precisely tunable or very wide emission spectra, respectively. An excellent overview on current LEDs and resources is given by Sato and Murthy, Imaging—Chap. 10 (29).

3. Methods

Calcium imaging is performed on visual interneurons expressing a genetically encoded fluorescent probe. Targeted expression of the indicator from an upstream activation sequence (*Gal4/UAS-system, 37*) enables indicator expression in large, functionally homogeneous populations of neurons, depending on the specificity of the used driver line. If different types of neurons are included in the expression pattern, they can still be analyzed separately given that the neurons coramify their neuritis at distances significantly larger than the point spread function of the used microscope. Activity in the labeled neurons as well as in the surrounding network of naive unlabeled neurons is elicited by presenting visual stimuli to the compound eyes of the fly. The visual stimuli are precisely synchronized to the scanning movements of the horizontal x-scanning mirror of the 2-Photon Laser Scanning Microscope (32) (see Sect. 2.4). This approach enables spatially resolved recordings of neuronal activity in the retinotopically organized lattice of visual interneurons and overcomes two difficult problems inherent to optical recording in the fly visual system. First, blinding of the fly can be prevented. Second, light from the visual stimulus is excluded from the recorded fluorescence signals given precise timing at the μs time-scale. The provided protocol was recently used to optically record visually evoked fluorescence changes exhibited by the genetically encoded calcium sensor TN-XXL expressed in L2 neurons (20).

3.1. Construction of the Recording Chamber and Selection of Flies

1. Prepare a custom-built metal (aluminum) holder with an iron/mild steel core (Fig. 1a, right). The iron core allows flexible and reversible attachment of the Plexiglas fly holder that has a magnet in its center (Fig. 1a, upper left).
2. A thin layer of medical grade silicone is used to glue a small piece of aluminum foil to the bottom side of the metal holder (Fig. 1a, c).
3. Use a sharp 25-gauge needle to cut a keyhole-shaped opening into the aluminum foil (Fig. 1c). The triangular shaped opening fits the thorax of the fly, and the round opening precisely fits

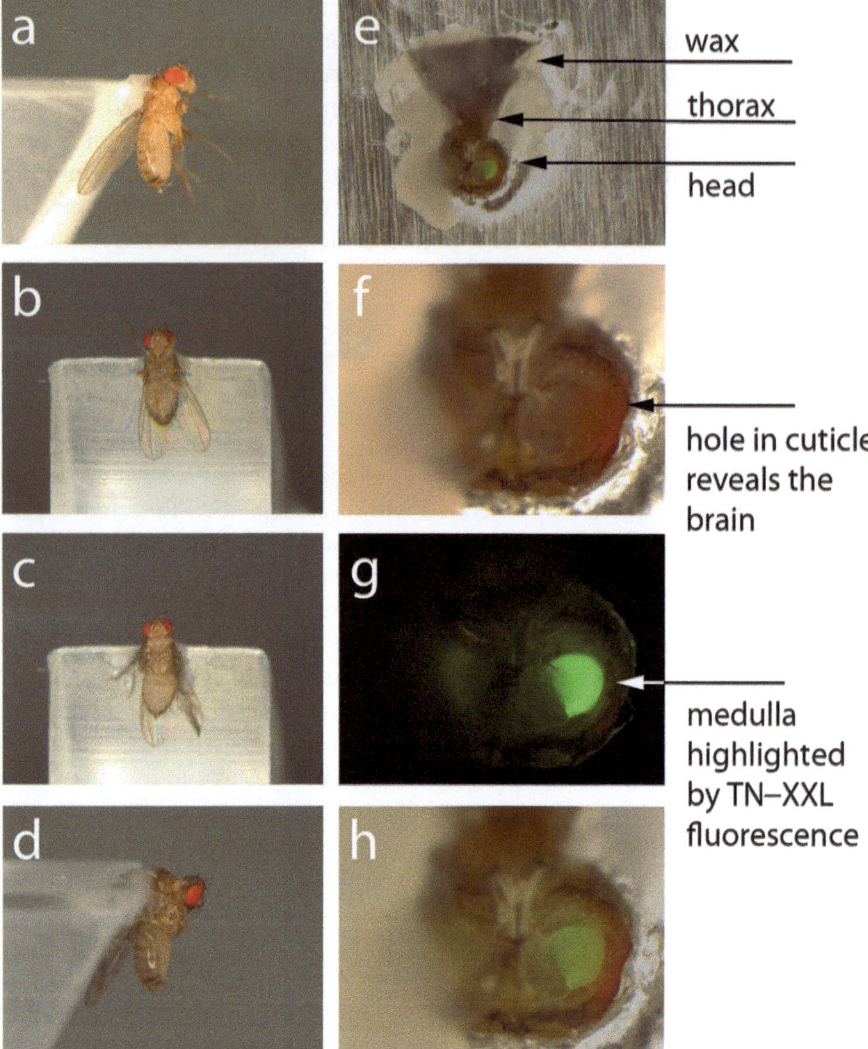

Fig. 2. Preparation of the fly. (**a**) Side view of a fly that has been glued (bee wax) to the beveled Plexiglas holder shown in Fig. 1a. The fly is alive and kicking. (**b**) Frontal view, the fly moves its legs in front of the eyes. (**c**) Frontal view, the legs have been affixed with wax. (**d**) The head has been bent down and affixed. (**e**) Composite image, a fly has been inserted into the whole in the aluminum foil. The thorax and the right side of the caudal surface of the fly's head have been affixed with bee wax. The cuticle, fat body, and trachea have been removed on the left side of the head. TN-XXL fluorescence highlights the medulla. (**f–h**) Close-ups of the same fly. The composite images in (**e**) and (**h**) were generated by combining a bright-field and a fluorescence image.

the head of the fly. This step requires some training. Placing a cut *Drosophila* head next to the opening might help.

4. Prepare a Plexiglas holder with a round magnet in its center (Fig. 1a). At a later stage, the fly will be fixed to the beveled side of this holder (Fig. 1a, bottom left and Fig. 2).
5. GECI expression in experimental flies might be checked using a fluorescence stereomicroscope (Fig. 1d). If you need to check your crossings, allow minimum 2 h recovery time. I routinely check the flies 1 day prior to the imaging session.

3.2. GECI Expression

The unique possibility to direct the expression of GECIs to particular cell types bypasses the need of dye injection, as the GECI molecules are synthesized in situ by the neurons themselves. Specific expression in particular neurons is routinely achieved by employing standard expression strategies. Promoter or enhancer elements that are active in a given cell type are used to drive the expression of a transcription factor typically Gal4 or LexA. The transcription factor in turn binds to the upstream activation DNA element of a second transgene containing the cDNA encoding the GECI. This routine approach is widely used in *Drosophila* to express any transgene (37, 38). We use it to bypass the loading of synthetic dyes and to specifically target an entire population of visual interneurons. The latter can be achieved by no other method. Moreover, most promoters have a fairly stable expression pattern, enabling highly reproducible optical recordings in single cells and large ensembles of neurons from one animal to the next. Finally, different types of neurons can be targeted by employing different promoters.

3.3. Fly Rearing and Crossing

Eight to ten virgins are crossed to 3–5 male flies. Flies are reared on standard cornmeal–agar medium at 60–70% humidity and typically 25°C. Make sure that the vials are not crowded; flip the vials daily or every second day. This ensures healthy flies growing to maximum size. Typically, all experimental animals are females several days after eclosion when the cuticle has hardened. In my hands, the elastic and flexible cuticle of freshly hatched flies makes the preparation harder. There is several ways to tune the expression level of the GECIs:

- The Gal4–UAS system (37) is strongly temperature dependent. At 20°C, the expression is much weaker compared with 25°C.
- The expression strongly depends on the number of copies for both enhancer-Gal4 and UAS-GECI (same for the LexA/LexAop or other expression systems).

A fluorescence stereomicroscope is very useful to routinely check the expression pattern and expression level of the flies (see Figs. 1d and 2).

3.4. Mounting and Preparation of the Fly

1. Place a single fly in a small flask and put on ice. After about 1 min, the fly is fully anesthetized and can be handled. Avoid the use of CO_2.
2. Take the wax melter and attach a small drop of bee wax at the upper edge of the beveled edge of the Plexiglas holder. This wax will be used to attach the fly with the back of its thorax (Fig. 2a).
3. Take forceps #3 and gently grab the fly without causing any damage. It is imperative to keep the fly fully intact. Place some obstacle between the prongs of the forceps to prevent complete closure.
4. Use one hand and melter to melt the drop of wax on the holder, use the other hand to hold the fly right next to the melting wax.

Quickly remove the melting device and attach the back of the thorax of the fly to the wax. The head of the fly is right above the edge/upper surface of the Plexiglas holder. Use very little wax to prevent heating of the fly (Fig. 2a, b). Ensure that the fly is alive. Check if the fly is moving its head to follow the movements of the forceps, prepare a new fly if the fly doesn't show head movements.

5. Attach the legs of the fly to the Plexiglas holder using bee wax (Fig. 2c). If you intend to perform imaging in the behaving animal, you will have to modify this protocol.
6. Gently bend down the head of the fly such that the backside of the fly head faces upward. Use a tiny amount of wax to keep the head fixed in this position (Fig. 2d). Make sure that no wax contaminates the compound eyes.
7. Attach the metal holder to (the magnet of) the Plexiglas holder. Adjust the position of the keyhole-shaped opening in the aluminum foil such that the thorax and head gently fit the opening (Fig. 2e). Use wax to cover and seal the thorax to the aluminum foil. Additional wax might be used to attach on side of the fly head to the aluminum. However, great care has to be taken not to damage the fly with the heat of the melting device and wax. Ensure that all gaps are sealed with wax.
8. Add ringer solution to the topside of the preparation.
9. Typically, the cuticle is removed from only one side of the fly head. Remove the cuticle and fat tissue from on hemisphere using a fresh 25-gauge needle and a sharpened forceps #5 (Fig. 2f). Gently remove trachea and air sacks. Make sure that all main tracheal branches and fat body has been removed. Remaining tissue diminishes the efficacy for successful 2-P excitation. During training, moving the prep to a fluorescence microscope might help to check the quality of the preparation. The structures of interest expressing the GECI must be optically accessible and clearly visible (Fig. 2g, h).
10. If muscle 16 is still intact, remove it. Muscle 16 connects the frontal pulsatile organ to the aortic funnel, its contractions cause massive movement of the brain. Grab one end of the muscle and tear it off. Now, the major pulsating movements of the brain should cease.
11. Wash the brain several times with fresh saline and let recover for several minutes. Now you can move the prep to the recording setup. There is plenty of time—the fly will be usable for several hours.

3.5. Optical Imaging: Getting Started

1. Attach the fly holder to the rig (Figs. 3 and 4a). Center the dissected hemisphere/optic lobe in the field of view of the objective by using the translational stage of the objective and a CCD camera/computer display.

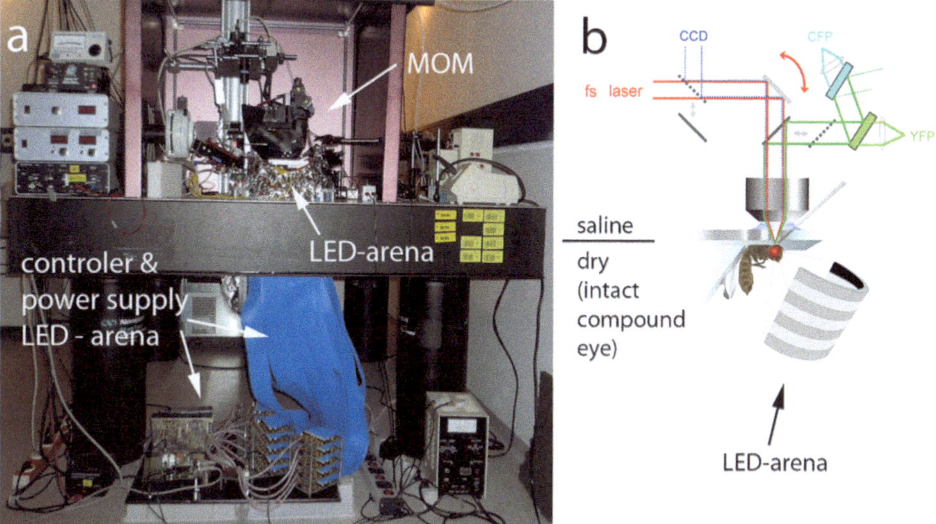

Fig. 3. Recording situation. (**a**) Recording setup. A custom-built movable optical microscope (MOM) is placed above an light-emitting diode (LED)-arena. The arena fits into a hole cut into the large vibration isolation table. The controllers and power supplies of the LED arena are located below the large vibration-free table and connected via the stream of blue cables. (**b**) Schematic of the recording situation. Saline has been put on the upper side of the aluminum foil and fly head. A water immersion objective is placed above the tiny hole in the foil and cuticle. The setup enables both wide field and two-photon fluorescence imaging. Wide-field excitation allows precise focusing and identification of the area of interest, however, the fly will be blinded by the excitation light. This problem is resolved by sliding two mirrors to enable 2PLSM. Two separate detectors (PMTs) are used to record the photons emitted by cyan- (CFP) and yellow fluorescent protein (YFP) in the 2PLSM mode if a ratiometric GECI is used (Fig. 3b has been reproduced from (20) with permission of from Nature Neuroscience).

2. You might use wide field fluorescence excitation to identify the GECI expressing neurons and to adjust the brain region of interest. Then, adjust the focal point of the objective about 100 μm above the area of interest.
3. Switch to 2PLSM (see light path of the microscope in Fig. 3b) and display the continuously acquired images (life mode) on your computer display. Carefully lower the objective while observing the fluorescence images.
4. I find the most effective technique to identify the region of interest is to briefly scan the expression pattern in its entire *x*, *y*, and *z* dimension.
5. Focus on the neuropile/region of interest and choose appropriate magnification and image specification. Now, you are ready to run your experiments.

3.6. Optical Imaging of Neural Activity During Visual Processing

1. The mounted fly is above the arena (Figs. 3 and 4a). The compound eyes of the fly must be fully intact, kept dry and looking at the LED arena as shown in the schematic drawing in Fig. 3b. In Fig. 4a, the microscope, objective, and LED arena are shown, the fly holder has been removed from its mount to better reveal the grating displayed on the LED arena. In the

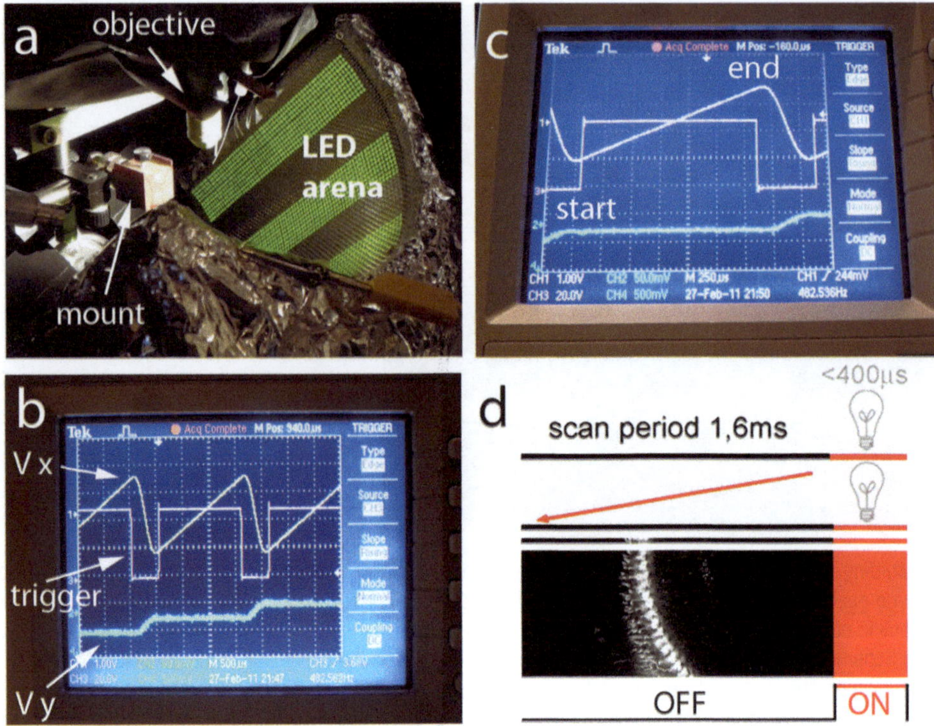

Fig. 4. Recording principle. (**a**) Recording objective, mount for the metal holder, and LED-arena. A horizontally drifting grating of vertical bars is presented on the circular LED arena (hemi-field). Due to the precise synchronization of the horizontal-scanning mirror and the LEDs, the 2PLSM can be operated and line scans or frames can be recorded without contamination by stimulus light. (**b**) Scope display of the voltage signals controlling the horizontal (V_x), the vertical scanning mirror (V_y), and the trigger used to switch the LEDs. About 2,5 cycles are displayed. (**c**) Same signals as in b but higher temporal resolution. During the scanning of a single line, V_y is constant and the *y*-mirror does not change its position, whereas V_x increases continuously until the end of the line-scan is reached. Now, the TTL-signal is pulled down and the LED arena is switched on. It remains on while V_x and the x-scanning mirror return to the starting condition which takes about 400 µs. At the beginning of the next round, i.e., the scanning of the next line, the trigger is pulled up again and the arena is immediately switched off. V_x ramps up for 1.6 ms and a new line is scanned without stimulus light contaminating the recording of the fluorescence signals emitted by the GECI. Thus, the stimulus is briefly flashed onto the LED arena at a rate of 500 Hz and a duty cycle of little less than 20%. The parameters that define the scan and the stimulus presentation can be varied, yet the arena has to be switched off during the fly-back of the x-scanning mirror (Fig. 4d has been reproduced from (20) with permission of from Nature Neuroscience).

recording situation, any light-source other than the LED arena is shut off. In addition, the entire rig is hidden in a metal shield housing blocking light from the computer monitors or other devices (Fig. 3a).

2. During an experiment, brightness and time-variable visual stimuli are displayed on the LED arena. We use Matlab (Mathworks Inc., USA) for programming and generation of the patterns as well as for sending the serial command sequences via RS-232 to the main controller board of the LED arena. The main board reads in the pattern information from a compact flash (CF) memory card as well as the TTL trigger signal. The main board controls and synchronizes the individual

microcontrollers of each dot-matrix display via an I2C interface (for details see materials). However, continuous stimulus presentation corrupts the recorded fluorescence signals as light emitted by the LEDs enters the detection pathway during the recording period (see also troubleshooting).

3. Exclude the visual stimulus from being recorded by precisely synchronizing the movement of the horizontal scanning mirror and the current flow to the LEDs. The TTL generated every 2 ms at the beginning of each line-scan of the horizontal scanning mirror (x-mirror) is used to trigger and to shut off the individual panels of the LED arena. The x-mirror scans a line and reaches its final position after 1.6 ms scanning. Now, with a delay of some 10 s of μs, the arena is switched on for about 330 μs. Then, some 10 s of μs prior returning to the new starting position, the arena is switched off again. This design enables the precise presentation of the visual stimulus during 330 μs of the 400 μs fly-back period of the horizontal scanning mirror. The time course of the individual signals is shown in Fig. 4b, c. You can adapt the timing to your experiment and microscope and chose other parameters.
4. Connect a copy of the voltage signals (V_x and V_y) controlling the x- and y-scanning mirrors as well as the TTL to a scope. Adjust the timescale of the scope appropriately (~250–500 μs) and trigger the display to the rising edge of the TTL. The displayed signals should be comparable to the ones shown in Fig. 4b, c. V_x continuously increases during the scanning of a line until the x-mirror reaches its final position. After having reached its final position, the x-mirror quickly returns into its starting position. During this so-called fly-back period, the trigger is pulled down and the y-mirror jumps to the next position if xy-images are scanned.
5. If the timing is correct, the LED arena is shut off for the entire period required for the scanning of a single line (here 1.6 ms scanning) and switched on during 330 μs of the 400 μs fly-back (close-up of voltage signals in Fig. 4c). This design enables artifact-free imaging and results in temporal switching at 500 Hz between fluorescence recording and stimulus presentation which is above the flicker-fusion frequency of the *Drosophila* eye (schematic in Fig. 4d).

4. Troubleshooting

4.1. Artifacts Because of Imprecise Timing of Scanning and Visual Stimulus Presentation

The complete elimination of stimulus light from the recorded GECI-fluorescence is key for the detection and proper analysis of subtle signals. Most people employ a combination of wavelength filtering and offline data correction. However, recording a clean

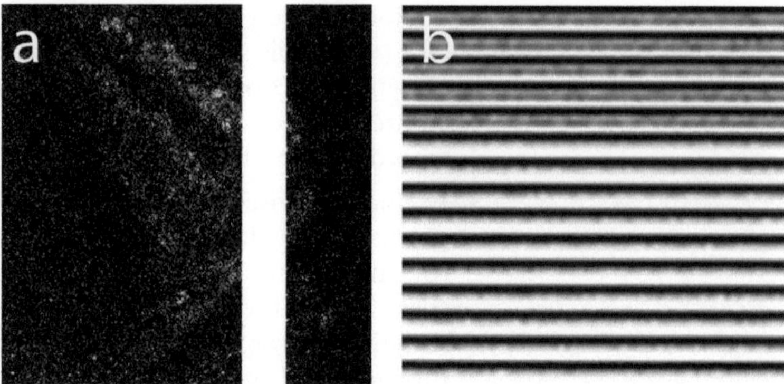

Fig. 5. Effects of imprecise timing. (**a**) An image (300 × 300 pixel) has been acquired at ~2.5 ms per line. The visual stimulus was presented 1.6 ms after the onset of each line-scan. There is no stimulus presentation during the fly-back of the x-scanning mirror. (**b**) Same timing of the visual stimulus (1.6 ms after the onset of each line-scan). However, the time for the scanning of each line was significantly smaller than 2 ms; thus, the TTL signal and the fixed delay of the LED arena interfere in time.

GECI signal is much preferred. The effect of imprecise presentation of the visual stimulus during the line scan period is shown in Fig. 5. Displaying the TTL and the voltage controlling the x-scanning mirror (Fig. 4) during the design and programming of the experiment helps avoiding these problems prior to the life experiment.

4.2. Recorded Fluorescence Signals, GECIs, and Expression Level of the GECI

The design of the GECI as well as the concentration at which a given GECI is expressed in vivo have a strong influence on the recorded fluorescence signal:

- The signal-to-noise ratio (SNR) of the recorded signal depends on the square root of the number of photons n that are detected during a given time Δt. Thereby, n depends almost linearly on the number of indicator molecules present in a given volume (voxel) that is excited by fs-pulsed laser light. Thus, a sufficient concentration of GECI molecules is required to obtain an SNR of 2 or higher. At the same time, laser energy has to be low in order to minimize bleaching of the indicator, thermal damage, and direct excitation of photoreceptors.
- The time course of the recorded fluorescence signal is dramatically slowed with increasing concentration of the indicator. This is because the time that is required for calcium to bind to the indicator and to reach steady-state increases with increasing GECI concentration.
- For the same reason, transient changes in calcium are reported with reduced amplitude of the fractional fluorescence change if the indicator is expressed at a high concentration. Steady-state of calcium-binding to the indicator is not reached by the

time calcium is cleared off from the cytosol by the efficient extrusion mechanisms and buffering in internal stores.

- Any calcium indicator will potentially interfere with intracellular signaling mechanisms by buffering calcium and by slowing down the time-course of the normal calcium changes. In addition, GECIs have been reported to directly interact with intracellular signaling molecules.

Ultimately, as low as possible concentration of the GECI is preferable that still enables optical recording of fluorescence signals with sufficient SNR. These issues have been extensively discussed in a number of excellent publications (39–42).

5. Conclusions

5.1. Strength of the Method

1. Biosynthesis of the GECI in situ bypasses the need for dye delivery and enables targeting of large functionally homogeneous populations of neurons by genetic methods. Ideally, only a single functional class of neurons is labeled, enabling assigning of the recorded signals to a specific cell type, even when recording from large areas and multiple cells.
2. 2PLSM enables excitation and recording of GECI fluorescence in visual interneurons in vivo while minimizing or preventing direct excitation of photoreceptors and blinding of the fly. The latter is fully prohibited by restricting the output power of the excitation laser to a maximum of about 5 mW in the specimen.
3. 2-Photon excitation further restricts fluorescence excitation to a small voxel of about 250 nm in *xy* and less than one micrometer in z. These values strongly depend on the characteristics and optics of the used microscope and tissue and allow for high optical resolution.
4. The very small number of photons emitted by the GECI molecules in the excited area prevents excitation of the photoreceptors by GECI-fluorescence.
5. So far, whole cell recordings in the *Drosophila* brain are limited to the soma. The spatially integrated voltage signal provides great temporal precision and sensitivity, yet very limited information is included about processing and signal transformation in different and distant compartments. Subcellular compartments can be analyzed using the described optical approach.
6. Optical recording enables the simultaneous observation of many neurons. Neurons in neighboring columns of the fly visual system process information from neighboring image points in space. The spatial–temporal pattern of the stimulus is preserved in the neuronal activity and the recorded fluorescence signals.

7. Optical imaging in the *Drosophila* visual system can be combined with molecular and genetic techniques for the manipulation of activity in neuronal circuits as well as behavioral essays.
8. The inherent difficulty of optical recording neuronal activity with GECIs in the fly visual system lies in presenting, i.e., drifting visual patterns to the eye of the fly while at the same time protecting the photomultiplier tubes in the 2PLSM from light that is not emitted by the fluorescent calcium indicator. This problem is solved by stimulating the compound eyes with light only during the return movement (fly-back) of the x-scanning mirror. The precise features of this method depend on the characteristics of the (custom-built) imaging setup and stimulus device.

Acknowledgements

This work was supported by the Max Planck Society.

References

1. Clifford CW, Ibbotson MR (2002) Fundamental mechanisms of visual motion detection: models, cells and functions. Prog Neurobiol 68:409–437
2. Borst A, Haag J, Reiff DF (2010) Fly motion vision. Annu Rev Neurosci 33:49–70
3. Hotta Y, Benzer S (1969) Abnormal electroretinograms in visual mutants of *Drosophila*. Nature 222:354–356
4. Hardie RC (1991) Voltage-sensitive potassium channels in *Drosophila* photoreceptors. J Neurosci 11:3079–3095
5. Hardie RC (1991) Whole-cell recordings of the light-induced current in dissociated *Drosophila* photoreceptors: evidence for feedback by calcium permeating the light-sensitive channels. Proc R Soc Lond B 245:203–210
6. Hardie RC (1989) A histamine-activated chloride channel involved in neurotransmission at a photoreceptor synapse. Nature 339:704–706
7. Kirschfeld K (1973) Das neuronale Superpositionsauge. Fortschr Zool 21:228–257
8. Meinertzhagen IA, O'Neil SD (1991) Synaptic organization of columnar elements in the lamina of the wild type in *Drosophila melanogaster*. J Comp Neurol 305:232–263
9. Meinertzhagen IA, Sorra KE (2001) Synaptic organization in the fly's optic lamina: few cells, many synapses and divergent microcircuits. Prog Brain Res 131:53–69
10. Gengs C, Leung HT, Skingsley DR, Iovchev MI, Yin Z, Semenov EP, Burg MG, Hardie RC, Pak WL (2002) The target of *Drosophila* photoreceptor synaptic transmission is a histamine-gated chloride channel encoded by ort (hclA). J Biol Chem 277:42113–42120
11. Heisenberg M, Buchner E (1977) The role of retinula cell types in visual behavior of *Drosophila melanogaster*. J Comp Physiol 117:127–162
12. Rister J, Pauls D, Schnell B, Ting CY, Lee CH, Sinakevitch I, Morante J, Strausfeld NJ, Ito K, Heisenberg M (2007) Dissection of the peripheral motion channel in the visual system of *Drosophila melanogaster*. Neuron 56:155–170
13. Yamaguchi S, Wolf R, Desplan C, Heisenberg M (2008) Motion vision is independent of color in *Drosophila*. Proc Natl Acad Sci U S A 105:4910–4915
14. Yamaguchi S, Desplan C, Heisenberg M (2010) Contribution of photoreceptor subtypes to spectral wavelength preference in *Drosophila*. Proc Natl Acad Sci U S A 107:5634–5639
15. Takemura SY, Lu Z, Meinertzhagen IA (2008) Synaptic circuits of the *Drosophila* optic lobe: the input terminals to the medulla. J Comp Neurol 509:493–513
16. Gao S, Takemura SY, Ting CY, Huang S, Lu Z, Luan H, Rister J, Thum AS, Yang M, Hong S-T, Wang JW, Odenwald WF, White BH,

Meinertzhagen IA, Lee C-H (2008) The neural substrate of spectral preference in *Drosophila*. Neuron 60:328–342

17. Katsov AY, Clandinin TR (2008) Motion processing streams in *Drosophila* are behaviorally specialized. Neuron 59:322–335
18. Joesch M, Plett J, Borst A, Reiff DF (2008) Response properties of motion-sensitive visual interneurons in the lobula plate of *Drosophila melanogaster*. Curr Biol 18:1–7
19. Joesch M, Schnell B, Raghu SV, Reiff DF, Borst A (2010) ON and OFF pathways in *Drosophila* motion vision. Nature 468:300–304
20. Reiff DF, Plett J, Mank M, Griesbeck O, Borst A (2010) Visualizing retinotopic half-wave rectified input to the motion detection circuitry of *Drosophila*. Nat Neurosci 13:973–978
21. Maimon G, Straw AD, Dickinson MH (2010) Active flight increases the gain of visual motion processing in *Drosophila*. Nat Neurosci 13: 393–399
22. Chiappe ME, Seelig JD, Reiser MB, Jayaraman V (2010) Walking modulates speed sensitivity in *Drosophila* motion vision. Curr Biol 16: 1470–1475
23. Schnell B, Joesch M, Forstner F, Raghu SV, Otsuna H, Ito K, Borst A, Reiff DF (2010) Processing of horizontal optic flow in three visual interneurons of the *Drosophila* brain. J Neurophysiol 103:1646–1657
24. Seelig JD, Chiappe ME, Lott GK, Dutta A, Osborne JE, Reiser MB, Jayaraman V (2010) Two-photon calcium imaging from head-fixed *Drosophila* during optomotor walking behavior. Nat Methods 7:535–540
25. Hassenstein B, Reichardt W (1956) Systemtheoretische Analyse der Zeit, Reihenfolgen und Vorzeichenauswertung Bei der Bewegungsperzeption des Russelkafers Chlorophanus. Zeitschrift fur Naturforschung 11:513–524
26. Reichardt W (1961) Autocorrelation, a principle for the evaluation of sensory information by the central nervous system. In: Rosenblith WA (ed) Sensory communication. Wiley, New York, pp 303–317
27. Götz KG (1964) Optomotorische Untersuchungen des visuellen Systems einiger Augenmutanten der Fruchtfliege *Drosophila*. Kybernetik 2:77–92
28. Eichner H, Joesch M, Schnell B, Reiff DF, Borst A (2011) Internal structure of the fly elementary motion detector. Neuron 70:1155–1164
29. Yuste R (2011) Imaging—a laboratory handbook. Cold Spring Harbor Laboratory Press, Cold Spring Harbor, NY
30. Zhang B, Freeman MR, Waddell S (2011) *Drosophila* neurobiology—a laboratory handbook, 2nd edn. Cold Spring Harbor Laboratory Press, Cold Spring Harbor, NY
31. Wilson RI, Turner GC, Laurent G (2004) Transformation of olfactory representations in the *Drosophila* antennal lobe. Science 303:366–370
32. Denk W, Strickler JH, Webb WW (1990) Two-photon laser scanning fluorescence microscopy. Science 248:73–76
33. Haag J, Denk W, Borst A (2004) Fly motion vision is based on Reichardt detectors regardless of the signal-to-noise ratio. Proc Natl Acad Sci U S A 101:16333–16338
34. Mank M, Santos AF, Direnberger S, Mrsic-Flogel TD, Hofer SB, Stein V, Hendel T, Reiff DF, Levelt C, Borst A, Bonhoeffer T, Hubener M, Griesbeck O (2008) A genetically encoded calcium indicator for chronic in vivo two-photon imaging. Nat Methods 5:805–811
35. Tian L, Hires SA, Mao T, Huber D, Chiappe ME, Chalasani SH, Petreanu L, Akerboom J, McKinney SA, Schreiter ER, Bargmann CI, Jayaraman V, Svoboda K, Looger LL (2009) Imaging neural activity in worms, flies and mice with improved GCaMP calcium indicators. Nat Methods 6:875–881
36. Reiser MB, Dickinson MH (2008) A modular display system for insect behavioral neuroscience. J Neurosci Methods 167:127–139
37. Brand AH, Perrimon N (1993) Targeted gene expression as a means of altering cell fates and generating dominant phenotypes. Development 118:401–415
38. Lai SL, Lee T (2006) Genetic mosaic with dual binary transcriptional systems in *Drosophila*. Nat Neurosci 9:703–709
39. Borst A, Abarbanel HD (2007) Relating a calcium indicator signal to the unperturbed calcium concentration time-course. Theor Biol Med Model 4:7
40. Yasuda R, Nimchinsky EA, Scheuss V, Pologruto TA, Oertner TG, Sabatini BL, Svoboda K (2004) Imaging calcium concentration dynamics in small neuronal compartments. Sci STKE 2004:l5
41. Pologruto TA, Yasuda R, Svoboda K (2004) Monitoring neural activity and [Ca2+] with genetically encoded Ca2+ indicators. J Neurosci 24:9572–9579
42. Hendel T, Mank M, Schnell B, Griesbeck O, Borst A, Reiff DF (2008) Fluorescence changes of genetic calcium indicators and OGB-1 correlated with neural activity and calcium in vivo and in vitro. J Neurosci 28:7399–7411

Part III

Behavior

Chapter 8

Behavioral Analysis of Navigation Behaviors in the *Drosophila* Larva

Matthieu Louis, Moraea Phillips, Mariana Lopez-Matas, and Simon Sprecher

Abstract

Functional and anatomical dissection of neural circuits is often hindered by the complexity of such systems. With only 10,000 neurons, the central nervous system of the *Drosophila* larva is at least one order of magnitude simpler than its adult counterpart. Despite this numerical simplicity, the behavioral repertoire of the larva contains a surprisingly diverse array of sophisticated behaviors. Larvae demonstrate robust orientation behavior toward light and odors (phototaxis and chemotaxis). The sensory organs and circuits underlying these behaviors are greatly reduced in comparison with the adult: the larval eye is composed of just 12 photoreceptor neurons, the nose of just 21 olfactory sensory neurons. While the larval olfactory pathway displays remarkable structural similarities with the adult system its numerical simplicity facilitates the analysis of individual, genetically identifiable neurons at anatomical and functional levels. The use of information arising from different modalities allows for investigation of the principles controlling multisensory integration. In this chapter, we review a series of assays to study light and odor-driven behaviors. The advent of high-resolution machine-vision algorithms to analyze behavior in real time is likely to revolutionize our knowledge of how organization of the larval brain mediates distinct behaviors. The simplicity of the larval sensory systems allows us to aim for a comprehensive and systems-level understanding of the relationships between circuit anatomy and function, from afferent sensory neurons through to higher brain centers where orientation decisions are made and communicated to efferent motor neurons.

Key words: Chemotaxis, Phototaxis, Orientation behavior, Behavioral analysis, Olfaction, Vision, *Drosophila* larva

1. Introduction

1.1. Development and Structure of the Larval Brain

The central nervous system (CNS) of the *Drosophila* larva develops during embryogenesis (1). It is therefore fully functional when the larva hatches. The larval CNS originates from an invariant set of neuronal stem cells, called neuroblasts, which delaminate during early embryonic stages from the neuroectoderm and subsequently

Bassem A. Hassan (ed.), *The Making and Un-Making of Neuronal Circuits in Drosophila*, Neuromethods, vol. 69, DOI 10.1007/978-1-61779-830-6_8, © Springer Science+Business Media, LLC 2012

divide to first produce the neurons of the larval nervous system (primary neurons), followed by adult specific secondary neurons which are born during larval stages (2). The CNS is divided into an anterior region, often referred to as the brain, and a posterior ventral nerve chord (VNC).

Developmentally the brain is derived from two ganglia, each consisting of three neuromeres: the supraesophageal and subesophageal ganglion (SOG). The supraesophageal ganglion consists of the protocerebrum, deuterocerebrum, and tritocerebrum; the SOG consists of the mandibular, maxillary, and labial neuromeres. The VNC develops from the ventral neurectoderm, the SOG from the neuroectoderm of gnathal segments, and the supraesophageal ganglion from the procephalic neuroectoderm (3, 4).

The fusion and condensation of the CNS in *Drosophila* makes recognition of segmental boundaries challenging. Subdivision into neuroectoderm and neuroblasts makes use of segmental boundary markers such as the pair-rule gene *engrailed* (5). Although an embryonic neuroblast map is available, a defined subdivision of the brain into individual neuromeres during larval stages has not yet been established. Thus, the terms referring to individual neuromeres are used loosely and do not correspond to developmentally defined terms.

Primary lineages expressing *engrailed* or derived from *engrailed* expressing neuroblasts can be identified in the larval brain and thereby provide potential landmarks for subdivision (6, 7). However, the three-dimensional structure, in particular the curved nature of the larval brain results in an ongoing debate regarding neuromeric subdivision.

Size differences between neuromeres are considerable. The most anterior part of the brain is the protocerebrum, which forms the greater part of the lobes. It encompasses several important and well-defined regions, such as the mushroom body and the larval optic neuropile (LON). The deuterocerebrum is less precisely defined and comprises structures such as the antennal lobes. Posterior brain domains have not yet been subdivided: they are referred to as SOG—this nomenclature is slightly misleading, since the so-called SOG also comprises the tritocerebrum. The SOG includes putative gustatory centers (8).

An alternative and more accessible subdivision relies on neuroanatomical markers. Here, the CNS can be subdivided into two compartments: a peripheral cortex formed primarily by cell bodies and a synapse dense central neuropile, where axonal and dendritic domains overlap. Each cell body extends a single neurite toward the neuropile (Fig. 1a). The neuropile is further divided into individual compartments, intersected by glia sheaths. In-depth analysis has revealed that these compartments first appear during mid to late embryonic stages when primary neurons extend their neurites into the neuropile (9). The compartments grow as the branching

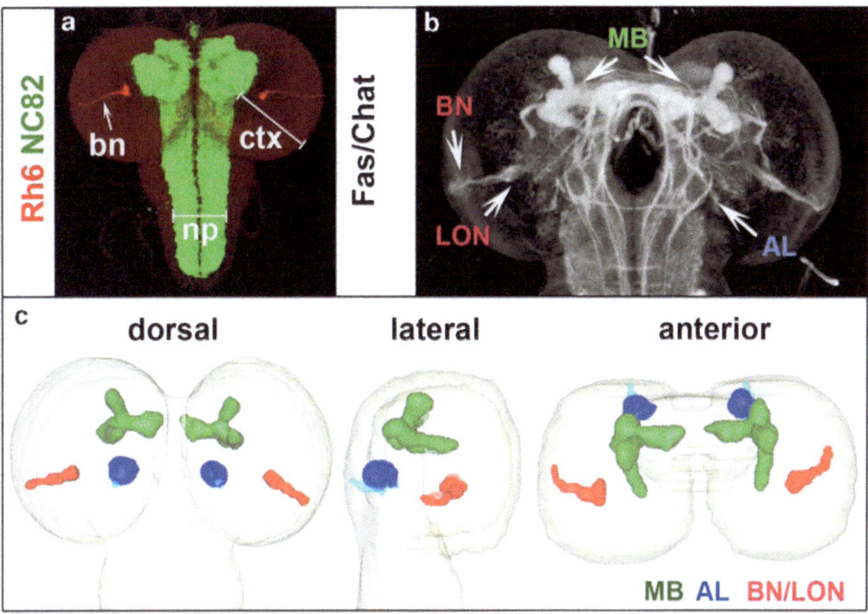

Fig. 1. Anatomical features of the larval brain. (**a**) Third instar larval brain stained with the neuropile marker NC82 (anti-Brp, *green*) and Rh6 (*red*) highlighting the Bolwig's nerve (bn). The brain is subdivided into the outer cortex (ctx) and central neuropile (np). (**b**) Third instar larval brain stained with a marker combination showing axon scaffolding (anti-FasII) and neuropile (Chat). Features of the visual and olfactory systems are highlighted as follows: Bolwig's nerve (BN, *red*), larval optic neuropile (LON, *red*), mushroom body (MB, *green*), and antennal lobe (AL, *blue*). (**c**) 3D model of the brain shown in (**b**) using the same color code to highlight structures in a dorsal, lateral and anterior view.

of dendrites and axons becomes denser and more refined. Still during larval stages, immature neurons of secondary lineages also project into preexisting neuropile compartments, resulting in substantial growth during this phase (10). As secondary neurons develop, the cortex expands dramatically in width. Major axon tracts and pioneer axons subdivide the neuropile; they can be viewed as prominent landmarks within the brain (Fig. 1b).

1.2. The Larval Olfactory and Visual Network

In the larva, organs mediating smell, taste, and vision are located in the head region. Several can be distinguished there: the dorsal organ (DO, larval equivalent of the "nose"), the terminal and ventral organs with putative gustatory functions, and the Bolwig's organ (larval equivalent of the "eye"). Both the DO and Bolwig's organ are paired structures situated symmetrically about the body axis.

Peripheral olfactory system: The Danish scientist Niels Bolwig reported already in 1946 that the pair of DOs represents the main olfactory organ in the house fly *Musca domestica*. In a series of remarkable experiments, Bolwig surgically ablated individual or pairs of DOs. He found that larvae with bilateral ablation where unable to respond to the presence of olfactory stimuli. In contrast, individuals with unilateral DO function were still able to smell (11). Since then, this conclusion has been corroborated in

Drosophila melanogaster, ablation and targeted loss of function being carried out by modern genetic techniques (12, 13). The DO is composed of a central dome perforated by thousands of pores thought to channel odorant molecules to odorant receptors (ORs) hosted on the membranes of the dendritic arbor formed by 21 olfactory sensory neurons (OSNs) (14–16). As a general rule, each OSN expresses one type of OR unique to itself, the "private" receptor, together with a "public" coreceptor Orco (formerly known as Or83b) (12, 17, 18). The total number of OR genes expressed at the larval stage is 25, indicating the existence of a few cases where two "private" OR genes are coexpressed in the same OSN (12). Expression of *Orco* is necessary for the trafficking of the private OR to the dendritic membrane in all larval OSNs (18, 19). *Orco* has also been shown to contribute actively to the odorant signaling pathway (20, 21). Consequently, the larval olfactory system is silenced in the *Orco* null background (18). Using Gal4 driver lines under the control of the *cis*-regulatory regions of a given OR gene, the expression of *Orco* can be restored in a single OSN, consequently restoring olfactory function in that OSN alone. This technique has been exploited in several studies to characterize the behavioral contribution and response profile of individual OSNs (12, 22–24).

As in the adult, larval OSNs project onto a primary olfactory center called the antennal lobe (AL), where each one innervates a distinct glomerular neuropile structure. The larval AL (LAL) is composed of 21 such glomeruli. The response property of a given glomerulus is largely determined by the receptive field of the OSN it is associated with. Within the LAL, two classes of neurons are targeted by the afferent OSNs: GABA-ergic local interneurons (LNs) and cholinergic projection neurons (PNs) (25, 26). LNs form an inhibitory network that establishes lateral connections within the LAL. LNs are thought to play a significant role in the processing of olfactory information proceeding from the OSNs. Functional and behavioral evidence suggests that they act as a gain control mechanism that tunes the dynamic range of the olfactory system (22). The potential redistribution of neural activity across the LAL may also contribute to a de-correlation of the incoming patterns of OSN activity (27, 28). Unlike in the adult fly (29), no excitatory local interneurons have been found in the larva.

An estimated total of between 25 and 30 PNs connect the LAL to higher-brain centers, namely, the lateral horn (LH) and the mushroom body (MB) (25, 26, 30–32). The dendrites of individual larval PNs are usually restricted to single LAL glomeruli (32). Cellular redundancy at the level of the PNs is low. The information carried by individual OSNs is modulated by the LNs and transmitted to the MBs and LH by the PNs. The MB calyx is composed of 34 glomeruli (25, 30, 31). Whereas most of the PNs target a single MB glomerulus, some establish connections with two or three glomeruli (30, 31, 33). Calcium imaging of odor-evoked activity at the level of OSN and PN axon terminals indicates that patterns of OSN activity are transmitted quite faithfully to the MB calyx.

The number of γ neurons in the MB calyx is in the order of 600. Clonal analysis based on the Flip-out and MARCM methods (see Chap. 4) have shown that each MB γ neuron innervates between 2 and 6 glomeruli in the calyx. The number of MB γ neurons connected by each PN is estimated to be between 30 and 180. Despite fundamental design similarities between larval and adult olfactory systems, the larval circuit is characterized by an approximate 1:1:1:1 connectivity pattern between OSNs, LAL glomeruli, PNs, and MB glomeruli, whereas in the adult OSNs converge on the LAL glomeruli in a 30:1 ratio and diverge out to the PNs in a 1:3 ratio. In the adult divergence from PNs to the MB γ neurons forming the calyces is approximately 1:15. In the larva it is 1:20 (30–32). Another important difference between the adult and larval olfactory systems is the unilateral projection of the larval OSNs onto the LAL (12, 13). Since no neurons connecting the two LALs have been found, bilateral comparison of signals is unlikely to occur in the LAL.

Visual system: The visual system of the larva is comparably simple. The larval eye consists of 12 photoreceptor neurons (PRs), which are subdivided into two types: four PRs express the blue-sensitive *rhodopsin5* (*rh5*), while the remaining eight express the green-sensitive *rhodopsin6* (*rh6*) (6, 34). Larval PRs project their axons to the LON: a small, distinct neuropile compartment located medio-laterally to the central brain (10). In contrast with the LAL comparably little is known about the downstream circuit of the LON. The best characterized target neurons are the main pacemakers of the clock circuit, termed lateral neurons (LNs) marked by the expression of the neuropeptide pigment-dispersing factor (Pdf) (35). The larval brain comprises four Pdf-expressing LNs per brain hemisphere. Clock neurons have been show to modulate the light-response of larvae in a circadian fashion. The fifth Pdf-negative LN is also part of the clock circuit and extends its dendritic arbors into the LON. The LON is further innervated by two serotonergic neurons. These serotonergic neurons are important for general responsiveness to light (36). How these neurons are connected to the larval PRs remains elusive. Likewise, their function is not yet known. Lastly, a set of three neurons known as optic lobe pioneers (OLPs) has been shown to project into the LON. The function of the OLPs is also unknown (37).

1.3. Orientation Behaviors

At the conclusion of the larval stage, individuals will have increased their body mass by two orders of magnitude in less than 7 days. To achieve this feat, larvae need to feed constantly. Olfaction is thought to play a key role in the localization of food sources (14, 38). Several studies have reported the ability of larvae to respond to a wide range of chemical odorant stimuli (12–14, 39–48). When exposed to an attractive odor source, larvae quickly migrate toward higher odorant concentrations. Even though the ecology of *Drosophila* larvae remains poorly understood, most of the larval life takes place in decaying fruits and other rich substrates. Whether they feed on cacti or rotting grapes, larvae are extremely sensitive to desiccation. It is therefore not surprising that *D. melanogaster* larvae demonstrate strong light-avoidance behavior (49, 50).

Orientation behaviors are usually classified into two groups: *kinesis* and *taxis* (51). Kinesis involves undirected changes in activity in function of stimulus intensity. A process is categorized as *klinokinesis* when the rate of turning is a function of stimulus intensity. In *orthokinesis*, the rate of movement (activity) is a function of stimulus intensity. For both orientation strategies, migration toward (or away from) the source of an attractive (or repulsive) stimulus results from a series of undirected reorientation events. A well-documented example of klino kinesis is bacterial chemotaxis. *Escherichia coli* navigate chemical gradients according to a biased random walk (a klinokinesis) (52). In short, bacterial locomotion is based on the alternation between two modes of motion: straight runs and random turns. When a bacterium experiences increasing concentrations of an attractant (e.g., sugar), it tends to suppress turning (positive chemotaxis: $\Delta C/\Delta t > 0$ leads to turn suppression). In all other conditions, turning occurs at a constant rate. The same principle governs negative chemotaxis: turning is suppressed when a concentration decrease is experienced (negative chemotaxis: $\Delta C/\Delta t < 0$ leads to turn suppression).

In contrast to the indirect orientation strategy featured by kinesis, taxis involves directed orientation based on spatiotemporal comparisons of stimulus intensities. In the case of *tropotaxis*, orientation results from the instantaneous comparison of sensory input transmitted by different sensors that are physically separated. For instance, adult flies, ants, and bees are capable of detecting concentration differences between their left and right antennae. This stereo-olfaction mechanism permits them to orient in the field of an odor gradient (53, 54). Not all taxis involve comparisons between the inputs of spatially separated sensors. *Klinotaxis* involves comparison between sensory inputs—snapshots—measured at different time points. For olfactory receptor and photoreceptor neurons located in the anterior part of the body, lateral motion of the head allows a larva to "canvass" its sensory conditions during forward locomotion (55). Through this temporal sampling mechanism, changes in direction are biased toward or away from the gradient. The same mechanism permits sharks to ascend odor trails (56). Notably, humans are also able to perform scent tracking by wavering their head either side of a scent trail (57)—the same mechanism observed in insects and other vertebrates (58, 59).

In response to light, larvae avoid regions of high stimulus intensities. For attractive odors, locomotion is directed toward high stimulus intensities. The navigation strategy controlling these two types of orientation behavior is currently the focus of much attention. Even though the orientation algorithms directing phototaxis and chemotaxis are not fully understood, they appear to involve a precise assessment of local gradients (55).

Chemotaxis: In *Musca*, single larvae respond to food odors by orienting toward the odor source and by staying in the vicinity of the odor (60). *Drosophila* larvae display a similar directed response (14, 39). Given that larvae have a pair of bilaterally symmetric DOs, it is reasonable to hypothesize that stereo-olfaction (comparison between left and right olfactory input, *tropotaxis*) is the strategy they use to orient in odor gradients. Initial evidence suggested that unilateral surgical ablation of the DOs led to circling behavior toward the side of the dysfunctional olfactory organ (61). Another series of experiments employed a probabilistic rescue strategy to obtain unilateral olfactory function (24). Using this technique, it was concluded that bilateral olfactory function is not necessary for larvae to chemotax. Notwithstanding, larvae with unilateral olfactory function showed reduced performance compared to individuals with bilateral function. This finding suggests that left-right comparisons enhance the signal-to-noise ratio when detecting changes in odor concentration. Having ruled out a mechanism based solely on stereo-olfaction (tropotaxis), what strategy might be used by larvae to chemotax? Our work indicates that they use an active sampling mechanism to navigate in odor gradients (47, 62). During forward locomotion, larvae appear to collect information about the odor gradient by sweeping their head from side to side. Such a mechanism would rely on decisions involving comparison between odor intensities measured at different points in space (klinotaxis). Orientation would result from temporally based decisions.

Phototaxis: During the feeding stage (up to late third instar), larvae are strongly photophobic. Upon sudden exposure to light, larvae stop moving and begin to sweep their head from side to side. This behavior was reported in *Drosophila* by Mast as early as 1911 (63). He also noted that they made larger movements when sweeping the head away from the light source than toward it. After a series of head sweeps, the larvae tend to orient in the opposite direction to the light source. About 30 years after Mast, Bolwig made the same observation in larvae of *Musca* (11). These findings have been confirmed in more recent experiments involving computer-aided tracking (36, 64, 65). As with chemotaxis, it is reasonable to hypothesize that larvae detect differences in light intensities during these lateral head sweeps. Turns would be directed toward the side where the PRs on the head receive less light. This model proposes klinotaxis as a mechanism for phototaxis.

1.4. Outstanding Questions in the Field

Despite our understanding of the functions carried out by neurons in peripheral sensory systems, the downstream circuits of the visual and olfactory pathways are completely unknown. During the coming 10 years, one focus of research will certainly be to establish relationships between circuits and function in the larval brain. Which neuronal subsets are responsible for the control of specific behaviors? Pioneering studies have investigated the function of genetically identifiable neurons labeled by Gal4 lines (66, 67).

The emergence of new genetic techniques to simultaneously label numerous neurons in the same brain with different colors using flybow or brainbow is likely to further improve the resolution of reconstructed circuits in the larval brain (68, 69). A major drawback when using light microscopy techniques to assess brain circuits is the diffraction limit of optical resolution. Electron microscopy (EM), however, allows us to determine the existence of a synaptic connection between two neurons. Advances in serial sectioning for EM scanning and subsequent reconstruction promise unique methodological possibilities to gain insight into brain circuitry (70). Serial reconstruction of brain compartments, entire neuromeres, or even the entire CNS will therefore not only provide us with a complete description of synaptic connectivity within a circuit but also allow us to determine the precise number of neurons and synapses. The numeric simplicity of the visual and olfactory circuits should permit the identification of all neurons and corresponding synaptic connections to higher brain centers. The same is likely true for other sensory modalities or motor systems. The analysis of an entire larval CNS will be a breakthrough in the field, and is likely to be forthcoming.

One limitation of EM analysis is the lack of genetic markers for individual neurons within a given circuit. Any comparison with Gal4 drivers therefore remains correlative. An alternative approach is provided by light microscopy techniques capable of overcoming the diffraction barrier such as STED, PALM, or STORM (71). It has been shown that the resolution achieved by STORM allows imaging of chemical synapses by specifically labeling pre- and postsynaptic proteins. A build-on from this breakthrough will be its application in the visualization and identification of synapses between genetically identifiable, labeled neurons.

At present, behavior analyses in larvae focus mainly on identifiable peripheral sensory neurons, neurotransmitter systems in the CNS, or individual neurons labeled with Gal4 drivers. The assessment of multiple neurons and dissection of circuits remains a major challenge in the coming years. Centers responsible for information processing or integration are largely undocumented in the larval brain. Exceptions are the antennal lobes and the mushroom bodies, which seem to maintain similar organization and function in the adult. Even though not all neurons innervating the antennal lobe have been identified and characterized structurally and functionally, the basic function of the glomerularly organized neuropile is as an elemental integration center of olfactory information processing (22). As with the adult fly, the larval mushroom bodies are centers required for olfactory learning and memory formation (72). Many details about higher centers for visual or gustatory information processing have been disregarded. An intriguing open question is how these distinct pathways converge on central pattern generators required for navigation. Identifying neurons or groups of neurons that function in similar manners will provide us with a preliminary link between higher brain centers and motor centers.

The resolution of commonly used behavioral assays remains crude. A big step toward a global understanding of larval behavior will be the development of computer-aided automated tracking. Various approaches have been initiated in this direction. Recently, several freeware programs have been developed to monitor the behavior of groups of adult flies at high resolution (73, 74, 109). Similar tracking software exists now for the larva (75), software created for *C. elegans* and zebra fish larvae should be adaptable to the *Drosophila* larva (see for instance (76, 77)). The Worm Tracker devised for *Caenorhabditis elegans* by the Schafer lab (73) could also be used as a powerful instrument to couple high-resolution behavioral tracking with calcium imaging. With these tools in hand, we should be equipped to answer several fundamental problems in sensory neuroscience:

1. How are changes in odor concentrations represented in the larval olfactory system? How are changes in light intensity represented by photoreceptors in the Bolwig's organ? How is this information interpreted by higher brain centers?
2. What are the exact orientation algorithms used by larvae to navigate odor and light gradients? Are they essentially the same? Do larvae rely on the use of side-to-side head sweeps to perform active sensing?
3. What circuits carry out the neural computation directing larval chemotaxis and phototaxis? Are any of these circuits or circuit components overlapping?
4. How relevant is our understanding of the circuit-function relationships controlling larval chemotaxis and phototaxis to other sensory modalities like thermosensation (77, 79)? Where is information from different sensory modalities combined?

As described above, the navigational algorithms allowing larvae to chemotax or phototax likely rely on temporally based decisions (klinotaxis). This process involves a comparison of odor intensities measured at different points in space. The contribution of spatial comparisons between paired olfactory or visual organs (tropotaxis) should not, however, be excluded. This type of process seems to involve some sort of memory (24, 50)—a hypothesis yet to be tested in the larva. Concerning the use of bilateral comparisons in the detection of olfactory signals, theoretical arguments predict that the typical concentration differences measured between the two DOs would fluctuate too much to be detected reliably (62). It is reasonable to speculate though that processing of the same signal by left and right sensors increases the signal-to-noise ratio when detecting minute changes in concentration over time. Such a process would have to involve cross talk between the left and right olfactory pathways. Circuits capable of implementing this operation have yet to be identified.

2. Materials and Reagents

2.1. Animal Rearing

Flies are crossed in a ratio of one male to two females and allowed to lay eggs in small food vials for the desired period: 4, 12, 24 h. After oviposition, parental flies are removed and the vials containing the eggs are incubated at 22°C on a 12 h–12 h light cycle. Under these conditions, 5.5 days are needed for larvae to reach the third instar stage of development, generally used in chemotaxis and phototaxis assays. Mature larvae can be separated from their food by immersing them in a 15% sucrose solution, which causes larvae to float to the surface. At this point, they can be decanted into a clean container with a small amount of the sucrose solution. It is best to limit the time for which larvae are left in sucrose to less than 2 h in order to avoid unaccounted for effects due to starvation (80). Alternatively if exposure to sucrose is undesirable larvae can be removed manually with a spatula and placed directly into a Petri dish. Noncalorific solutions, such as water, PBS or polyethylene glycol, are sometimes used to wash larvae from food as an alternative to sucrose (81).

Drosophila larvae show a significant circadian modulation of their naïve light response: light responses are stronger during late night and early day phases (19). The strongest decrease in light response is observed within the first 2 h of the day (after the light has been turned on, Zeitgeber time ZT 0-2). The light-preference is stable during the remaining 10 h of light. For this reason, it is important to avoid testing larvae for the first 2 h following the light break. The 12 h–12 h light cycle should be timed to suit the experimenter. To compare results across experiments, it is important to perform experiments at equivalent points of the circadian cycle, typically during the same period of the day. In addition, it is best to minimize thermal fluctuations, which are known to affect larval behavior. Ideally, larvae should be reared at temperature close to the conditions in which experiments are carried out.

Although larvae often survive harsh treatment, they can also show acute nociceptive behavior (82). In reaction to pain, larvae are known to freeze, roll, and/or engage in rapid forward locomotion. At present, it is not known for how long such responses last. To avoid these undesirable effects on the study of chemotaxis and phototaxis, one should minimize physical stresses while preparing the larvae for an experiment. It is recommended to manipulate groups of larvae with bristles (4–5 mm bristle length) or fine-point brushes. As opposed to a passive wash in a sucrose solution, larvae can be actively cleaned with PBS. Although the latter alternative reduces the time elapsed before the onset of an experiment (thereby reducing the duration of starvation), it is also stressful to the larva.

2.2. Chemotaxis Assays

Odor dilutions: Paraffin oil is a common solvent used to dilute odors. The desired concentrations can be obtained through serial dilutions. Start, for instance, with a 1.0 M solution and perform a 1:2 dilution to obtain a solution of 0.5 M. By repeating the process one consecutively reduces the concentration. To enhance the precision of the dilutions, it is advisable to use a digital scale for each preparation.

Since many organic odors react with plastic, it is preferable to prepare and store odor dilutions in glass vials with Teflon caps. For instance, Agilent Technologies offers a series of 1.5-mL glass vials with Teflon caps. We recommend preparing fresh odors on a daily, or at least weekly, basis. This is particularly true when using chemicals with high vapor pressures. To avoid confusion about the identity of stimuli used in an experiment, it is good practice to make note of the CAS (Chemical Abstracts Service) registration number associated with each odor.

Petri dish assay: A commonly used assay to study odor-driven behavior is carried out in a 100 × 15 mm Petri dish containing solidified agarose. Several options exist to provide an odor source. The odor dilution can be pipetted onto a small piece of filter paper in direct contact with the agarose. To avoid potential diffusion of the chemical in the agarose matrix, the odor dilution can be placed in a small plastic container. The top part of an Eppendorf tube (or any other cap) covered with a small piece of filter paper can be used as a container (12). The main drawback of these disposable caps is that they cannot be efficiently closed. In an en masse assay, larvae which enter in direct contact with the chemical cannot be excluded from the experiment. This problem been solved by the availability of customized Teflon cups with small removable lids perforated by a few holes to allow for odor diffusion (83). While these Teflon (PTFE) cups ensure cleaner experiments, their manufacturing cost requires that they be reused. Thorough cleaning is necessary (see below). Whatever odor container is used, larvae will spend significant time exploring it. Therefore, it is important to control for these nonolfactory effects by placing one odor container at each extremity of the plate. For the 1-odor paradigm (sect. 3.1), the extra container is filled with the solvent (Paraffin oil).

Larvae are sensitive to humidity. Their hygrotaxis responses (84) are correlated with a high sensitivity to desiccation. After short periods of dehydration larvae stop moving thus it is essential to maintain them sufficiently moisturized throughout the experiment. It is also important to avoid humidity gradients within the behavioral arena. It is with that aim in mind that Petri dishes are coated with a surface of agarose. To prevent larvae from digging into the surface, the gel should be at least 1.5% agarose, preferably up to 3%. Ensure that the dishes are leveled before pouring and remove any bubbles that form on the surface. Small differences in the concentration of agarose are not known to affect sensory behavior. Agarose dishes should be freshly prepared, preferably on the day of the experiment. Older dishes can become dehydrated, resulting in shrinkage of the agarose layer. Any gaps or textural irregularities promote digging behavior, which represents an uncontrollable source of variation across experiments.

Odor responses are quantified by the fraction of larvae found on each side of the plate (preference index). To prevent the inclusion of larvae suffering from locomotor defects or those showing no preference, a neutral zone equidistant from the two odor cups is defined, dividing the Petri dish into two zones. The exact width of this zone does not seem to be critical it is known to vary between 3 and 14 mm across laboratories. To facilitate counting of larvae, it is helpful to print a template of the Petri dish defining the neutral zone. The printout can be slid under the dish before counting.

96-Well plate assay: For high-resolution analyses of chemotaxis, controlled odor gradients can be created by using multiple odor sources. As detailed in ref. (85), a versatile assay can be set up by stacking three lids of rectangular 96-well microplates, which bear condensation rings corresponding to the wells. The bottom lid is recommended to isolate the rest of the system from the light pad and reduce convection in the arena. The second lid, coated with 25 mL of a 3% agarose gel, serves as a stage on which the larvae respond to the odor gradient. The top lid is inverted to close the system, the odor droplet suspended from its inner surface. The droplet is pipetted directly into one of the condensation rings (wells) of the lid, thus regulating its position. To prevent the odor from spilling over the rings, droplets of 10 μL are recommended. Single or multiple odor sources can be used to generate gradients with distinct geometries (22, 24). When loading multiple odor droplets it is important to minimize the time spent between introduction of the first droplet and the last one. It is preferable to load droplets from high to low concentrations, since lower concentrations are likely to be depleted faster.

Tracking setup: The configuration of the tracking setup will be highly dependent on the software used. Commercial products usually come with a CCD camera and frame grabber video card system. Customized tracking software leaves more flexibility in the choice of cameras and acquisition systems. For online tracking analysis at a temporal resolution lower than 5 Hz (5 frames analyzed per second), or for post hoc analysis, Matlab (The MathWorks) is often recommended. Matlab offers an extensive library of functions specifically dedicated to image acquisition and image processing. Different computer-vision algorithms can be implemented to capture the behavior of moving objects (86, 87). Even though a description of these algorithms is beyond the scope of this book chapter, the quality and contrast of the images will strongly influence the amount of information that can be extracted from the image processing. It is advisable to acquire good quality digital CCD cameras and optics. Basler (Basler Vision Technologies, DE) offers competitive products. To facilitate data transfer between the camera and PC it is best to purchase a model that makes use of the FireWire (IEEE 1394) interface. The A622f of Basler represents a good quality/cost trade-off with a $1,024 \times 1,280$ pixel resolution. Communication between FireWire cameras and Matlab will necessitate the installation of a specific driver. For this purpose, the Carnegie Mellon University has developed an excellent freeware driver (CMU 1394 Digital Camera Driver, http://www.cs.cmu.edu/~iwan/1394/).

Lenses can be purchased from many suppliers. Edmund optics, for instance, offers a wide collection of high quality products. When selecting a lens, one should pay attention to three parameters: the focal length, which determines the field of view; the working distance and the sensor diagonal. The sensor diagonal is determined by the specifics of the camera (e.g., the Basler A622f is a type 2/3″). To maximize the number of pixels covering the body of a larva, the focal length should be chosen to fit the field of view of the assay. Flexibility in this respect can be gained by using a varifocal system that allows the experimenter to adjust the field of view. A lens with a focal length ranging between 12 and 36 mm (Edmund Optics part number NT57-680) has been successfully made use of. Finally, it is important to make sure that a working distance is left between the camera and behavioral setup. A minimum of 20 cm should be considered.

The camera should be mounted on a solid stand, such as those used for picture development. Stages can be purchased from suppliers of professional photography material. Again, platforms, stands, and adapters can be purchased from Edmund Optics and Thorlabs. To avoid light interferences, it is best to enclose the tracking setup in a cabinet or an ensemble of black curtains. Unlike experiments involving vision, it is not critical to test odor-driven behavior in dark conditions, but rigorous attention should be paid to the homogeneity of the illumination system.

The functioning of many computer-vision algorithms is based on the principle of background subtraction. To maintain a good contrast between the light-absorbing larva (dark object) and a homogeneous bright background, it is convenient to illuminate the arena from underneath. A nonheating light source is recommended. Given the high sensitivity of larvae to light, the homogeneity of the surface is critical. For white illumination, the light boxes manufactured to visualize transparencies represent cheap and well-suited options (e.g., "Slim edge" light pad manufactured by Logan Electric). To conduct experiments in a red background invisible to the larva, more advanced backlight systems can be acquired from Advanced Illumination (USA). Finally, the reader is encouraged to keep a thermometer and humidity recorder near the behavioral setup in order to monitor the atmospheric conditions during behavioral tests (optional).

Room lighting: It is important to avoid exposure to stimuli other than those of interest. Ideally, light-driven behavioral experiments should be conducted in a darkened room with constant temperature and humidity conditions. The known visual pigments in *Drosophila* larvae are not sensitive to red light (wavelength higher than 650 nm), and it is not known to induce phototactic responses. If any additional lighting is required in the behavior room, it should also therefore be red. Beware that not all commercially available red-light bulbs are restricted to emitting only red light. It is crucial to verify manufacturer specifications. The light spectrum emitted by a source of light can be determined empirically using a photodiode system. Different apparatus are commercially available (e.g., Thorlabs offers a benchtop photodiode amplifier).

2.3. Phototaxis Assay

Adaptation of the Petri dish assay: For phototaxis experiments, we used regular Petri dishes coated with 2–3% agarose (see Sect. 2.2). Most assays are run on standard 100-mm Petri dishes (88). Smaller plates, such as 16-mm Petri dishes, can be used to study the behavior of first instar larvae.

Petri dish lighting and light-protection: Different lighting systems have been used for light-preference assays. Typically the light source is located at least 30–40 cm above the illuminated arena. Keeping the source at a distance from the arena is important to minimize temperature changes. Ideally, a cold-light source is used. LEDs or LED-bulbs (assemblies of several LEDs fitting a standard light bulb socket such as OSARAM LED, 80012 White) offer two major advantages: first, they do not emit much heat; second, they emit a defined spectrum of light. If other light-sources are used, effects due to temperature changes upon light exposure should be controlled for. Lighting of the behavioral arena must be uniform, which can be achieved by increasing the distance between light-source and arena. When using photodiodes, it is important to ensure that the light intensity is homogeneous throughout the stage on which larvae move.

Tight control of the light conditions in the illuminated vs. shaded areas in the Petri dish is crucial for the behavioral setup. Since behavior of larvae in light-gradients has not been assessed in detail, nor has the behavior at the interface between light and dark, we cannot conclude how they "choose" shaded areas over light-exposed areas. Thus, the assay can provide an index for light-preference but not explain the phenomenon. Assuming that the decision between light and dark occurs at the interface, one should avoid diffuse areas of light and create clear-cut edges. To that end, the light-intercepting material should be located close to the surface of the agar plate. For most assays black duct tape on the cover of the Petri dish will suffice (for optimal results, the duct tape can doubled with aluminum foil). In addition, the internal surface of the upper lid should be optically isolated by duct tape. To decrease reflection of light by the agarose surface, a black surface can be placed underneath.

Room lighting: Light in the behavioral arena should stem exclusively from the light source used to illuminate the agarose plate. Thus, it is important to perform behavioral experiments in a dark room with no white light source and constant temperature and humidity conditions. Since *Drosophila* does not have photopigments sensitive to red light (wavelength higher than 650 nm) and does not seem to display phototactic responses when exposed to red light, additional red or infrared LED lighting can be used.

3. Protocol and Expected Results

3.1. Olfactory Assay: Petri Dish, 1-Odor Paradigm

Rationale: Odor-driven behaviors have been studied and documented in the larva throughout the past century (14). Early observations consisted of manually tracking individual larvae exposed to an odor source. The directional nature of locomotion mediating larval chemotaxis was already reported by Hafez (60). To quantify sensory-locomotor responses elicited by odors, researchers have adopted a statistical method based on en masse assays. Typically, 30–100 larvae are loaded in a Petri dish. The dish contains an odor source and a visually identical control loaded with solvent on the opposite side of the dish. After a few minutes, the fraction of animals found on the side of the odor source is calculated. Significant departure from homogeneous distribution in the plate (50% on left, 50% on right) indicates the existence of attraction to or repulsion by the odor. Despite its apparent simplicity, this assay has been used in numerous labs and has led to significant findings (14, 88). It has the advantage of being robust, cheap and high-throughput. The main drawback is that the user has very limited control over the olfactory conditions in the dish: the only parameter that can be manipulated in a controlled manner is the concentration of the stimulus. Furthermore, the distribution and stability of the odor in the plate is not known.

Procedure: 10 μL of the test odor is pipetted into an odor cup and 10 μL of paraffin oil (solvent) into another. Excess humidity is removed from the surface of the Petri dish and the odor cups are placed on either side of the agarose surface, approximately 7 mm from the edge of the dish. Odor cups should be manipulated with the help of forceps.

Leaving the Petri dish to stand covered for 1 min before introducing the larvae allows for a gradient to be established in the plate. Meanwhile, larvae can be transferred to an agarose plate using a damp fine-point brush or a bristle (see Sect. 2.1). Once larvae have dispersed a little, in doing so removing excess sucrose solution from their cuticle, they can be transferred to the experimental Petri dish.

The larvae are timed as they move freely for 5 min in the covered dish. After this period, the experimenter counts the number of larvae (their position defined by the mouth hooks) in each predefined zone (west, neutral, and east). It is helpful to mark the position of each larva on the lid of the Petri dish as it is counted. Larvae on the lid or those that have dug into the agarose are not usually included in the count. Systematically alter the relative positions of the odor cups between experiments.

The tendency of larvae to accumulate on or away from the odor source is typically calculated with a preference index:

$$\begin{aligned}\text{Preference index(PREF)} = {} & (\text{number of animals on the odor side} \\ & - \text{number of animals on the solvent side}) \\ & / (\text{total number of animals in all three zones})\end{aligned}$$

This normalized score ranges from +1 (complete attraction) to −1 (complete repulsion).

The result of the 1-odor paradigm can be sensitive to the length of time larvae are left to move before assessing their distribution. The maximum preference index is also dependent on the odor. This point is illustrated in Fig. 2a. The evolution of the preference index in time is compared between two odors: isoamyl acetate (IAA, CAS: 123-92-2) and acetone (ACE, CAS: 67-64-1). IAA has a vapor pressure of 5 mmHg at 25°C. The vapor pressure of acetone is about 40 times higher: 200 mmHg at 25°C. For both odors, the concentration tested was 1:50. Despite this difference, we observe that the time course of the preference index is very similar. A "steady state" distribution is reached after approximately 5 min. In the case of IAA, the preference index tends to decrease at around 10 min. The maximum preference index is 5 times higher for IAA than ACE. This result may be due to several reasons: first, the innate attraction to IAA may be higher than to ACE. Second, the sensitivity of the larval olfactory system may be more acute for

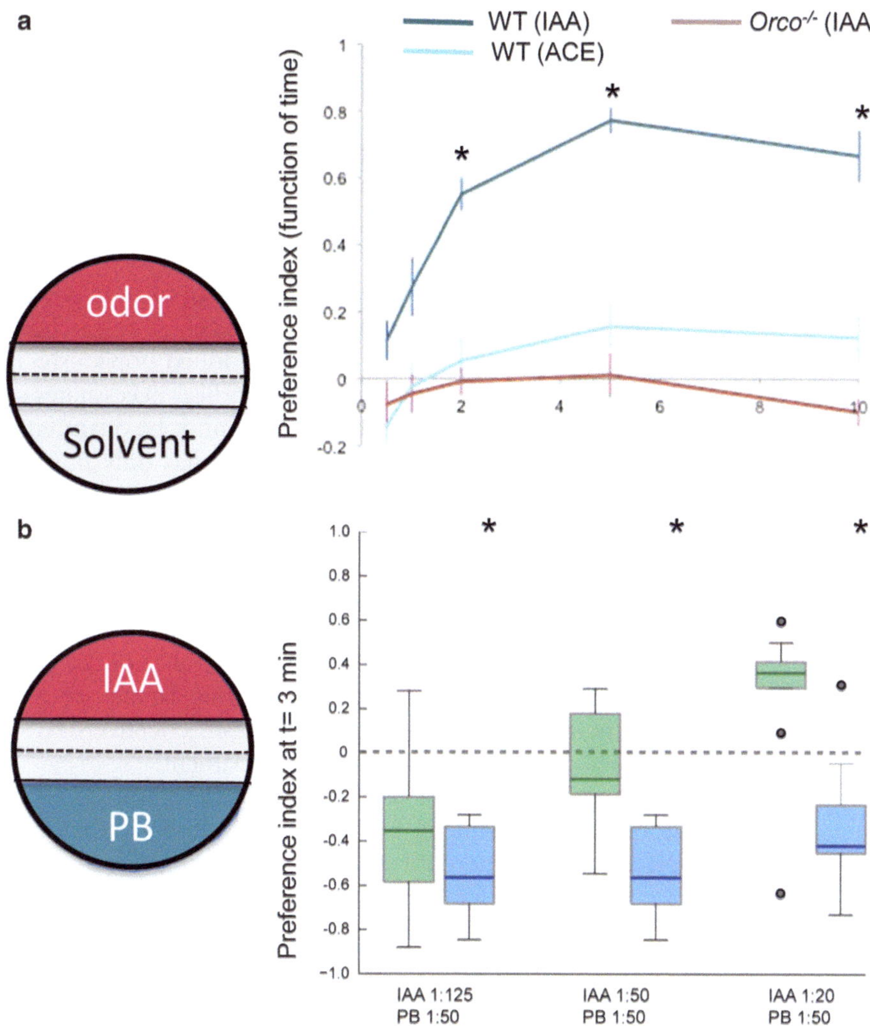

Fig. 2. Illustration of the Petri dish assay for assessment of olfactory behavior (manual counting). (**a**) 1-Odor paradigm where the behavior of larvae is tested against a single odor source. Attraction to the odor is quantified by a preference index measuring the number of animals on the odor side minus those on the nonodor side divided by the total number of animals (in both odor zones and the neutral zone). The preference index is measured in function of the time elapsed after the introduction of the larvae in the plate. Wild type (WT) are tested for two odors at a concentration of 1/50: isoamyl acetate (IAA, *dark blue*) and acetone (ACE, *light blue*). The graph reports preference index as mean values over ten experiments. Error bars: SEM. Upon application of a *t*-test, only points noted with a star (*) are significantly different from 0 ($p<0.05$ after Bonferroni correction). The behavior of wild type larvae is controlled by larvae with an *Orco*$^{-/-}$ null background (17). Preference indices are measured on groups of 20–30 larvae. (**b**) 2-Odor preference paradigm. The preference for 1 odor, propyl butyrate (PB), over another odor (IAA), is tested with successive increments in the concentration of IAA. Odor preferences are tested for two genotypes: wild type (*green*) and *Or42a*-functional larvae (*blue*) (11). Samples noted with a *star* (*) are statistically different from 0 (sign test, $p<0.05$ after Bonferroni correction).

IAA. Third, the high volatility of ACE will lead more evaporation of the odor. As a result, the gradient may be shallower, and thereby more difficult to navigate than that of IAA. It is verified that the anosmic larvae *Orco*$^{-/-}$ do not show a significant bias toward the odor side. In conclusion, Fig. 2a highlights that larvae do not immediately accumulate on the side of odor preference. It is not advisable to measure the preference index before 3 min have passed (see Fig. 2a). Ideally, the point at which a "steady state" distribution is obtained should be determined for each odor. For odors that are highly volatile, one should also check that the preference index does not begin to decline in under the allocated time (14, 83)—an effect which can be due to a gradual flattening of the gradient in the arena.

General considerations: Even though a neutral zone is not included in the evaluation of the preference index by all researchers, the use of one is a prudent measure. As mentioned previously, the interpretation of differences in preference indices necessitates caution as the olfactory conditions in the arena cannot be quantitatively controlled or measured. The experimenter has only limited control over the initial conditions of the experiment: it is usually assumed that larvae start oriented in random directions. At the onset of the experiment, larvae heading down-gradient may take several seconds (or even minutes) to reorient. While enough time must be given for the majority to locate and accumulate around the source, waiting too long can lead to the disappearance of the gradient and apparent habituation effects. Over time, larvae are more likely investigate the walls and lid of the dish, effectively reducing the number of larvae included in the experiment.

Another important consideration is the consistent use of the lid during the experiment, be it present or absent. When the lid is in place, most of the odor diffusing from the source remains in the arena. The absence of turbulent airflow makes the gradient more static. In this regime, it is reasonable to assume in a first approximation that the odor is distributed by a purely diffusive process (Fig. 4b). Without the lid, odor distribution in the dish cannot be predicted on the basis of simple physical principles rendering behavioral correlations unfeasible. Finally, one should keep in mind that the preference index captures a complex multistep behavioral process: accumulation of the larvae on one side of the dish results from odor detection, odor processing, integration of the signal with other cues, an orientation decision, and a motor response. It is determined by a series of molecular, cellular, and behavioral events.

It is recommended to draw conclusions from sample sizes of minimum 20 experiments. As illustrated in Fig. 2, smaller sample size (e.g., $N=10$) can lead to apparently clear trends without statistical significance.

3.2. Olfactory Assay: Petri Dish, 2-Odor Paradigm

Rationale: The 1-odor paradigm addresses the following question: can a larva detect a particular odor at a given concentration? This assay is useful to determine the sensory threshold of an odor (the lowest concentration leading to a significant behavioral response) and assess potential changes in sensitivity resulting from alterations in the olfactory circuits (89).

Another question that one may want to address is whether one odor is preferred over another. In other words, is a larva more attracted to odor X than odor Y? A 2-odor paradigm where each cup contains a different stimulus is suitable to address this problem. This assay is widely used in associative conditioning experiments (15).

Procedure: As with the 1-odor paradigm, replacing the solvent in the control cup with the second odor in this case.

In Fig. 2b, we illustrate the 2-odor assay in WT larvae and larvae with olfactory function restricted to a single OSN that expresses the *Or42a* receptor gene (12). We ask whether the preference for one odor can be altered by increasing the concentration of the competing odor. The two odors used were IAA and propyl butyrate (PB, CAS: 105-66-8). Since both odors have a similar vapor pressure (IAA: 5 mmHg at 25°C, PB: 5.45 mmHg at 25°C), they are likely to generate similar gradients with similar concentration ranges and geometries (assuming that differences in other aspects like solubility do not exert as much of an influence). When tested at the same concentration (a dilution of 1:50), WT larvae show equal preference for both odors. When the concentration of IAA at the source is increased to 1:20, WT show a clear preference for IAA (positive preference index). In contrast, a clear preference for PB is observed when the concentration of IAA is decreased to 1:125. Similar observations have been recently reported by Gerber and coworkers (90). Interestingly, this concentration-dependent reversal of preference is not observed for *Or42a*-functional larvae within same range of concentrations. A clear preference for PB 1:50 is shown even over higher concentrations of IAA. This finding suggests that the affinity of the Or42a receptor is higher for PB than for IAA.

General considerations: Although the analogy may be tempting at first, one should not forget that a 1-odor Petri dish assay is not equivalent to the 2-alternative forced-choice assay used for higher-order organism like rats (91). The accumulation of larvae on one side of the plate results from a series of behavioral decisions controlled by the chemotaxis algorithm. Furthermore, the larval olfactory system may not adapt and/or habituate to distinct odors in the same way. Such effects may affect the evolution of the preference index over time.

The results shown in Fig. 2b draw attention to the concentration-dependence of odor preferences. When testing multiple odors, establishing a hierarchy in the preference between odor pairs is only valid for specific concentrations. Conclusions drawn from a given set of concentrations cannot be readily generalized to others. As shown in Fig. 2b, modest changes in concentration can invert the preference for one odor over another. This observation raises the question of which concentrations should be tested for pairs of odors with different vapor pressures. If it is assumed that the level of attraction is proportional to the stimulus intensity, one approach entails identifying the concentrations of each odor necessary to elicit equal naïve attraction. This is routinely achieved in human psychophysics experiments where the experimenter defines the equivalence point at which all stimuli are of the same perceived intensity (92). Finding the equivalence point is achieved by trial and error according to an intensity-matching procedure. It is reasonable to start the matching procedure with a moderately higher concentration of the odor with lowest vapor pressure. Once matched concentrations have been defined for each odor of a set, shifts in relative preferences due to cross-adaptation or alterations in neural activity can be studied.

3.3. Visual Assay: Petri Dish

Rational: Two distinct plate assays have been used to measure light preferences: the half-plate assay and the quarter-plate assay. Both behavioral setups provide similar results that can be displayed and assessed by distinct parameters. Often used as the dark preference is the percentage of larvae found in darkness over light (Fig. 3a). The value is strongly dependent on light intensity, circadian time, and genetic background.

An alternative display is the performance index, which depicts to what degree an animal prefers darkness to light. The performance index for darkness preference is calculated as follows: $PI_{(darkness)}$ = (number of animals in dark – number of animals in light)/ total number of animals. At moderately high light intensities (750–800 lux), the performance index varies between 0.4 and 0.6. This implies that 65–80% of the larvae prefer darkness (Fig. 3b).

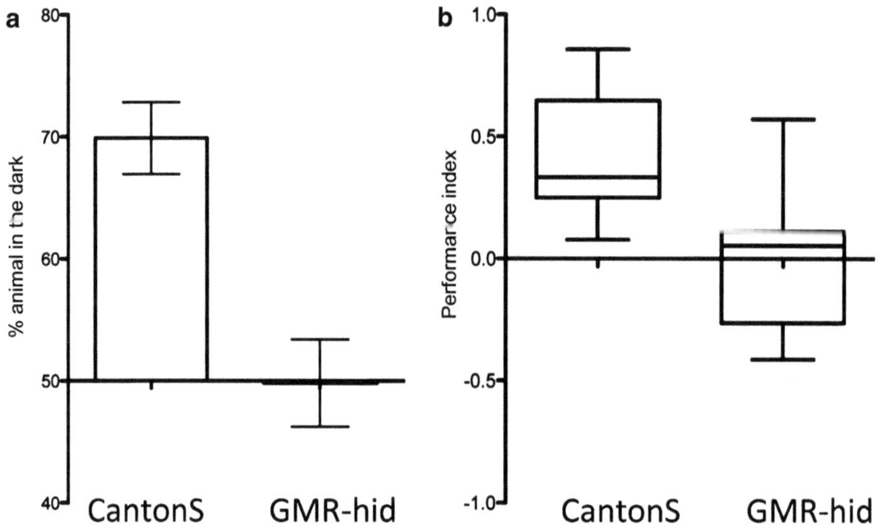

Fig. 3. Illustration of the Petri dish assay for assessment of visual behavior (manual counting). (**a**) Mean percentage of animals on the dark half is displayed. Control animals (Canton-S) score 69.9%, while eyeless larvae (GMR-hid) score 49.89%. GMR-hid do not display a preference for either side (N= 15). (**b**) Preference indices from (**a**) displayed as box plots. The mean preference index of Canton-S is 0.39, while that of GMR-hid it is 0.003. Error bars: SEM.

Experimental procedure: Larvae should be transferred from the incubator to the behavior room immediately before performing the experiment. To avoid thermal inconsistency, the behavior room and the incubator should be kept at same temperature. Experiments are usually carried out at 25°C. It is important to wash larvae thoroughly as any residual food transferred to the agar plate will act as a gustatory attractant, which will likely affect the behavior observed. Light preference is typically measured after 5, 10, or 15 min. As with the olfactory assay, the distribution of larvae on the dark and light exposed quadrants of the plate can be found to be adjusted as early as 1 min into the experiment. Mutants with partial locomotor deficiencies may require more time to distribute over the plate. It is advisable to wait a minimum of 5 min before quantifying larvae in each quadrant. A semiautomated version of this assay consists in taking snapshots of the plate every 5 min. Quantification of the number of larvae on the light-exposed quadrants can be carried out when the experiment is complete. The calculation of performance indices requires foreknowledge of the number of larvae introduced in the plate.

3.4. Automated Tracking: Revisited Petri Dish Assay

Rationale: When using a Petri dish, the distribution of larvae is a function of the past and present odor gradients in the plate. As the gradient decreases in slope, larvae tend to experience more difficulties locating the position of the odor source. As a result, their distribution will be more diffuse around it. Another potential factor influencing odor responses is habituation. Upon prolonged exposure to a given odor, larvae demonstrate a loss of interest, or even avoidance responses. The performance index refers to the spatial distribution of the larvae, a measure, which, as discussed above, evolves in time (Fig. 2a). To avoid time-based artifacts influencing conclusions, measurements can be repeated at various intervals during the experiment. Aside from being cumbersome and tedious, manual counting is prone to errors when repeated numerous times on large numbers of animals. Therefore, it is advantageous to delegate counting tasks to a computer.

Procedure: Several options exist to monitor the behavior of larvae (individuals or groups). To analyze the distribution of larvae in a Petri dish (irrespective of their identity), one can develop or adapt computer-vision algorithms based on object recognition. Matlab once again provides an excellent tool for this type of analysis recognition (93). This approach has been recently used to analyze the olfactory behavior of mosquito larvae (94).

The reader less inclined toward programming will be happy to learn that several commercial software products have been specially designed for behavioral analysis. Author experience extends to Ethovision (Noldus Information Technology) (12, 24). Similar commercial software products exist, such as Trackit 2D (BiobServe) and DIAS developed by Soll Technologies. DIAS has been successfully used by several groups to study larval locomotor

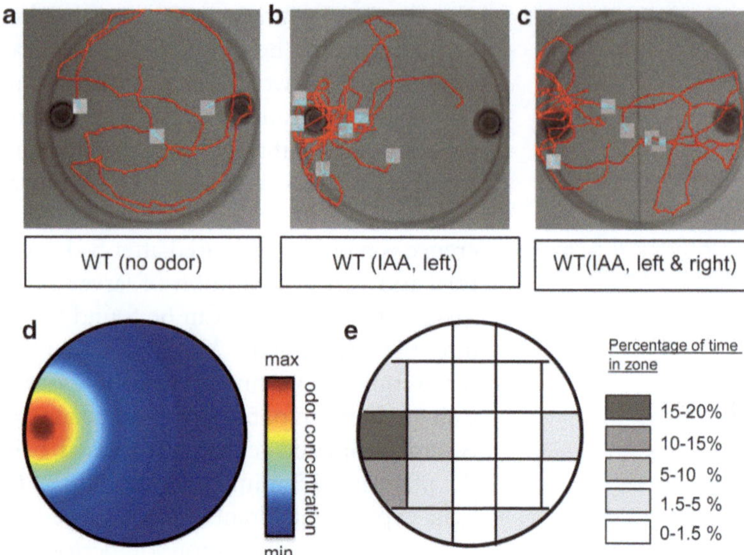

Fig. 4. Illustration of the Petri dish assay for olfactory behavior with automated tracking of larvae. Trajectories were generated with the tracking software Ethovision (Noldus, The Netherlands). Each recording was made in a separate Petri dish and later superimposed. (**a**) Wild type larvae exposed to the odorless solvent. (**b**) Wild type larvae exposed to 2 μL of pure IAA on one side of the plate (*left*). (**c**) Wild type larvae exposed to 2 μL of pure IAA on both sides of the dish. (**d**) Simulated distribution of the odor in the Petri dish. (**e**) Sector plot analysis of the progression of 72 larval trajectories in response to 2 μL of pure IAA on the *left hand side* of the dish. The mean fraction of time observed in each sector is reported by the grey scale on the *right*. It is evident that in this case larvae spend about 20% of their time on the sector corresponding to the odor zone.

behaviors (36, 65, 95). The advantage of programs like these is that they track the position of single larvae at relatively high temporal resolution (typically 5 Hz) without storing large images. However, most of the existing commercial software is unable to monitor groups of larvae in the same plate. At present time, several groups have undertaken to develop freeware capable of tracking numerous larvae at the same time. For instance, the Multi-worm Tracker created by the Kerr lab at Janelia Farm (HHMI) for *C. elegans* offers promising possibilities for the *Drosophila* larva. Another software adapted specifically to larvae has been created by the Samuel lab (75).

Figure 4 illustrates trajectories obtained with commercial software (Noldus) for the 1-odor Petri dish assay. Figure 4a displays the foraging behavior of three larvae with no odor stimulus. Note that the three trajectories displayed in the figure were acquired during separate experiments. Upon exposure to a high concentration of IAA, larvae accumulate close to the odor source (Fig. 4b, source on left side of plate). When the same concentration of IAA is introduced on both sides of the plate, larvae choose either odor source and remain in its vicinity (Fig. 4c). If it is assumed that the odor originating from a single odor source leads to an exponential distribution profile (Fig. 4d), basic correlation between stimulus and behavior can be carried out. As shown in Fig. 4e, the odor plate can be subdivided into small sectors. The average fraction of time spent in each sector can be measured over several trajectories. In the example of Fig. 4e, the fractions of time spent in each sector were measured for a single source of IAA (72 trajectories). Not surprisingly, it is observed that larvae tend to accumulate in the three sectors surrounding the sector containing the source (middle left). The sector plot is compatible with the existence of an exponential gradient such as that shown in Fig. 4d.

In Fig. 5, the information extracted by the tracking software is further exploited to address the following question: are wild type (WT) larvae more accurate chemotaxing to a given odor (IAA) than larvae with a single functional OSN (*Or42a*-functional)?

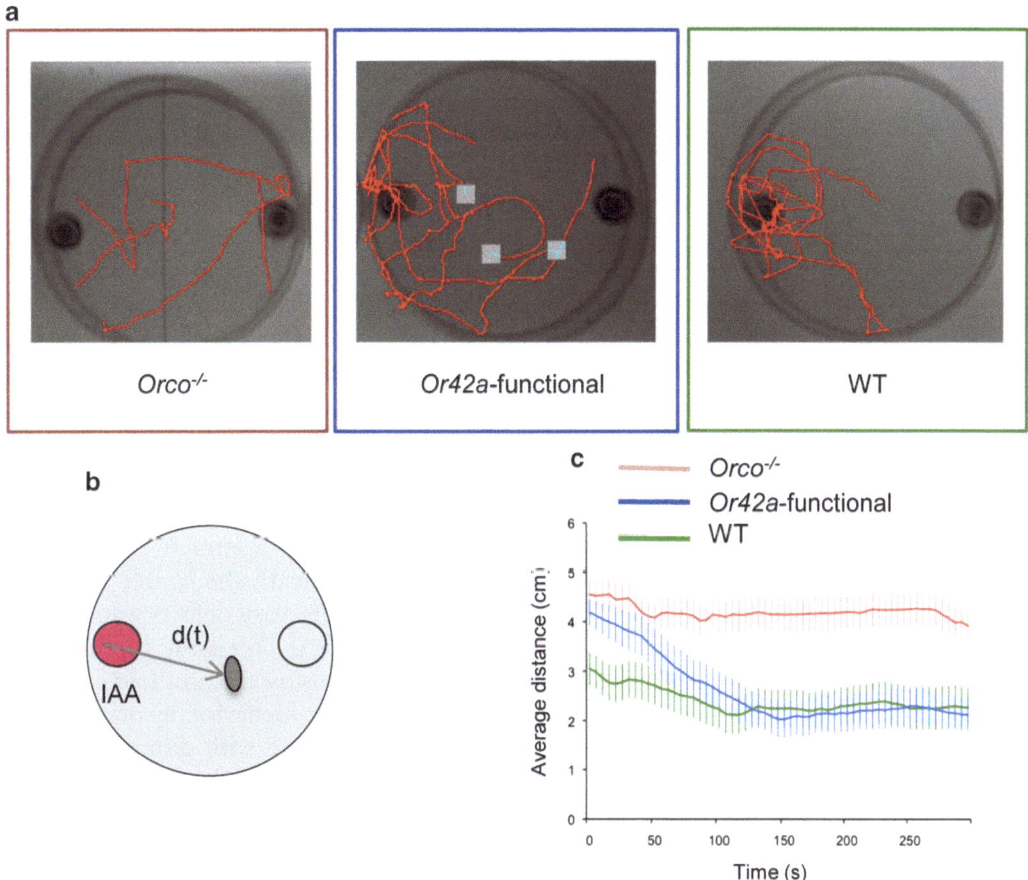

Fig. 5. Time course of olfactory response in 1-odor paradigm. Attraction to an odor source vs. solvent is measured over time. (**a**) Representative trajectories of *Orco*$^{-/-}$ (*red*), *Or42a*-functional (*blue*) and wild type (*green*) larvae in response to 2 μL of pure IAA on the *left hand side* of the dish. Top and bottom lids display different experiments. In all experiments the odor source contains 2 μL of pure IAA. (**b**) Distance between the larva and the odor source is measured in reference to the larval center of mass. (**c**) Average distance to the odor source calculated over 30 trajectories. Error bars: SEM.

To address this question, we monitored the behavior of 30 larvae in response to a single source of IAA. We quantified the distance to the odor source as a function of time (Fig. 5b). Figure 5a illustrates representative trajectories. Visual inspection of these trajectories conveys the impression that WT are more focused on the odor source than *Or42a*-functional larvae. This impression is corroborated by the time course of the mean distance to the odor source (Fig. 5c). We find that *Or42a*-functional larvae require more time to reach the vicinity of the source. As expected, we find that the anosmic control $Orco^{-/-}$ stays at a constant average distance from the source (interestingly, this value is approximately equal to the radius of the plate).

3.5. Olfactory Assay: 96-Well Plate, Single and Multiple Odor Source(s)

Rationale: A major disadvantage of the Petri dish assay is the inability to control the odor gradient generated by a single odor source. A second drawback is the difficulty to measure the topology of the gradient enclosed in the arena. Both limitations have been solved by the development of a system based on 96-well plates (as described in Sect. 2.2). Using a spectroscopy-based technique, the stability and geometry of the gradient can be quantified (24). For odor distributions with one axis of symmetry, the rectangular geometry of these plates permits accurate reconstruction of the gradient in a noninvasive manner. The use of a single odor source centered in the middle of plate leads to a radially symmetric gradient. Such assay represents a controlled alternative to the 1-odor Petri dish. As described below, the geometry of the gradient can be tailored by using multiple odor sources. This feature is specific to the 96-well plate assay.

Procedure for single-odor source assay: As detailed in Fig. 6a and (85), a single odor droplet (10 μL) is pipetted into ring #E7. A single larva is introduced under the odor source (or close vicinity). The movement of the larva is recorded for 3–5 min, or until it makes contact with the wall of the plate. For attractive stimuli (e.g. IAA), wild type larvae accumulate beneath the source (Fig. 6b). Anosmic larvae or larvae subject to olfactory defects wander away from the droplet quite soon (Fig. 6c). This assay represents an efficient tool to reliably determine the sensory threshold of a given genotype to certain odors (22). The behavior associated with groups of trajectories can be characterized with different types of parameters. As in Fig. 4e, one can quantify the amount of time spent in zones of interest. For instance, rings of equal width, but increasing radius, can be defined around the source (22). The distribution of average time spent in each ring is informative of the attraction to the source. Also, the space for distribution allows the experimenter to evaluate the precision with which a given genotype is able to pinpoint the source and stay in its close vicinity. Following this approach, we discovered that *Or42a*-functional larvae accumulate at a short distance from a high-concentration

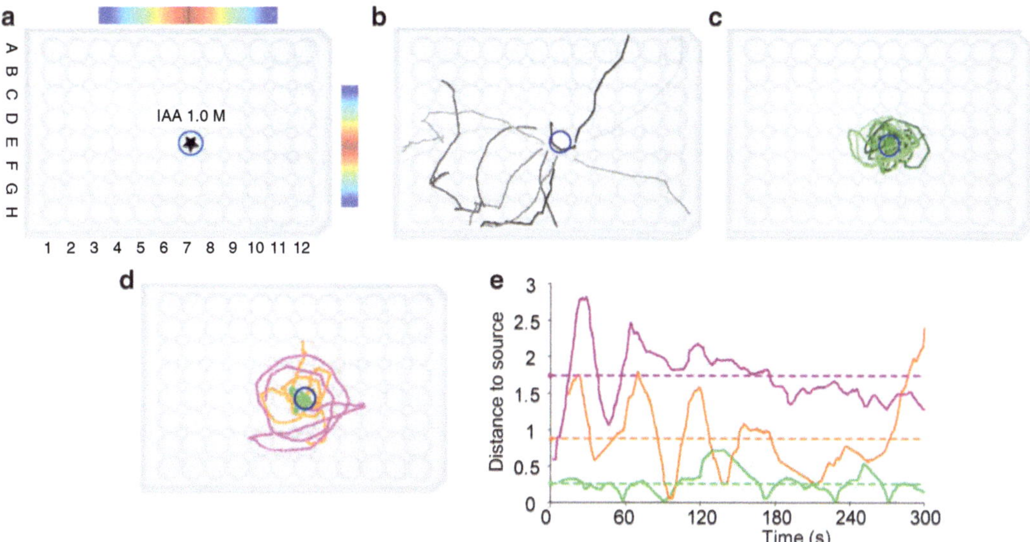

Fig. 6. Illustration of the single-odor source assay in a rectangular arena. (**a**) Schematic of the odor delivery setup. A 10 μL droplet 1.0 M IAA, is suspended from the inside of the lid at position #E7 (*blue circle*). As detailed in ref. (23), the resultant odor gradient can be approximated by a Gaussian distribution. Individual animals are monitored after being placed on the agarose surface directly under the odor source. (**b**) Trajectories of *Orco*$^{-/-}$ larvae ($N = 10$). (**c**) Trajectories of wild type larvae ($N = 10$). Trajectories remain clustered under the odor source. (**d**) Trajectories generated in response to different odors reflect different behavior in relation to the source. The trajectory represented in green stays tightly centered under the odor. The trajectory in orange departs significantly from the source, making use of sharp turns to return to it. The *pink* trajectory remains at a fixed distance as the larva circles around the source. (**e**) Trajectories from (**d**) represented as distance from the source over time. The mean for each one is represented as a *dashed line*.

source of ethyl butyrate (21). Instead of staying beneath the odor source, *Or42a*-functional larvae follow a concentration isocline. This variation is illustrated in Fig. 5d where three distinct behavioral responses are represented: one trajectory remains under the source (green). Reorientation toward the source occurs almost immediately after the source is overshot. A second involves movements back and forth under the source (orange). This trajectory may reflect a reduced precision in the ability of the larva to detect concentration decrease. As a consequence, the onset of a turn is delayed and the trajectory is less tightly clustered under the source. A third trajectory (magenta) corresponds to a "circling" response like that of *Or42a*-functional larvae, where the larva keeps a distance from the source. This behavior may be induced by an avoidance response elicited at high concentrations (93). Note that the richness of these phenotypes would have been overlooked if the behavior had been quantified as a percentage of time in a unique odor zone centered on the source. In addition, the odor response can evolve in time. Trends can be captured throughout the time course in relation to the distance from the odor source (Fig. 5b). After several back-and-forth movements, the orange trajectory leaves the source ($t > 270$ s). The magenta trajectory evolves at a distance from the source that decreases over time. One may speculate that this effect is due to a flattening of the gradient which transposes a given concentration isocline closer to the odor source. In contrast, the green trajectory remains near the source at all times. The point we aim to illustrate here is that separating the data into arbitrary zones may not always be the most appropriate way to characterize a given phenotype. One should also consider the evolution of continuous variables such as the distance from the odor source, the speed, etc.

Procedure for multiple odor sources: To create gradients with specific topological features along the width and length of the arena, several odor droplets are pipetted in the rings inside the lid. As shown in Fig. 7a, an odorant trail can be created by aligning six droplets in the central row E (rings #E2, #E4, #E6, #E8, #E10, and #E12, reference system shown in Fig. 6a). We have used this assay to characterize the searching strategy utilized by *Drosophila* larvae to navigate odor gradients. The multiple odor-source assay is illustrated for two gradients of IAA where the concentration increases either exponentially (Fig. 7b) or linearly (Fig. 7c) along the length of the plate. These two different geometries are obtained by using geometric or arithmetic series respectively for the concentrations of the six droplets laid in row E.

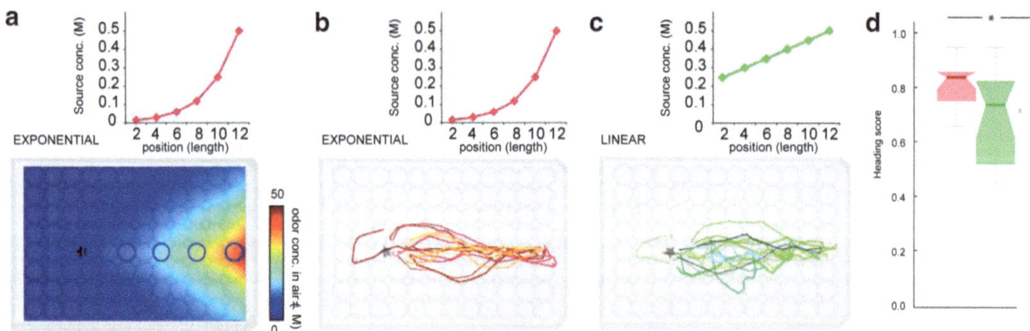

Fig. 7. Illustration of the multiple-odor source assay in a rectangular arena. (**a**) Schematic of the odor delivery setup. Six 10 μL droplets of IAA are suspended along the inside of the lid in row E (*blue circles*). To create an exponential gradient, the concentration of each droplet is increased according to the geometric series represented in the graph above the figure: #E2: 0.03 M, #E4: 0.06 M, #E6: 0.12 M, #E8: 0.25 M, #E10: 0.5 M, #E12: 1.0 M. The odor gradient in the arena was reconstructed using an infrared-spectroscopy-based technique described in ref. (23). The absolute concentration of the stimulus in gaseous phase is reported in the color scale on the right. Individual animals are monitored after being placed on the agarose surface between #E3 and #E4. (**b**) Trajectories of ten wild type larvae following the exponential gradient described in (**a**). (**c**) Trajectories of ten wild type larvae following a linear gradient. The concentrations of the droplets used to generate this gradient follow the arithmetic series shown in graph above the figure: #E2: 0.25 M, #E4: 0.30 M, #E6: 0.35 M, #E8: 0.40 M, #E10: 0.45 M, #E12: 0.50 M. (**d**) Quantification of the average alignment of the motion with the odor gradient. As explained in ref. (23), we define a heading angle between the instantaneous direction of motion and the local direction of the odor gradient (direction of steepest concentration change). Heading angles are scored by a cosine function. The mean cosine of the heading angles is then calculated for all positions of a given trajectory to give a heading score. The boxplots report the distribution of the heading scores obtained for the exponential and linear gradients ($N = 10$). Heading scores are significantly lower for the linear gradient (Wilcoxon test, $p < 0.05$), suggesting that larvae have more difficulty orienting in gradients with reduced slopes.

At the beginning of the experiment, single larvae are introduced under the odorant line (row E) between rings #E3 and #E4. The top lid containing the odorant droplets is lifted just enough to allow for loading of the larva. It is important to ensure that the initial conditions correspond to a position sufficiently immersed in the gradient; or else many larvae fail to detect the signal from the start. In addition, it is preferable to orient the newly introduced larvae in the direction of the gradient, ensuring that the body and head face the highest concentration. Initial drift away from the odorant line leads to many false negatives as experiments are stopped as soon as an animal contacts any wall of the arena. The duration of each trial is limited to 3 min, as larvae tend to lose interest and abandon the gradient after that. The recording can be stopped sooner if the larva reaches a target area or makes contact with the wall.

Healthy larvae start crawling immediately after their introduction into the arena. If the experimenter is slow at beginning the recording, initial data points may be lost. At the end of the recording, lift the lid and remove the larva with a paintbrush. We advise the experiment to not reuse larvae. If multiple animals are tested in a given arena, do not delay the loading of the next larva.

Expected results and general considerations: While anosmic larvae wander at random from the starting point (24), wild-type larvae show robust chemotaxis behavior along the odorant trail. Navigation in the direction of increasing odor concentration is evident for both exponential and linear gradients (Fig. 7b, c). However, the average alignment of the trajectory with the odorant trail is slightly better for the exponential gradient than for the linear one. This trend can be quantified by measuring the heading angle between instantaneous direction of the motion and the direction of the local odor gradient. We defined a heading score as the average cosine of the heading angle observed for all positions of a trajectory (see Supp Fig. 12 of (24)). The heading score measured for a linear gradient was lower than that for an exponential gradient. Navigating a linear gradient is more challenging than navigating an exponential gradient. The accuracy with which a larva ascends the odor trail is correlated with its ability to detect temporal variations in the odor stimulus. Convoluted paths are often caused by defects in the sensory system (reduced sensitivity) or in central processing of olfactory inputs. Other possible parameters for behavioral quantification have been discussed elsewhere (24, 47). They include percentage of time spent by a larva under the odorant trail, mean heading angle with respect to the local direction of the odor gradient, and a combined chemotaxis score measuring the global tendency of a larva to follow the odorant line. Many other options are available to the experimenter (96).

Gradient profile verification with IR-spectroscopy: Before testing larvae in a particular gradient, it is important to determine the topology and stability over time. To measure odor concentration in gaseous phase, we have developed a noninvasive method based on IR spectroscopy (24). In this manner, the experimenter can adjust the duration of each trial and decide whether several successive trials can be conducted in the same arena.

4. Notes and Troubleshooting

4.1. Influence of Internal States and Overcrowding Effects

Predisposition and internal states of an animal or a group of experimental animals will have a profound impact on the behavior being observed. Even though the influence of the internal states has not been studied in depth, the notion of motivation to perform a specific behavior has been discussed at various points (97). Adult flies have to be starved before they display certain behaviors, which constitutes another variable necessitating control. Internal states can affect the value allocated to external sensory information from potential food sources (98). Currently, the question of whether olfactory and visual behaviors are altered by starvation remains unexplored.

On the other hand, overcrowding during development is likely to influence the outcome of any behavioral experiments. Even though stampede effects have not been observed in the larva (44, 99), Godoy-Herrera et al. recently reported the existence of attraction and repulsion between different *Drosophila* species at the larval stage (100). Chemosensory communication across individual larvae is then possible (see also ref. (99)). Stringent schedules for egg collection and small crosses of progenitor flies (about 50) help control the amount of larvae in a vial.

In experiments involving Petri dishes, it is not recommendable to test more than 50 larvae at once. Larvae should not be clumped when they are loaded into the dish but spread out along the neutral zone with the brush or bristle. As described earlier in this chapter, *Drosophila* larvae contain a molecular clock in defined brain neurons. This clock modulates specific behavior in a circadian-dependent manner. Until now only visual behaviors have been shown to undergo circadian modulation but it seems likely that other behaviors are modulated in the same way. For this reason the use of light-dark incubators and performing experiments at the same time of the day should be routine practice.

4.2. Age-Specific Changes

The increase in body mass from first instar to the beginning of pupation is substantial: within 6 days, larvae increase their weight a hundredfold through constant feeding. Yet, feeding behavior is drastically altered prior to pupation. Near the end of the third instar larvae stop feeding and search for dry surfaces, which represent adequate pupation sites. At this point, larvae make a transition from the "feeding" stage to the "wandering" stage. During the wandering stage, exposure to light does not trigger an avoidance response: negative phototaxis turns to indifference or even positive phototaxis (100). The neural mechanisms explaining this reversal from negative to positive phototaxis are still unknown. Similarly, it is likely that the loss of interest in food is correlated with transformations in olfactory behaviors. That larvae undergo a change from positive to negative chemotaxis at the onset of the wandering stage seems plausible. While this problem represents an excellent insight into the tuning of sensorimotor transformation (50), it can also affect conclusions pertaining to innate behavior at the feeding stage.

For many behavioral assays, experimenters use early third instar larvae. This choice is mainly motivated by practical considerations: third instar larvae are simple to handle and less fragile than first and second instar animals. Very few labs work with first instar larvae, but see (101). If a different developmental stage is analyzed, it is advisable to first perform in-depth control experiments to verify that the behavior occurs reproducibly. In any case, it is not advisable to work with late third instar larvae, unless the intent is to study changes in behavioral responses prior to pupation. In particular, larvae found crawling on the walls of food vials should be removed before making use of the remaining larvae. To have reasonable control over the age of larvae, staging can be achieved by restricting egg collection to 2–4 h periods. Other methods to stage larvae based on morphological markers also exist (102).

4.3. Adaptation and Plasticity of Olfactory Behaviors

Preexposure to an odor modifies subsequent olfactory responses. When a larva is stimulated by a given odor for several minutes, its naïve attraction to that odor can decrease or be even abolished (103). Until now, the nature of this behavior has remained under debate. On one hand, odor exposure could induce peripheral adaptation. Accordingly, the OSNs activated by the odor would become unresponsive during prolonged stimulation. This effect would lead in turn to insensitivity to the odor.

On the other hand, odor exposure could also involve habituation through the remodeling of neural circuits. This effect could turn naïve attraction to an avoidance response. Recently, odor exposure has been shown to reverse behavior from attraction to avoidance (104). Using a combination of behavioral and genetic manipulations, light has been shed on the nature of odor-exposure effects. Behavioral habituation arises, at least in part, from central synaptic plasticity mechanisms at the level of the antennal lobe (105). Activation of peripheral OSNs is not necessary for olfactory habituation. Habituation results from the activity of two classes of neurons: local interneuron (LNs) and projection neurons (PNs). A model has been put forward whereby habituation occurs as the result of a pathway in which projection neurons signal depression on OSN-PN synapses and/or facilitation of LN-PN synapses. No equivalent effect has been reported upon preexposure to light. Nonetheless, peripheral adaptation or habituation to light seems highly plausible. Therefore, the recycling of larvae across successive experiments is ill-advised. Larvae tested for chemotaxis or phototaxis should not have undergone any form of prestimulation other than that associated with regular rearing. Most sensory responses are clearly not invariant: they are to a great extent dependent on the history of the individual in question.

As illustrated in Fig. 2b, one should keep in mind that innate odor preferences are concentration dependent (90). Classifying odors as attractive and repulsive can be misleading. The vast majority of odors are attractive to larvae at low to mid-range concentrations and many are repulsive at excessively high concentrations. Moreover, variations in olfactory behavior are expected across different species of *Drosophila* (106). Major evolutionary shifts have been documented between *D. melanogaster* and *D. sechellia* where high concentrations of one odor (methyl hexanoate) can lead to attraction or repulsion behavior in different adult flies of different species (107). Similar shifts are likely to exist across different geographical isolates of the same species as well as different lab strains. Thus, care should be taken in selecting the lab strain used as a positive control. Canton-S is usually accepted as a standard control. For experiments involving genetic manipulations, the control strain should have the same genetic background as the transgenic lines. Here, one should remember that the *yw* (yellow white) and w^{1118} strains are associated with visual defects (lack of pigments) and potential cognitive impairments. Interestingly, no obvious results of these visual defects have been observed in *w* or *yw* mutant larvae up until now.

4.4. Miscellaneous Tips to Avoid Odor Contaminations

Following are listed various manners of avoiding undesirable sources of contamination in olfactory experiments:

First, use agarose as opposed to agar. While significantly cheaper, agar is less refined, hence the reduced production costs. Agar contains odorants. Variations across batches may create differences in the experimental odorant background.

Second, plastic dishes should not be recycled across experiments. Many odors react with the polystyrene of Petri dishes and 96-well plates—pipetting a couple of drops of ethyl butyrate on the lid of a 96-well plate quickly melts the plastic. Plate recycling could be done with glass Petri dishes, but thorough cleaning would need to be carried out with ethanol (or hexane) followed by a rinse with soap and water. Ideally, the plate should dry overnight at high temperatures (50°C). When using the 1 or 2-odor Petri dish assays, plastic or Teflon cups are a preferable vessel for the odor than filter paper in direct contact with the agarose (12, 83, 108). Finally, we propose the following protocol to decontaminate Teflon (PTFE) cups used as odor sources.

Decontamination of Telfon (PTFE) odor cups: The size of the odor cups and oily nature of the solvent make it difficult to remove odor solutions from one experiment to another. To help counteract this, the cups are cleaned with hexane. Due to its toxic nature, cleaning should be carried out in the fume hood with the use of nitrile gloves.

Ideally, odor cups used for different odors should be stored and cleaned separately to minimize cross-contamination by residues. Lids are removed from cups with forceps and both parts placed in a small glass container filled with hexane past the level of the cups. The flasks are manually agitated to remove air bubbles. Leaving them on a mechanical agitator aids the perfusion of the hexane.

After being left to soak for 5 min, the cups are strained and transferred to a fresh glass container filled with warm water, odorless soap, and ethanol, which speeds drying later. These are left on a mechanical agitator for half an hour, taking care that there is sufficient agitation to move the cups, but not damage them. After straining they are placed upright on a fresh open Petri dish in the fume hood to air dry overnight, and then stored away from possible sources of odor contamination.

5. Conclusions and Perspectives

In this chapter, we present a series of assays to study orientation behavior in the *Drosophila* larva in response to visual and olfactory stimuli. We review coarse and high-throughput paradigms to assess the ability of larvae to smell specific odors and their level of preference in a two-alternative olfactory assay (Fig. 2). Similar paradigms exist to assess the ability of larvae to respond to light (Fig. 3). The advantage of automated tracking has been highlighted in the analysis of behavioral responses over time (Figs. 4 and 5). The same procedure should be applicable to study light responses if tracking is carried out in the infrared range with a low-pass filter blocking the light stimulus. We also describe a manner in which to carry out behavioral analyses at higher spatial and temporal resolution (Figs. 6 and 7). While high-throughput assays are convenient to conduct large-scale behavioral screens, a detailed analysis of elementary behavioral responses is necessary to establish circuit–function relationships. The numerical simplicity and ongoing anatomical characterization of visual and olfactory systems (Fig. 1) provides an excellent opportunity to progress in our understanding of the molecular and cellular basis of orientation behavior. Major progress is expected to be made in this direction by combining behavioral analysis with the ever-expanding molecular toolkit to manipulate neural functions (109, 110) and the creation of large collections of driver lines expressed in defined subsets of neurons in the larval brain (66, 67). During the coming few years, the study of neurobiology in the larva is likely to be increasingly influenced by fields that are already gaining momentum: high-resolution behavioral analysis combined with optogenetics, calcium imaging, electrophysiology, and neuroethology (55).

The behavioral analysis presented in Figs. 6 and 7 permits correlations between behavioral responses and sensory inputs (24). However, the description it offers of larval behavior is too simplistic, as it only captures positional information for the center of mass. Considerable advances have already been made in order to monitor the evolution of various body points (e.g., the head and the tail) to ultimately capture the movement of individual segments (101). This level of description will be necessary to fully capture the complexity of behavioral responses. In particular, it will be important to characterize the elementary behavior underlying chemotaxis and phototaxis. Thus far, Connolly and coworkers have established the "ethogram" associated with foraging behavior (47). Having defined four elementary behaviors (feeding, locomotion, retreating, bending), they calculated the probability of each behavioral transition. The probabilities served as a quantitative basis to compare variations in foraging patterns across strains, species or environmental conditions (47, 48). In the future, it will be important to perform this type of analysis with automated algorithm for tracking, identification and classification of elementary behavior (73). To fully account for larval responses to odors and light, one may need to capture the head movement in three dimensions. Sudden increase in stimulus intensities is usually accompanied by vigorous head movements on either side of the sagittal plane (lateral head casting). Head lifting ("rearing") is likely to contribute to active sampling of the environment (47).

To characterize the encoding of sensory information in the peripheral olfactory circuits and the processing of this information in higher-order brain centers, it will become critical to develop new methods to conduct in vivo electrophysiology and calcium imaging in the *Drosophila* larva. Such techniques have become standard in the adult fly. Asahina et al. (22) have developed a useful preparation to record from the peripheral olfactory system which is likely to offer a solid basis for future work. The pioneering work of Oppliger and Cobb is also paving the way for extracellular recordings from larval sensory neurons (23, 45). Here, the next challenge will consist in conducting calcium imaging in freely moving larvae. This should be greatly facilitated by the recent advent of new tracking setups devised for optogenetic use in *C. elegans* (79). Bioluminescence represents another promising tool for semitransparent organisms like zebra fish larvae (111).

Finally, though works such as those by Raul Godoy-Herrera have laid a foundation for our understanding of naturalistic behaviors in the *Drosophila* larva, knowledge about the neuroethology of this organism is largely limited. To investigate the neural logic of behavior, it seems essential to better understand its application in the wild. Despite their great capacity to chemotax, the extent to which larvae actually engage in odor-detection during their development remains unclear, though the ability to do so likely confers some competitive advantage (38).

Acknowledgment

ML is thankful to the Vosshall lab where most of data for Figs. 4, 5, 6, and 7 were generated. The Louis lab acknowledges funding from the Spanish Ministry of Science and Innovation (MICINN, BFU2008-00362), and the EMBL-CRG Systems Biology Program.

Financial support to S.G.S. was provided by the Swiss National Science Foundation grant number PP00P3_123339.

References

1. Campos-Ortega JA, Hartenstein V (1997) The embryonic development of *Drosophila melanogaster*, vol xvii, 2nd edn. Springer, Berlin, p 405
2. Hartenstein V et al (2008) The development of the *Drosophila* larval brain. Adv Exp Med Biol 628:1–31
3. Sprecher SG, Reichert H (2003) The urbilaterian brain: developmental insights into the evolutionary origin of the brain in insects and vertebrates. Arthropod Struct Dev 32(1): 141–156
4. Urbach R, Technau GM (2004) Neuroblast formation and patterning during early brain development in *Drosophila*. Bioessays 26(7): 739–751
5. Kumar A et al (2009) Arborization pattern of engrailed-positive neural lineages reveal neuromere boundaries in the *Drosophila* brain neuropil. J Comp Neurol 517(1):87–104
6. Sprecher SG, Reichert H, Hartenstein V (2007) Gene expression patterns in primary neuronal clusters of the *Drosophila* embryonic brain. Gene Expr Patterns 7(5):584–595
7. Urbach R, Technau GM (2003) Segment polarity and DV patterning gene expression reveals segmental organization of the *Drosophila* brain. Development 130(16): 3607–3620
8. Kwon JY, Dahanukar A, Weiss LA, Carlson JR (2011) Molecular and cellular organization of the taste system in the Drosophila larva. J Neurosci 31(43):15300–15309
9. Younossi-Hartenstein A et al (2006) Embryonic origin of the *Drosophila* brain neuropile. J Comp Neurol 497(6):981–998
10. Pereanu W et al (2010) Development-based compartmentalization of the *Drosophila* central brain. J Comp Neurol 518(15): 2996–3023
11. Bolwig N (1946) Senses and sense organs of the anterior end of the house fly larvae. Vidensk Medd Naturhist Foren 109: 81–2017
12. Fishilevich E et al (2005) Chemotaxis behavior mediated by single larval olfactory neurons in *Drosophila*. Curr Biol 15(23): 2086–2096
13. Heimbeck G et al (1999) Smell and taste perception in *Drosophila melanogaster* larva: toxin expression studies in chemosensory neurons. J Neurosci 19(15):6599–6609
14. Cobb M (1999) What and how do maggots smell? Biol Rev 74:425–459
15. Gerber B, Stocker RF (2007) The *Drosophila* larva as a model for studying chemosensation and chemosensory learning: a review. Chem Senses 32(1):65–89

16. Singh RN, Singh K (1984) Fine structure of the sensory organs of *Drosophila melanogaster* Meigen larva (Diptera: Drosophilidae). Int J Insect Morphol Embryol 13(4):255–273
17. Kreher SA, Kwon JY, Carlson JR (2005) The molecular basis of odor coding in the *Drosophila* larva. Neuron 46(3):445–456
18. Larsson MC et al (2004) Or83b encodes a broadly expressed odorant receptor essential for *Drosophila* olfaction. Neuron 43(5):703–714
19. Benton R et al (2006) Atypical membrane topology and heteromeric function of *Drosophila* odorant receptors in vivo. PLoS Biol 4(2):e20
20. Sato K et al (2008) Insect olfactory receptors are heteromeric ligand-gated ion channels. Nature 452(7190):1002–1006
21. Wicher D et al (2008) *Drosophila* odorant receptors are both ligand-gated and cyclic-nucleotide-activated cation channels. Nature 452(7190):1007–1011
22. Asahina K et al (2009) A circuit supporting concentration-invariant odor perception in *Drosophila*. J Biol 8(1):9
23. Hoare DJ, McCrohan CR, Cobb M (2008) Precise and fuzzy coding by olfactory sensory neurons. J Neurosci 28(39):9710–9722
24. Louis M et al (2008) Bilateral olfactory sensory input enhances chemotaxis behavior. Nat Neurosci 11(2):187–199
25. Marin EC et al (2005) Developmentally programmed remodeling of the *Drosophila* olfactory circuit. Development 132(4):725–737
26. Python F, Stocker RF (2002) Adult-like complexity of the larval antennal lobe of *D. melanogaster* despite markedly low numbers of odorant receptor neurons. J Comp Neurol 445(4):374–387
27. Niessing J, Friedrich RW (2010) Olfactory pattern classification by discrete neuronal network states. Nature 465(7294):47–52
28. Olsen SR, Bhandawat V, Wilson RI (2010) Divisive normalization in olfactory population codes. Neuron 66(2):287–299
29. Shang Y et al (2007) Excitatory local circuits and their implications for olfactory processing in the fly antennal lobe. Cell 128(3): 601–612
30. Masuda-Nakagawa LM et al (2009) Localized olfactory representation in mushroom bodies of *Drosophila* larvae. Proc Natl Acad Sci U S A 106(25):10314–10319
31. Masuda-Nakagawa LM, Tanaka NK, O'Kane CJ (2005) Stereotypic and random patterns of connectivity in the larval mushroom body calyx of *Drosophila*. Proc Natl Acad Sci U S A 102(52):19027–19032
32. Ramaekers A et al (2005) Glomerular maps without cellular redundancy at successive levels of the *Drosophila* larval olfactory circuit. Curr Biol 15(11):982–992
33. Masuda-Nakagawa LM et al (2010) Targeting expression to projection neurons that innervate specific mushroom body calyx and antennal lobe glomeruli in larval *Drosophila*. Gene Expr Patterns 10(7–8):328–337
34. Sprecher SG, Desplan C (2008) Switch of rhodopsin expression in terminally differentiated *Drosophila* sensory neurons. Nature 454(7203):533–537
35. Kaneko M, Hall JC (2000) Neuroanatomy of cells expressing clock genes in *Drosophila*: transgenic manipulation of the period and timeless genes to mark the perikarya of circadian pacemaker neurons and their projections. J Comp Neurol 422(1):66–94
36. Rodriguez Moncalvo VG, Campos AR (2009) Role of serotonergic neurons in the *Drosophila* larval response to light. BMC Neurosci 10:66
37. Tix S, Minden JS, Technau GM (1989) Pre-existing neuronal pathways in the developing optic lobes of *Drosophila*. Development 105(4):739–746
38. Asahina K, Pavlenkovich V, Vosshall LB (2008) The survival advantage of olfaction in a competitive environment. Curr Biol 18(15):1153–1155
39. Aceves-Pina EO, Quinn WG (1979) Learning in normal and mutant *Drosophila* larvae. Science 206(4414):93–96
40. Ayyub C et al (1990) Genetics of olfactory behavior in *Drosophila melanogaster*. J Neurogenet 6(4):243–262
41. Cobb M (1996) Genotypic and phenotypic characterization of the *Drosophila melanogaster* olfactory mutation Indifferent. Genetics 144(4):1577–1587
42. Cobb M, Bruneau S, Jallon JM (1992) Genetic and developmental factors in the olfactory response of *Drosophila melanogaster* larvae to alcohols. Proc Biol Sci 248(1322): 103–109
43. Cobb M, Dannet F (1994) Multiple genetic control of acetate-induced olfactory responses in *Drosophila melanogaster* larvae. Heredity 73(pt 4):444–455
44. Monte P et al (1989) Characterization of the larval olfactory response in *Drosophila* and its genetic basis. Behav Genet 19(2):267–283

45. Oppliger FY, Guerin P, Vlimant M (2000) Neurophysiological and behavioural evidence for an olfactory function for the dorsal organ and a gustatory one for the terminal organ in *Drosophila melanogaster* larvae. J Insect Physiol 46(2):135–144
46. Siddiqi O (1980) Development and neurobiology of *Drosophila*. In: Siddiqi O, Babu P, Hall LM, Hall JC (eds) Basic life sciences. Plenum Press, New York, p 496
47. Green CH, Burnet B, Connolly KJ (1983) Organization and patterns of inter- and intraspecific variation in the behaviour of *Drosophila* larvae. Anim Behav 31(1): 282–291
48. Godoy-Herrera R, Connolly K (2007) Organization of foraging behavior in larvae of cosmopolitan, widespread, and endemic *Drosophila* species. Behav Genet 37(4): 595–603
49. Mazzoni EO, Desplan C, Blau J (2005) Circadian pacemaker neurons transmit and modulate visual information to control a rapid behavioral response. Neuron 45(2): 293–300
50. Sawin-McCormack EP, Sokolowski MB, Campos AR (1995) Characterization and genetic analysis of *Drosophila melanogaster* photobehavior during larval development. J Neurogenet 10(2):119–135
51. Fraenkel GS, Gunn DL (1961) The orientation of animals. Dover Publications, New York
52. Berg HC (2000) Motile behavior of bacteria. Phys Today 53:24–29
53. Borst AA, Heisenberg M (1982) Osmotropotaxis in *Drosophila melanogaster*. J Comp Physiol A Neuroethol Sens Neural Behav Physiol 147(4):479–484
54. Martin H (1965) Osmotropotaxis in the honey-bee. Nature 208(5005):59 63
55. Gom-Marin A, Louis M (2011) Active sensation during orientation behavior in the Drosophila larva: more sense than luck. Curr Opin Neurobiol 22(2):208–215
56. Mathewson RF, Hodgson ES (1972) Klinotaxis and rheotaxis in orientation of sharks toward chemical stimuli. Comp Biochem Physiol A Comp Physiol 42(1): 79–84
57. Porter J et al (2007) Mechanisms of scent-tracking in humans. Nat Neurosci 10(1):27–29
58. Gibbons B (1986) Magazine Article Nat Geogr Mag 221(170):324–361
59. Schöne H (1984) Spatial orientation: the spatial control of behavior in animals and man, Princeton series in neurobiology and behaviour. Princeton University Press, Princeton, NJ, p 347
60. Hafez M (1950) On the behaviour and sensory physiology of the house-fly larva, *Musca domestica L. I.* Feeding stage. Parasitology 40(3–4):215–236
61. Bala ADS, Panchal P, Siddiqi O (1998) Osmotropotaxis in larvae of *Drosophila melanogaster*. Curr Sci 75:48–51
62. Gomez-Marin A et al (2010) Mechanisms of odor-tracking: multiple sensors for enhanced perception and behavior. Front Cell Neurosci 4:6
63. Mast SO (1911) Light and the behavior of organisms, vol xi, 1st edn. Wiley, New York, p 410
64. Busto M, Iyengar B, Campos AR (1999) Genetic dissection of behavior: modulation of locomotion by light in the *Drosophila melanogaster* larva requires genetically distinct visual system functions. J Neurosci 19(9): 3337–3344
65. Scantlebury N, Sajic R, Campos AR (2007) Kinematic analysis of *Drosophila* larval locomotion in response to intermittent light pulses. Behav Genet 37(3):513–524
66. Pfeiffer BD et al (2008) Tools for neuroanatomy and neurogenetics in *Drosophila*. Proc Natl Acad Sci U S A 105(28):9715–9720
67. Shinomiya K et al (2010) Flybrain neuron database: a comprehensive database system of the *Drosophila* brain neurons. J Comp Neurol 519(5):807–833
68. Hadjieconomou D et al (2011) Flybow: genetic multicolor cell labeling for neural circuit analysis in *Drosophila melanogaster*. Nat Methods 8(3):260–266
69. Hampel S et al (2011) *Drosophila* Brainbow: a recombinase-based fluorescence labeling technique to subdivide neural expression patterns. Nat Methods 8(3):253–259
70. Cardona A et al (2010) An integrated micro- and macroarchitectural analysis of the *Drosophila* brain by computer-assisted serial section electron microscopy. PLoS Biol 8(10): e1000502
71. Huang B, Babcock H, Zhuang X (2010) Breaking the diffraction barrier: super-resolution imaging of cells. Cell 143(7):1047–1058
72. Pauls D et al (2010) *Drosophila* larvae establish appetitive olfactory memories via mushroom body neurons of embryonic origin. J Neurosci 30(32):10655–10666
73. Branson K et al (2009) High-throughput ethomics in large groups of *Drosophila*. Nat Methods 6(6):451–457
74. Dankert H et al (2009) Automated monitoring and analysis of social behavior in *Drosophila*. Nat Methods 6(4):297–303

75. Gershow M, Berck M, Mathew D, Luo L, Kane EA, Carlson JR, Samuel AD (2012) Controlling airborne cues to study small animal navigation. Nat Methods 9(3):290–296.
76. Ramot D et al (2008) The Parallel Worm Tracker: a platform for measuring average speed and drug-induced paralysis in nematodes. PLoS One 3(5):e2208
77. Cronin CJ, Feng Z, Schafer WR (2006) Automated imaging of *C. elegans* behavior. Methods Mol Biol 351:241–251
78. Ben Arous J et al (2010) Automated imaging of neuronal activity in freely behaving *Caenorhabditis elegans*. J Neurosci Methods 187(2):229–234
79. Leifer AM et al (2011) Optogenetic manipulation of neural activity in freely moving *Caenorhabditis elegans*. Nat Methods 8(2):147–152
80. Zinke I et al (1999) Suppression of food intake and growth by amino acids in *Drosophila*: the role of pumpless, a fat body expressed gene with homology to vertebrate glycine cleavage system. Development 126(23):5275–5284
81. Khurana S, Abu Baker MB, Siddiqi O (2009) Odour avoidance learning in the larva of *Drosophila melanogaster*. J Biosci 34(4): 621–631
82. Tracey WD Jr et al (2003) Painless, a *Drosophila* gene essential for nociception. Cell 113(2):261–273
83. Neuser K et al (2005) Appetitive olfactory learning in *Drosophila* larvae: effects of repetition, reward strength, age, gender, assay type and memory span. Anim Behav 69(4):891–898
84. Benz G (1956) Der Trockenheitssinn bei Larven von *Drosophila melanogaster*. Experientia 12:297
85. Louis M, Piccinotti S, Vosshall LB (2008) High-resolution measurement of odor-driven behavior in *Drosophila* larvae. J Vis Exp 11:638
86. Gilat A (2005) MATLAB: an introduction with applications, vol Viii, 2nd edn. Wiley, Hoboken, NJ, p 343
87. Gonzalez RC, Woods RE (2008) Digital image processing, vol xxii, 3rd edn. Pearson/ Prentice Hall, Upper Saddle River, NJ, p 954
88. Lilly M, Carlson J (1990) Smellblind: a gene required for *Drosophila* olfaction. Genetics 124(2):293–302
89. Kreher SA et al (2008) Translation of sensory input into behavioral output via an olfactory system. Neuron 59(1):110–124
90. Saumweber T, Husse J, Gerber B (2010) Innate attractiveness and associative learnability of odors can be dissociated in larval *Drosophila*. Chem Senses 36(3):223–235
91. Uchida N, Mainen ZF (2003) Speed and accuracy of olfactory discrimination in the rat. Nat Neurosci 6(11):1224–1229
92. Khan RM et al (2007) Predicting odor pleasantness from odorant structure: pleasantness as a reflection of the physical world. J Neurosci 27(37):10015–10023
93. Gomez-Marin A, Stephens GJ, Louis M (2011) Active sampling and decision making in Drosophila chemotaxis. Nat Commun 2:441.
94. Liu C et al (2010) Distinct olfactory signaling mechanisms in the malaria vector mosquito *Anopheles gambiae*. PLoS Biol 8(8):e1000467
95. Suster ML et al (2003) Targeted expression of tetanus toxin reveals sets of neurons involved in larval locomotion in *Drosophila*. J Neurobiol 55(2):233–246
96. Benhamou S (2004) How to reliably estimate the tortuosity of an animal's path: straightness, sinuosity, or fractal dimension? J Theor Biol 229(2):209–220
97. Schleyer M, Saumweber T, Nahrendorf W, Fischer B, von Alpen D, Pauls D, Thum A, Gerber B (2011) A behavior-based circuit model of how outcome expectations organize learned behavior in larval Drosophila. Learn Mem 18:639
98. Ribeiro C, Dickson BJ (2010) Sex peptide receptor and neuronal TOR/S6K signaling modulate nutrient balancing in *Drosophila*. Curr Biol 20(11):1000–1005
99. Kaiser M, Cobb M (2008) The behaviour of *Drosophila melanogaster* maggots is affected by social, physiological and temporal factors. Anim Behav 75(5):1619–1628
100. Godoy-Herrera R et al (1992) The development of the photoresponse in *Drosophila melanogaster* larva. Rev Chil Hist Nat 65:91–101
101. Luo L et al (2010) Navigational decision making in *Drosophila thermotaxis*. J Neurosci 30(12):4261–4272
102. Roberts DB (1998) *Drosophila:* a practical approach, 2nd edn. Oxford University Press, Oxford
103. Boyle J, Cobb M (2005) Olfactory coding in *Drosophila* larvae investigated by cross-adaptation. J Exp Biol 208(pt 18):3483–3491
104. Colomb J et al (2007) Complex behavioural changes after odour exposure in *Drosophila* larvae. Anim Behav 73(4):587–594

105. Larkin A et al (2010) Central synaptic mechanisms underlie short-term olfactory habituation in *Drosophila* larvae. Learn Mem 17(12):645–653
106. Ibba I (2010) Neuroethology of olfaction in *Drosophila*. In: Department of plant protection biology. The Swedish University of Agricultural Sciences, Alnarp
107. Dekker T et al (2006) Olfactory shifts parallel superspecialism for toxic fruit in *Drosophila melanogaster* sibling, *D. sechellia*. Curr Biol 16(1):101–109
108. Michels B et al (2005) A role for synapsin in associative learning: the *Drosophila* larva as a study case. Learn Mem 12(3):224–231
109. Luo L, Callaway EM, Svoboda K (2008) Genetic dissection of neural circuits. Neuron 57(5):634–660
110. Simpson JH (2009) Mapping and manipulating neural circuits in the fly brain. Adv Genet 65:79–143
111. Naumann EA et al (2010) Monitoring neural activity with bioluminescence during natural behavior. Nat Neurosci 13(4):513–520

Chapter 9

Drosophila as a Genetic Model to Investigate Motion Vision

Daryl M. Gohl, Marion A. Silies, and Thomas R. Clandinin

Abstract

The neural circuits that underlie motion vision in *Drosophila* provide an excellent model system for studying the logic of neural computation. A rich history of quantitative behavioral analysis has provided a detailed theoretical framework for investigating the neuronal basis of motion detection. Here, we describe the many inventive methods that have been used to quantify optomotor responses in *Drosophila*. Next, we discuss stimuli that can probe the computational structure of motion detection circuitry. Finally, we highlight some of the sophisticated genetic tools for targeting and manipulating neurons that are being applied to the visual system. Studies combining quantitative behavioral assays with this ever-expanding genetic toolkit are beginning to uncover the computational roles played by individual neurons. The circuits that process visual motion in *Drosophila* provide an exciting opportunity to understand a complete neural circuit linking perception to behavior.

Key words: Motion vision, Quantitative behavior, Single fly assays, Population assays, Visual stimuli, Neurogenetics, Enhancer traps, Intersectional strategies

1. Motion Vision: A Paradigmatic Computation

Vision provides a wealth of information about the world and is a critical sensory modality for many animals. In every multicellular organism that can see, photons are converted into voltage changes in photoreceptors; downstream circuits reconstruct elements of the visual scene from processing these signals. By implementing various computations, such circuits can extract information about the spatial, temporal, and spectral pattern of photons reaching the eye, and can use this information to guide behavior. Amongst this plethora of potential signals, visual motion cues are particularly important, providing salient information about the relative movement of the organism and its environment. Indeed, every organism that possesses an eye that is, in principle, capable of detecting motion has the neural circuitry necessary to do so. Importantly, because motion signals can only be deduced by comparing photoreceptor voltage

Bassem A. Hassan (ed.), *The Making and Un-Making of Neuronal Circuits in Drosophila*, Neuromethods, vol. 69,
DOI 10.1007/978-1-61779-830-6_9, © Springer Science+Business Media, LLC 2012

changes across both space and time, motion vision has long been thought of as a paradigmatic example of a well-constrained neural computation. However, in spite of considerable effort in many experimental systems, the fundamental mechanisms by which motion is detected in the brain remain unclear, and no complete motion-sensitive circuit has been described that would link a retinal input to a direction-selective behavioral output. Thus, many labs are presently taking advantage of the stereotyped nervous system, genetic tools, and behavioral assays to unravel motion-detecting circuits in the fruit fly. In this review, we summarize the behavioral assays, stimuli, and genetic tools that are presently available in this system to unravel the circuit mechanisms of motion vision.

2. Behavioral Assays for Motion Vision in *Drosophila*

How does one know what a fly can see? As with any other animal, behavioral responses to visual stimuli provide a key entry point to understanding circuit function. Flies have robust innate responses to visual motion that can be readily exploited to indirectly measure the properties of the visual system, by presenting defined stimuli and monitoring behavioral response. However, it is important to recognize that flies likely perceive many stimuli that do not elicit instinctive responses under the experimental conditions that have so far been tested, and thus that these assays do not give a full picture of the visual abilities of the fly. Moreover, while it is possible to probe additional visual faculties by training flies to associate specific visual cues with aversive or attractive stimuli, responses to visual motion are dominated by innate reflexes. Therefore, conditioning assays have generally not been used in this context. Thus, while the extant behavioral assays do provide a rich armamentarium for investigating motion vision in fruit flies, it is likely that additional, more sophisticated assays will further illuminate our understanding of motion computation.

The assays described below rely on optomotor responses, direction-selective changes in head or body movement. The optomotor response is usually thought of as a stabilizing reflex, as flies will attempt to turn their heads or bodies to minimize visual slip when presented with a rotating grating (1, 2). These reflexes can be incredibly robust, with little diminution of responses reported even after 48 h of exposure to a saturating stimulus (3). However, the phrase "optomotor response" has also been also applied to other nonrotary responses in *Drosophila*, such as the tendency of freely walking flies to orient and walk against the direction of visual motion (4). This directional response of walking flies to motion would be expected to increase, rather than reduce, relative motion on the retina, suggesting that either the optomotor response is not purely a stabilizing reflex or that distinct reflexes operate under different behavioral conditions. These observations may be reconciled by the fact that a more complex response was observed when flies were presented a combination of rotational and translational optic flow patterns (which is what freely moving flies would be expected to encounter) (5). In this case, responses to translational flow were stronger than rotational responses, with flies strongly avoiding image expansions caused by linear translation. Whatever the function of optomotor responses in *Drosophila*, they provide extremely useful readouts for visual system function.

In an ideal behavioral assay, a fly would be able to engage in its full behavioral repertoire, either walking or flying, while an arbitrary visual stimulus of any intensity, contrast, wavelength, and temporal frequency was precisely mapped at high spatial resolution onto the curved ommatidial axes of the eye. Behavioral data would be collected such that the fly's responses could be quantitatively related to the stimulus presentation at high temporal resolution. Furthermore, experiments could be performed under either open-loop conditions, where the stimulus presentation is independent of the fly's movements, or closed-loop conditions, where the fly's movement feeds back on the stimulus. Ideally, one would also be able to simultaneously monitor and/or modulate neural activity, and the assay would be high-throughput to allow for many replicates or genotypes to be easily tested and to make forward genetic screens feasible.

In practice, such an ideal assay does not exist and, indeed, is difficult to even imagine. However, a number of assays have been developed which either fix or ignore several of the above parameters in order to tightly control a limited number of inputs and/or readouts. The different classes of assays differ widely in the degree of behavioral freedom the animal is given, the precision of stimulus delivery, and the throughput of the assay. Thus, the choice of an optimal motion vision assay requires careful consideration of the trade-offs inherent in the different experimental paradigms and ultimately depends upon the experimental goals.

2.1. Trade-Offs Between Different Behavioral Assays

The experimental systems that have been used for studying motion vision can be broadly segregated into two major types: single fly assays and population assays. Single fly assays have the benefit that the stimulus delivery can be precisely controlled, allowing one to define what each fly is seeing. Such experiments can be performed either under open-loop conditions, where the stimulus is presented without regard to the fly's intended movements, or under closed-loop conditions, where the intended movements of the fly are fed back into the stimulus presentation. Population assays have the advantage of increased throughput, allowing for better statistical power and the potential for performing forward genetic screens. However, because such population assays generally involve freely moving flies, it is not possible to present the stimulus in precisely the same way to each individual, although it is possible to subsequently calculate what each fly saw (6). In addition, freely moving flies can receive vestibular/proprioreceptive feedback from the halteres created by self-motion, a signal that is clearly altered in tethered flies (7, 8).

While studying freely moving flies causes some difficulties in stimulus presentation, several recent studies have shown that neuronal responses to visual stimuli are affected by the behavioral state of the animal (9, 10). Specifically, flies that are walking or flying have increased gain in the responses of HS or VS lobula plate tangential cells. Furthermore, while these studies described only changes in neuronal gain, it is possible that other neuronal tuning properties might be sensitive to immobilization of the animal. It is thus formally possible that immobilizing the fly causes the visual system to compute motion signals in an entirely different mode from that seen in a freely moving preparation. These findings should be kept in mind when choosing and interpreting the results of behavioral assays, as many assays require that the movement of the fly be at least partially restricted.

2.2. Single Fly Motion Vision Assays

The first measurements of *Drosophila's* behavioral responses to visual motion were done by Hecht and Wald (4). In these assays, a fly was placed in a glass cell, and shown moving stripes of different widths and intensities. In response to the moving stripes, the fly would stop and then orient opposite to the direction of motion. This robust behavior allowed measurements of the visual acuity of *Drosophila* and the fly's ability to distinguish different intensities as a function of pattern contrast. This work was followed a decade later by tests of a number of eye color mutants and other mutant strains (11). These studies thus provided the conceptual foundation for using the fruit fly for dissecting the genetic underpinnings of visual behavior.

2.2.1. Tethered Single Fly Assays

Three main classes of assays have been used for measuring the behavioral responses of tethered flies to visual motion: flight simulators, walking "fly-on-a-ball" assays, and measurements of optokinetic head movements. The flight simulator was first developed by Götz who used a torque meter to measure the yaw (rotational) torque movements produced by flying *Drosophila* in response to rotating patterns (Fig. 1a1) (1). Variants of this device have been used with great success to measure many properties of *Drosophila's* behavioral response to motion, under both open- and closed-loop conditions (12, 13). Modifications to the torque-based flight simulator have allowed measurements of translational movements (14) and of roll and pitch movements (3, 15). In addition, analysis of wing beat movements has been used, in place of a torque meter, to infer the behavioral responses of flying, tethered flies (16, 17).

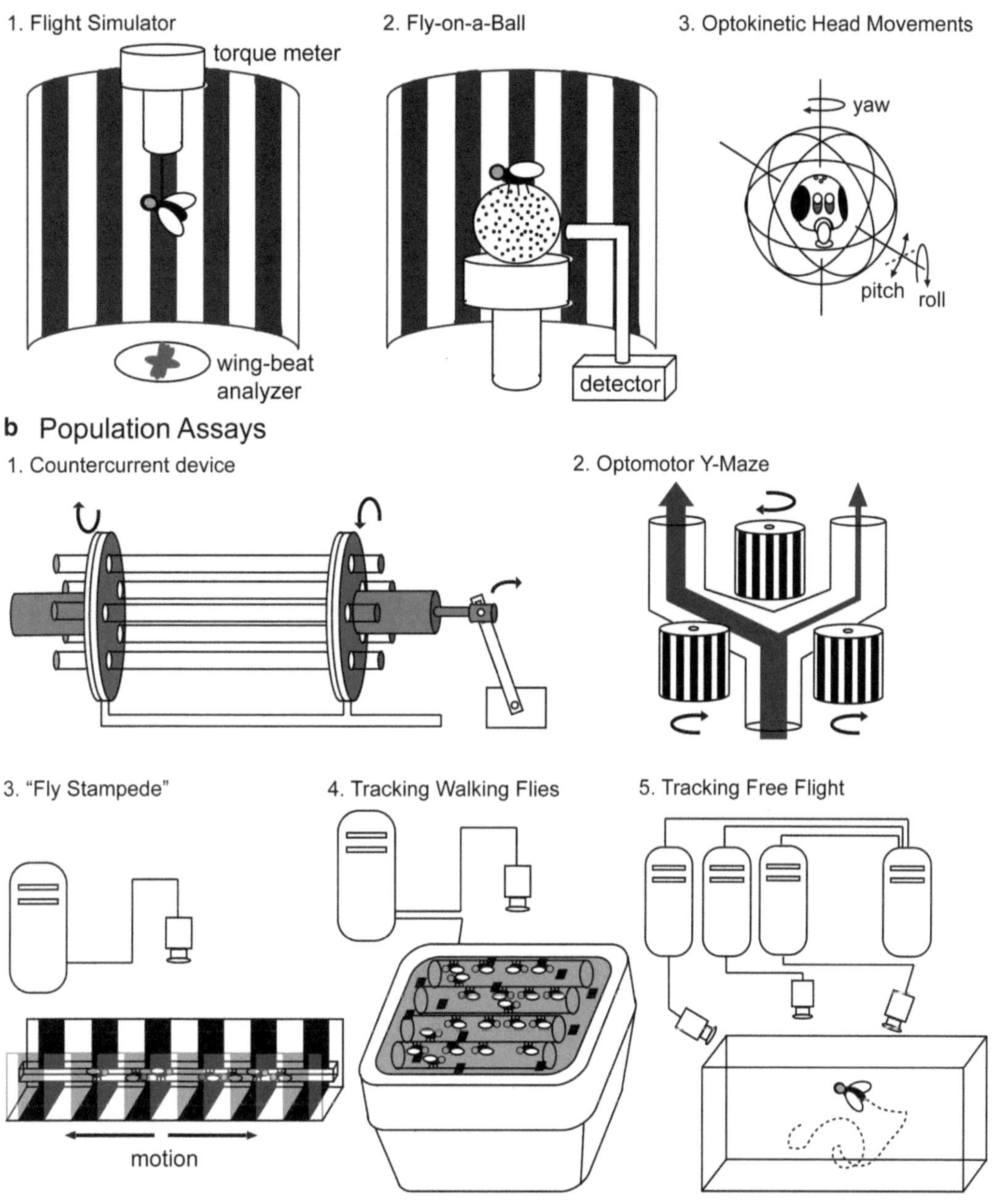

Fig. 1. Behavioral assays used to study motion vision. (**a**) Single fly assays. *1.* Flight simulator used to measure yaw torque or wing beat amplitudes of flying *Drosophila*. Adapted from refs. (14–16). *2.* Fly-on-a-ball assay used to measure intended movements of walking flies. Adapted from ref. (22). *3.* Measuring optokinetic head movements. Adapted from ref. (3). (**b**) Population assays. *1.* Countercurrent device for fractionating flies based on optomotor response. Adapted from ref. (34). *2.* Optomotor Y-maze. Adapted from ref. (35). *3.* Fly stampede assay. Adapted from ref. (29). *4.* Assay for real-time tracking of populations of walking flies. Adapted from ref. (36). *5.* System for tracking freely flying *Drosophila*. Adapted from ref. (37).

Finally, tethers that allow yaw rotation such as a string or more recently a magnetic field have been used to allow visual stimuli to be presented while the movements of flies in the yaw plane are recorded (18–20).

Another choice that must be made is whether to study flies that are walking or flying. While much of the work on motion vision has concentrated on flight assays, *Drosophila* spends much of its time walking, suggesting that this behavior is also ethologically relevant (21). The tethered fly-on-a-ball assay allows measurements of the intended movements of walking flies (Fig. 1a2) (2, 22). In this assay, a fly is placed atop an air-supported ball that is marked to allow the displacement of the ball to be captured by external detectors (22, 23). From these recordings, the intended trajectory of the fly in response to a visual stimulus can be determined. Because the fly-on-a-ball assay, like the flight simulator, uses stationary (often head-immobilized) flies, it also allows for very precise stimulus presentation. For instance, using such a system, it was possible to probe the structure of the *Drosophila* elementary movement detectors (EMDs) (22). In addition, as with the flight simulator, visual responses can be studied under both open- and closed-loop conditions. Another advantage of tethered single fly assays, especially those in which the head is fixed in place, is that they can allow measurement of behavioral responses to be paired with electrophysiological measurements or functional imaging of neuronal responses (9, 10, 23).

The final class of single fly motion vision assay that is performed on stationary subjects is the measurement of optokinetic head movements (Fig. 1a3). Such measurements were first performed on larger flies, such as *Calliphora* (24–26). More recently, similar measurements have been made in *Drosophila* to assess the contribution of different lamina monopolar cells to motion vision (27–29).

2.2.2. Freely Moving Single Fly Assays

The advent of video recording and tracking software, which allows the trajectory of a fly to be followed with high precision over the course of an experiment, sparked renewed interest in extending motion vision assays to freely moving animals. The first description of a high-resolution freely walking fly optomotor assay used a modified version of a tracking assay originally developed to study object fixation responses (30). In this assay, a fly with clipped wings was allowed to walk on a disk which floated in a water moat (to prevent escape) (31). The fly was surrounded by an arena comprising LED arrays, on which a variety of visual stimuli could be presented. The arena was illuminated with infrared light, which the fly cannot see, and tracked with an infrared-sensitive camera. This set-up allowed comparison of results between tethered and freely walking flies, under open-loop or "virtual open-loop" conditions, in which the locomotion-induced visual feedback of the fly was suppressed through compensatory movements in the visual surround. Subsequently, tracking programs and visual display arenas have also been developed to study freely flying flies (6, 32, 33) (Fig. 1b5).

2.3. Population Motion Vision Assays

A number of population optomotor assays have been based on fractionating flies by their differential responses to motion stimuli. A countercurrent device, surrounded by a rotating "barber pole" cylinder, was used to fractionate different mutants (Fig. 1b1) (34). In addition, an optomotor Y-maze, consisting of a series of choice points, surrounded by rotating striped drums was used to separate wild type and mutant lines and to screen for new mutants affecting motion vision (Fig. 1b2) (35). Heisenberg and Götz also developed a "hybrid" assay, in which single flies were placed in toroidal channels and the movements of the fly in response to a pattern of rotating stripes was recorded using a directionally selective light gate, which recorded the numbers of passages in either direction. While each chamber contained only a single fly, the device contained 24 assay chambers, allowing a relatively large number of flies to be tested in parallel (35).

As with the freely moving single fly assays, the development of a number of fly tracking software platforms for both walking and flying animals has made population assays much more sophisticated and sensitive in their readouts (36–38). The different tracking platforms differ substantially in their implementation. For instance, C-trax can be used to analyze video files and can track flies with high accuracy over long experiments (38). However, such analysis is computationally intensive, often requiring substantial computer time for analysis, a limiting factor in large-scale experiments such as genetic screens. A faster, live tracking platform has also been developed, in which trajectories are not extracted from videos, but acquired online (36). This algorithm is substantially faster, but is not suited for experiments in which individual flies need to be followed over long timescales.

Two studies have demonstrated the feasibility of combining such population assays with genetic tools for inactivating specific groups of neurons to implicate specific neuronal populations in motion vision. Zhu et al. used an assay in which groups of flies were placed in a transparent tube in an elongated track surrounded on three sides by LED displays, which displayed moving bars that caused the flies to "stampede" against the direction of motion (Fig. 1b3) (29). Images of the flies were captured by video tracking of infrared reflections, and from these data, changes in the probability histograms of fly positions were measured. A collection of enhancer trap lines associated with expression in various visual interneurons were then used to drive tetanus neurotoxin light chain (TNT) and these flies were screened for optomotor defects, identifying a subset of lamina neurons that were required motion detection but left phototactic behaviors intact. A second high-throughput motion vision assay allows the tracking of groups of flies in tubes, which receive a visual stimulus from a high refresh rate CRT monitor positioned below them (Fig. 1b4) (36). In this study, a collection of enhancer trap lines driving a temperature-sensitive *dynamin* mutant (*shibire*^{ts}), which blocks synaptic vesicle recycling at the nonpermissive temperature was screened for optomotor defects, identifying a small group of neurons that preferentially affected the rotational responses of flies to different random dot stimuli.

2.4. Choosing a Motion Vision Assay

While many of the above assays have provided detailed characterization of the visual responses of wild type and mutant flies, in attempting to crack the neuronal circuits underlying motion vision the advantages and disadvantages of particular behavioral paradigms are important to consider. For instance, the first step in dissecting a neuronal circuit is to identify the cellular constituents of the circuit. This can be accomplished by forward genetic screening using *Gal4* lines to manipulate circuit components, or by screening large numbers of candidate *Gal4* lines identified first by their expression patterns. Such screens have the advantage that they simultaneously allow identification of the neurons required for a given behavior, and provide reagents to manipulate the implicated neurons. To carry out forward genetic screens, high-throughput population assays are probably the best choice. However, once neurons have been implicated in aspects of motion vision, single fly assays, which allow precise delivery of stimuli, are necessary for detailed characterization of neuronal function.

3. Visual Stimuli

A wealth of visual stimuli has been used to examine behavioral and physiological responses to motion. The earliest behavioral experiments used vertically striped plates or cylinders that were moved manually to create a pattern of moving edges (4, 11). Götz developed a more quantitative analysis of optomotor behavior by presenting fixed stripes on a back-illuminated Plexiglas drum to head-fixed flying flies (1). The drum could be rotated at different velocities, allowing behavior to be examined as a function of pattern wavelength and angular velocity. These experiments also demonstrated that the optomotor response varies with stimulus contrast (1, 2, 22). Similar stimuli are also used for freely walking flies, or in "virtual reality environments," where the displayed pattern is dynamically updated with the fly's behavior (6, 31, 39). Finally, while these stimuli were initially generated by illumination projections of mechanically moving patterns (often drawn on paper), more recent work has used LED arrays (31, 40), CRT (36) or LCD screens (33, 41), to allow computer-controlled dynamic stimulus presentations that are more sophisticated than those that can be generated mechanically. In these latter cases, computer-generated visual stimuli must be adapted to the relatively poor spatial resolution (on average 4.6° for motion signals (1)), and high temporal resolution of the fly eye (with a flicker fusion rate approaching 200 Hz at room temperature (42)).

In addition to these simple stimuli, work in vertebrate models has made use of a number of more complex stimuli for probing motion vision. These stimuli include moving plaid patterns, dynamic random dot stimuli, white noise stimuli, or naturalistic scenes (43). Using random dot stimuli in flies, spatiotemporal frequency patterns different from those found in periodic stimuli can be tested (36). These stimuli interleave noise periods with motion pulses, in which the dots obtain high coherence and thus present a motion pulse into a certain direction. Still more complex stimuli, so-called white noise stimuli, which support a variety of systems identification approaches (44–46) can also be used to evoke responses to motion. These stimuli have been used in seminal work on adaptation in blowflies (47, 48), and are just beginning to be used in studying *Drosophila* optomotor responses (49). Finally, artificial stimulus patterns have been replaced by naturalistic images, which generally contain strong correlations and are closest to physiological conditions. The challenge here is to later extract the behaviorally relevant stimulus features. Again, these stimulus types have mainly been used in larger flies to describe coding properties of motion sensitive neurons, and have only recently been used in *Drosophila* (50–54).

All of the motion stimuli described so far are accompanied by local luminance changes in the presented images. Motion cues that are characterized by a spatiotemporal change in luminance are defined as first-order motion, or Fourier motion. However, a class of motion illusions, so-called second-order motion cues, can be also seen by higher organisms including humans (55–57). In these stimuli, correlated changes in other stimulus features, for example in contrast polarity or texture, define the motion signal. Such second order motion stimuli lack the luminance cues that define first-order motion and have also been applied to flies. Remarkably, moving objects without correlated luminance signals could in fact elicit optomotor responses in flies in a flight simulator (58). If contrasting first and second order motion stimuli are applied (theta stimuli), flies can actually follow the second order component of the stimulus, at least under some conditions (49, 58). Thus, the simplified nervous system of the fruit fly is capable of direction-selective responses to motion signals that are intrinsically complex.

4. Genetic Tools

Motion-evoked behavioral responses are, of course, dependent on the appropriate patterns of activity of the specific cells that comprise the motion detection circuitry. These are, in turn, dependent on the functions of specific genes in each cell. One of the biggest challenges in sensory neuroscience is to describe animal behavior at the cellular and molecular level, and genetic approaches to unraveling visual behavior have variously focused on either cells or genes. Initial approaches using mutations in genes to study the underpinnings of behavior took off in the Benzer laboratory and were followed by other labs using different elegant and inventive assays (35, 59–61). However, it quickly became apparent from these studies that the mapping of genetic activities onto behavioral functions is generally not straightforward, with genes playing roles in the shaping the activities of many neurons, affecting their differentiation, connectivity, and function at different stages in the life cycle. Thus, neurons, not genes, are the functional units that can provide an understanding of motion computation. The current approaches to dissect behavior are based on disrupting cellular function instead of gene function, without perturbing the connectivity of the circuit. Once the cellular composition of a behaviorally relevant circuit is known, one might then be able to return to study neuronal function at a molecular level, using forward genetic approaches, with temporal and spatial specificity.

In spite of this switch toward studying cells, not genes, the underlying logic of modern circuit dissection borrows heavily from the logic underpinning studies of gene function. By analogy, there are now genetic tools that allow the experimenter to test necessity and sufficiency for a cell in a circuit. In particular, by blocking either changes in membrane potential, or by disrupting synaptic transmission, it is possible to determine whether the activity of a given neuron, or group of neurons is necessary for a certain behavioral response. In addition, ectopic activation of a neuron can be used to test whether increasing the activity of a given neuron is sufficient to alter behavior. Finally, combining these manipulations of activity with direct measurements of neuronal activity during stimulus presentation (and ideally behavioral response) allows one to correlate changes in neural activity with the specific sensorimotor transformation under study. The key to all these genetic manipulations is the ability to target single neurons or classes of neurons specifically, ideally manipulating their activity only acutely. As we briefly describe, new genetic tools that will allow this level of specificity are now being established.

4.1. Mutant Analysis

The first approaches to dissecting *Drosophila* visual behaviors involved the analysis of mutants or genetic mosaic animals in UV vs. visible light choice assays, optomotor assays, or phototaxis tests (35, 59–61). Unfortunately, the first attempt to dissect motion vision, testing EMS mutagenized flies in the elegant Y-maze optomotor assay described above, largely identified mutants that were blind and had also been identified in nonmotion visual behavior screens (59, 60, 62, 63). The screens led to the identification of many of the key components of the phototransduction cascade (reviewed in ref. (3, 64)). However, the vast majority of these mutants affected the very peripheral units of visual processing, and despite these successes, these early screens failed in their original goals of identifying genes specific for certain behaviors. A key limitation of these initial efforts were limitations imposed by the genetic tools available, as these screens could only examine random somatic mosaic animals, or were limited to screening for viable mutations. Subsequently, similar behavioral assays were used in conjunction with eye-specific somatic mosaics in order to understand development of the visual system, including photoreceptor cell targeting (65, 66). A broader overview about visual behavior mutant screens, early and recent, has been given elsewhere (67, 68).

In terms of understanding motion vision, mutants helped to assign motion sensitivity to the outer photoreceptor cells, R1-R6 (69). Postsynaptic to these cells are the lamina neurons L1-L3, and analysis of *vacuolar medulla* mutants first suggested that the L1 and L2 monopolar cells of the lamina contribute to motion processing (70), a finding which has been confirmed by many subsequent studies (28, 29, 36, 71). One mutation identified in an initial screen turned out not to be generally blind, but was strongly defective in motion-evoked behavioral responses (72). This mutation, the *optomotor blind* allele omb^{H31} induces severe structural defects in the lobula plate and affects development of some giant neurons of the lobula plate (35, 72). These studies assign the respective lobula plate neurons to the optomotor turning response pathway. This result, however, remains an exception and is based on the fact that omb^{H31} represents a tissue-specific allele of the *optomotor blind* gene, a gene with broad expression and developmental roles in many tissues (73–75).

Under some circumstances, mutants affecting certain genes can provide useful tools for studying visual behavior. As examples, *norpA* or *ninaE* can be used as a control for blind or motion blind flies, respectively (29, 65). Another elegant strategy involving mutants was performed by Rister et al. (28), which took advantage of the fact that histamine is the neurotransmitter of all photoreceptor (R) cells, but other than that is not used broadly (76). This facilitated the use of mutations in the histamine receptor *outer rhabdomeres transientless* (*ort*) in a targeted rescue experiment. A *UAS-ort* transgene was expressed in only subsets of photoreceptor target neurons in the lamina. Thus, individual neuronal pathways could be individually rescued, complementing neuronal silencing results.

4.2. Targeting Single Neurons

In order to identify neurons that mediate a certain behavior in an unbiased fashion, the *Gal4/UAS* system can be used to block neuronal activity and then screen for a behavioral phenotype. Such *Gal4* collections consist of either randomly generated *Gal4* enhancer traps, or collections of enhancer fusion constructs, in which transgenic flies carry regulatory elements that are cloned in front of the genetic effector. In addition to the *Gal4/UAS* system, other binary or ternary expression systems exist in *Drosophila* and can be used together to allow simultaneous, independent manipulations of different cell types (77–79). As mentioned above, expression of a gene is generally not restricted to a single neuron or single neuronal class, nor is the expression of commonly generated *Gal4* enhancer traps, or enhancer fusion constructs. If silencing neuronal activity in a given *Gal4* pattern causes interesting behavioral phenotypes, but its expression is not confined to a single cell or cell type, different approaches can be taken to refine the expression pattern. Generally, either intersectional or subtractive strategies between existing *Gal4* lines with partially overlapping expression patterns are used to produce more specific expression patterns, or specific enhancer/promoter subelements are employed to narrow down expression of a *Gal4* line. The underlying principle behind the two strategies is that *Gal4* mediated transcriptional activation can be silenced by binding of the Gal80 protein (80) or the Gal4 protein can be split into two halves and reconstituted by adding a protein–protein interaction domain (81).

In order to achieve a reduction of given expression patterns, one can generate and test new collections of *Gal80* lines, or generate enhancer-*Gal80* fusions based on known (or hypothesized) enhancer expression patterns (82, 83). Furthermore, other binary or ternary expression systems, namely, *LexA/LexOp* and the *Q* system can be used in conjunction with Flp recombinase to create various logical gates and reduce expression (78, 79, 84). Flip-out strategies have been used in other systems to stochastically narrow down expression patterns and then correlate a behavioral phenotype with the respective expression pattern (85) (for a detailed description see Chap. 4). These approaches are powerful, but can only be implemented using robust single fly assays, and are labor intensive. Furthermore, such stochastic approaches are not amenable in the visual system, where visual information is transmitted in many parallel units implementing identical computations on different parts of visual space.

P elements insert preferentially into the 5′ UTRs of genes in *Drosophila* (86), and might thus be prone to detecting broader enhancer activity patterns than randomly inserted transposable elements. Since different transposons have distinct insertion spectra, different patterns might be achieved by generating enhancer traps from transposable elements other than P elements, such as the *piggyBac* element (87–89) or *Minos* elements (90). Another way to subdivide patterns of neuronal gene expression is based on molecular dissection of gene regulatory elements. Here, potential regulatory elements are taken outside out of their normal chromosomal context and used to generate enhancer fusion constructs. Pfeiffer et al. (82) reasoned that such enhancer fusion constructs would recapitulate a distinct subset of a gene's full expression pattern and demonstrated that patterns driven by individual fragments contain fewer cells than are typically targeted in enhancer trap lines. Known enhancer fragments can then be recloned to drive expression of *Gal80*, split-*Gal4*, etc. in order to refine expression patterns (Fig. 2a). Alternately, subdivision of the enhancer DNA sequence can be used to further reduce the pattern of expression. A large library of such enhancer *Gal4* fusion constructs is currently being generated (82). This enhancer fusion strategy allows one to

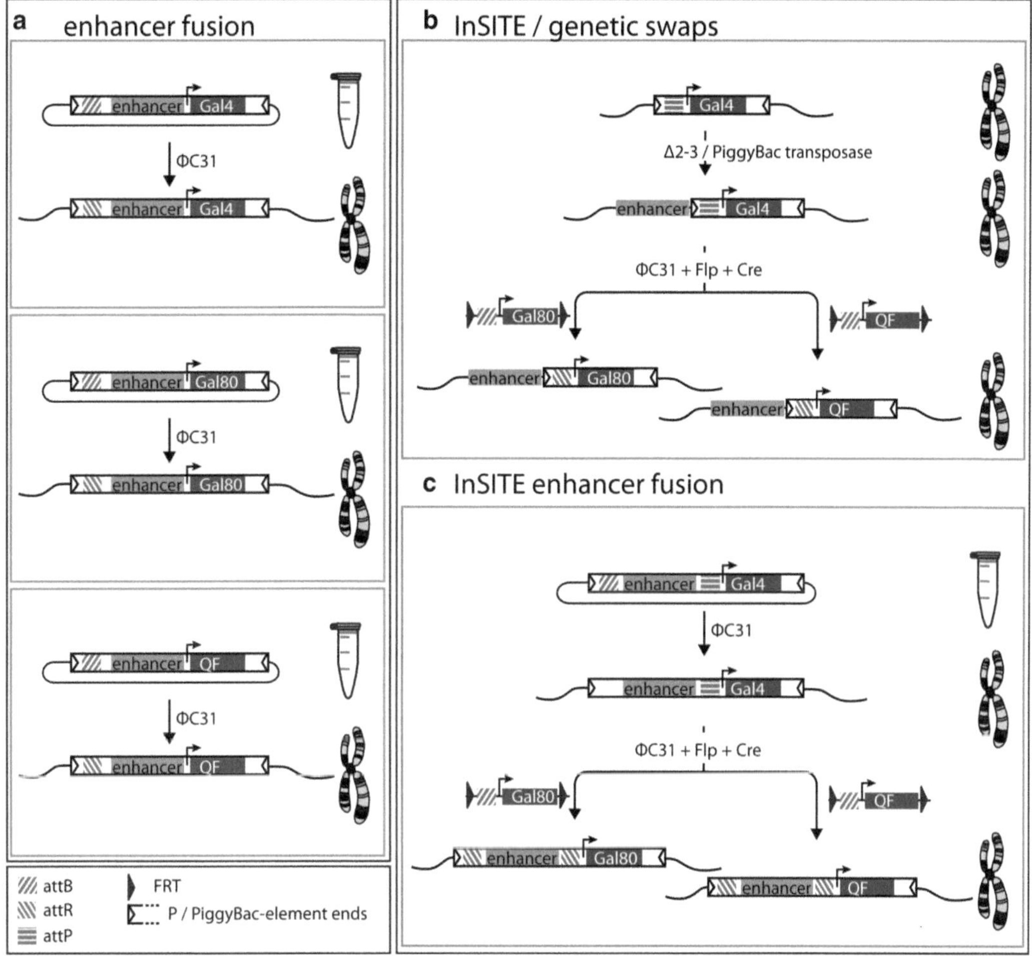

Fig. 2. Genetic approaches for repurposing expression patterns. (**a**) Enhancer fusions. Enhancer fusion *Gal4* constructs are generated *in vitro* and injected into flies via PhiC31 mediated transgenesis (*upper panel*) (82). Useful expression patterns can be repurposed by recloning the corresponding enhancer in front of other activator molecules *in vitro* and regeneration of transgenes. *Gal80* (*middle panel*) and QF (*lower panel*) are shown as examples. (**b**) InSITE system based *in vivo* swapping strategy (89). InSITE enhancer trap are based on either *P element* or *piggyBac* starter lines, which can be mobilized to generate diverse driver lines. The corresponding expression patterns can be repurposed by swapping *Gal4* to another sequence entirely *in vivo*, using a combination of the PhiC31, Flp, and Cre recombinases. *Gal80* and *QF* swaps are shown as examples. (**c**) InSITE compatible enhancer fusion conversion. An InSITE compatible enhancer fusion *Gal4* construct can be constructed *in vitro*. The corresponding *Gal4* can then be swapped *in vivo* in order to reuse the enhancer expression pattern, employing the same strategy as in (**b**).

repurpose expression patterns and design intersectional strategies; however, every new intersection requires *in vitro* cloning of a construct and generation of a transgenic fly line.

Recent work has described an alternate strategy by which a given expression pattern can be repurposed by converting *Gal4* into *LexA*, or any other DNA sequence of interest (89, 91). Yagi et al. developed a system in which a microinjected *LexA* plasmid can be inserted via PhiC31-mediated transgenesis at the 3′ end of a *Gal4* enhancer trap. Cre recombinase is then used to excise the original *Gal4*. Whereas it has been mainly validated to create genetic *Gal4-LexA* mosaics, this system can also be used repurpose a *Gal4* expression pattern to express the *LexA* transcription activator (91). Gohl et al. (89) developed a different strategy, which does not require microinjection of the donor, but instead can be done entirely *in vivo*. Swapping of *Gal4* to another sequence can be done in several efficient genetic steps, and does not involve recloning of enhancers, generation of transgenes, or labor-intensive screening for rare genetic events (Fig. 2b). The InSITE (integrase swappable *in vivo* targeting element) system utilizes a recombinase-mediated cassette exchange (RMCE) strategy to replace *Gal4* with another sequence. A genomic donor molecule can be excised by Flp recombinase and inserted into the InSITE *Gal4* enhancer trap with PhiC31 integrase, which directs site-specific integration of the *attB* site-containing donor into an *attP* site of the InSITE enhancer trap. The original *Gal4* insert can then be excised via a Cre recombinase, leaving only the donor molecule at the genomic position that was previously occupied by the *Gal4* transcription factor. InSITE-compatible enhancer fusion constructs can also be made (Fig. 2c). This system is presently compatible with several common genetic regulators, namely *Gal80*, *Gal4AD*, *VP16AD*, *LexA*, and *QF*. Recently, many of the existing activator molecules have been modified for greater efficiency and new technologies are constantly emerging, including stronger *Gal80*-repressible or nonrepressible *LexA* transcription activators, stronger *Gal4* or split-*Gal4* transcription factors, *Gal80* repressors, and a codon-optimized Flp recombinase ((84, 91, 92) and C. Potter, personal communication). These tools, as well as unanticipated future technologies, can easily be integrated into the existing InSITE framework, or used in conjunction with enhancer fusion constructs (82, 89).

4.3. Effectors

The ability to specifically target single or small subsets of cells within the nervous system ideally allows the spatially and temporally restricted expression of downstream effector molecules. Over the past few years, a variety of tools have been generated that allow one to identify, modulate, or image neurons of interest. Several excellent recent reviews discuss the existing strategies and available genetic tools (93–95). In the following sections, we briefly summarize the strategies that are used or might be most applicable to study visual processing.

4.3.1. Inhibiting Neuronal Activity

Neuronal activity can generally be inhibited by the expression of molecules that disrupt neurotransmitter release, or by expressing channels that reduce the membrane potential. The most commonly used effector molecules are the temperature-sensitive dominant negative version of *Drosophila* Dynamin, *shibire* (UAS-shi^{ts1} (96)), tetanus neurotoxin light chain (*UAS-TNT* (97)), or the inward-rectifying potassium channel *UAS-Kir2.1* (98, 99). All three of these reagents act to reduce neurotransmitter release, either by blocking endocytic vesicle recycling (shi^{ts1}) or exocytic vesicle fusion (through cleavage of Synaptobrevin by TNT), or by membrane hyperpolarization (*Kir 2.1*). Several criteria can influence the choice of effector. The most commonly used effector, *UAS*-shi^{ts1}, has the advantage that its effect is rapid (operating on the timescale of a few minutes), inducible by shifting to the nonpermissive temperature, and can be reversible (96). As *TNT* or *Kir2.1* are not intrinsically temperature sensitive, and would thereby alter synaptic transmission throughout the life of the animal, their expression must be controlled both in time and space. This can be trivial if a driver line happens to be expressed only in adult animals, but is more commonly ensured by using the temperature-sensitive $Gal80^{ts}$ repressor (100). Alternatively, *UAS> STOP> TNT* flies have been generated that, if expressed in photoreceptor cells together with *hs-Flp* induced expression via heat shock on three successive days, showed the identical *TNT* expression pattern to that of *UAS-TNT* flies, and displayed no optomotor behavior (101). This approach might depend more on the strength of the *Gal4* line, is less practical for large-scale analyses of big collections of *Gal4* lines, and in certain systems might allow for compensatory mechanisms to take place.

Importantly, different effector molecules might affect behavior differently, depending on the connectivity of the circuitry. For example, effector molecules that act at chemical synapses do not disrupt information processing via electrically coupled circuit elements, whereas hyperpolarizing currents generated by *Kir2.1* expression can pass gap junctions and thus disrupt neuronal activity in electrically coupled neighbors, if there is sufficiently large input. Such a strong coupling has been shown in the visual system between the lamina neurons L1 and L2 (71). The use of various effectors could hence be informative about the nature of the coupling mechanisms in the circuit.

Interestingly, different effector molecules sometimes give dissimilar phenotypes upon expression in the same cell type, suggesting that TNT and $shibire^{ts1}$ function differently in various cell types. TNT has, for example, been shown to be ineffective in adult photoreceptor cells (28, 29, 102). These results highlight the importance of comparing the effects of multiple manipulations on each circuit element.

4.3.2. Activating Neuronal Activity

Once a certain neuron or class of neurons has been implicated in a distinct behavior through loss of function studies, one ideally also wants to know whether stimulus-independent increases in neuronal activity are sufficient to drive the behavior of interest. Effector molecules exist that increase electrical excitability, generally by depolarizing the neuronal cell membrane. These effectors have so far not been used to study visual system function in adult flies, but several transgenic tools have been generated and used in other systems. For example, the voltage-gated bacterial sodium channel NaChBac further increases the electrical excitability upon depolarization (103). One can also optogenetically stimulate neuronal populations using the rapidly-gated light activated cation channel Channelrhodopsin-2 (ChR2) or a modified version of it (Chr2-H134R) (104–106). Its cofactor retinal is not present in flies but can be supplied by feeding. ChR2 activation is fast and reversible, but has to be activated by high intensity blue light, an input that also activates photoreceptors, creating challenges associated with using optogenetics to study visual processing. There might be some ways to circumvent this problem in certain contexts, such as using mutations that render flies blind; however, these have not yet been demonstrated. Furthermore, fly rhodopsins show no absorbance above ~600 nm (107) and there are efforts underway to create fluorescent proteins that span the near-infrared range of the visual spectrum (108, 109). Incorporation of these red-shifted fluorescent proteins into actuators that trigger or inhibit neuronal activity, or reporters of neuronal activity (see below) could be immensely useful for studying visual system function.

Light-activated channels are also not ideal for structures deep in the fly brain. In analogy to *UAS-shi^ts1^*, the warmth-activated dTrpA1 cation channel can be conditionally activated by shifting to warm temperatures (around 25°C (106, 110, 111). Endogenous dTrpA1 appears to be expressed only in very few cells of the adult brain (110), although a recent study suggested that dTrpA1 can be of functional importance even if its expression level is below immunodetection (112). In this study, TrpA1 was shown to be involved in a light transduction pathway in class IV dendritic arborization neurons of *Drosophila* larvae. Alternatively, the cold activated rat TRPM8 channels can be used in flies at a restrictive temperature of 15–18°C (113). However, optomotor behavior is generally worse at such low temperatures, and *Gal4* transgene expression also greatly depends on temperature. Ideally, one wants to use a modality for activation or inactivation that does not interfere with fly activity (as temperature and light do). One interesting modification to the use of Trp channels as effectors is the application of magnetic fields in conjunction with magnetic nanoparticles to heat-activate them. Although this technique looks promising, activation of a transgene *in vivo* has not yet been demonstrated (114).

In summary, many tools to manipulate neuronal activity exist, but "ideal" effector molecules that can be controlled by a modality other than light, which can be rapidly induced to affect populations of neurons, can be applied to tissues or neurons at any depth, act on a millisecond time scale, are reversible, and do not depend on the presence of cofactors, do not yet exist. No matter the approach taken to modulate neuronal activity, maintaining a uniform genetic background is an important consideration. Many visual behaviors are sensitive to changes in genetic background, an observation that was made originally with the *sevenless* mutant which was described to have lower optomotor responses, a phenotype that disappeared after backcrossing to a wild type strain (69). A reasonable standard practice is to backcross all transgenes to an isogenized wild type strain for five generations, followed by replacement of all non-transgene-containing chromosomes with isogenized chromosomes. New tools or collections of transgenes are thus ideally generated in an isogenized wild-type background, whose behavior has been extensively characterized.

4.3.3. Visualizing Neuronal Activity

Once the neurons have been identified that mediate a certain behavior, the question arises how signals are processed and computed in the key cells. One approach is to measure intracellular changes in calcium concentrations evoked by visual stimuli. The currently most commonly used genetically encoded calcium indicators in flies are GCamP3.0 and TN-XXL (115, 116) (for further details see Chap. 7). In the visual system, lobula plate tangential cells or the axon terminals of lamina neurons in the medulla have been imaged in head-fixed flies that were presented with motion stimuli from an LED arena (117, 118). Sophisticated set-ups have even allowed simultaneous measurement of calcium transients and optomotor walking behavior (10, 23).

5. The Future

The study of motion vision in *Drosophila* has seen dramatic methodological change over the past decade. While the field has long benefited from quantitative behavioral assays, combined with a rich theoretical framework, recent innovations have added new assays, stimuli, and genetic tools. As a result of these advances, the next decade will undoubtedly see enormous progress in our understanding of motion computation at the cellular and circuit levels.

References

1. Gotz KG (1964) Optomotorische Untersuchung des visuellen Systems einiger Augenmutanten der Fruchtfliege *Drosophila*. Kybernetik 2(2):77–92
2. Gotz KG, Wenking H (1973) Visual control of locomotion in the walking fruitfly *Drosophila*. J Comp Physiol A 85(3): 235–266

3. Heisenberg M, Wolf R (1984) Vision in *Drosophila*: genetics of microbehavior. In: Braitenberg V, Barlow HB, Bullock TH, Florey E, Grüsser OJ, Peters A (eds) Studies of brain function. Springer, Berlin
4. Hecht S, Wald G (1934) The visual acuity and intensity discrimination of *Drosophila*. J Gen Physiol 17(4):517–547
5. Tammero LF, Frye MA, Dickinson MH (2004) Spatial organization of visuomotor reflexes in *Drosophila*. J Exp Biol 207(pt 1): 113–122
6. Tammero LF, Dickinson MH (2002) The influence of visual landscape on the free flight behavior of the fruit fly *Drosophila melanogaster*. J Exp Biol 205(pt 3):327–343
7. Sherman A, Dickinson MH (2003) A comparison of visual and haltere-mediated equilibrium reflexes in the fruit fly *Drosophila melanogaster*. J Exp Biol 206(pt 2):295–302
8. Bender JA, Dickinson MH (2006) A comparison of visual and haltere-mediated feedback in the control of body saccades in *Drosophila melanogaster*. J Exp Biol 209(pt 23):4597–4606
9. Maimon G, Straw AD, Dickinson MH (2010) Active flight increases the gain of visual motion processing in *Drosophila*. Nat Neurosci 13(3):393–399
10. Chiappe ME et al (2010) Walking modulates speed sensitivity in *Drosophila* motion vision. Curr Biol 20(16):1470–1475
11. Kalmus H (1943) The optomotor responses of some eye mutants of *Drosophila*. J Genet 45(2):206–213
12. Heisenberg M, Wolf R (1979) On the fine structure of yaw torque in visual flight orientation of *Drosophila melanogaster*. J Comp Physiol A 130(2):113–130
13. Wolf R, Heisenberg M (1980) On the fine structure of yaw torque in visual flight orientation of *Drosophila melanogaster*. J Comp Physiol A 140(1):69–80
14. Gotz KG (1968) Flight control in *Drosophila* by visual perception of motion. Kybernetik 4(6):199–208
15. Blondeau J, Heisenberg M (1982) The three-dimensional optomotor torque system of *Drosophila melanogaster*. J Comp Physiol A 145(3):321–329
16. Gotz KG (1987) Course-control, metabolism, and wing interference during ultralong tethered flight in *Drosophila melanogaster*. J Exp Biol 128:35–46
17. Lehmann FO, Dickinson MH (1997) The changes in power requirements and muscle efficiency during elevated force production in the fruit fly *Drosophila melanogaster*. J Exp Biol 200(pt 7):1133–1143
18. Mayer M, Vogtmann K, Bausenwein B, Wolf R, Heisenberg M (1988) Flight control during 'free yaw turns' in *Drosophila melanogaster*. J Comp Physiol A 163(3):389–399
19. Duistermars BJ et al (2007) Dynamic properties of large-field and small-field optomotor flight responses in *Drosophila*. J Comp Physiol A Neuroethol Sens Neural Behav Physiol 193(7):787–799
20. Bender JA, Dickinson MH (2006) Visual stimulation of saccades in magnetically tethered *Drosophila*. J Exp Biol 209(pt 16): 3170–3182
21. Carey JR et al (2006) Age-specific and lifetime behavior patterns in *Drosophila melanogaster* and the Mediterranean fruit fly, *Ceratitis capitata*. Exp Gerontol 41(1):93–97
22. Buchner E (1976) Elementary movement detectors in an insect visual system. Biol Cybern 24:85–101
23. Seelig JD et al (2010) Two-photon calcium imaging from head-fixed *Drosophila* during optomotor walking behavior. Nat Methods 7(7):535–540
24. Land MF (1973) Head movements in flies during visually guided flight. Nature 243:299–300
25. Hengstenberg R (1993) Multisensory control of insect oculomotor systems. In: Miles FA, Wallman J (eds) Visual motion and its role in the stabilization of gaze. Elsevier, Amsterdam
26. Strange G, Hengstenberg R (1996) Tri-axial, real-time logging of fly head movements. J Neurosci Methods 64:209–218
27. Heimbeck G et al (2001) A central neural circuit for experience-independent olfactory and courtship behavior in *Drosophila melanogaster*. Proc Natl Acad Sci U S A 98(26): 15336–15341
28. Rister J et al (2007) Dissection of the peripheral motion channel in the visual system of *Drosophila melanogaster*. Neuron 56(1):155–170
29. Zhu Y et al (2009) Peripheral visual circuits functionally segregate motion and phototaxis behaviors in the fly. Curr Biol 19(7):613–619
30. Bulthoff H, Gotz KG, Herre M (1982) Recurrent inversion of visual orientation in the walking fly, *Drosophila melanogaster*. J Comp Physiol A 148(4):471–481
31. Strauss R, Schuster S, Gotz KG (1997) Processing of artificial visual feedback in the walking fruit fly *Drosophila melanogaster*. J Exp Biol 200(pt 9):1281–1296
32. Straw AD, Lee S, Dickinson MH (2010) Visual control of altitude in flying *Drosophila*. Curr Biol 20(17):1550–1556
33. Fry SN, Rohrseitz N, Straw AD, Dickinson MH (2008) TrackFly: virtual reality for a

behavioral system analysis in free-flying fruit flies. J Neurosci Methods 171:110–117
34. Gotz KG (1970) Fractionation of *Drosophila* populations according to optomotor traits. J Exp Biol 52(2):419–436
35. Heisenberg M, Gotz KG (1975) The use of mutations for the partial degradation of vision in *Drosophila melanogaster*. J Comp Physiol A 98(3):217–241
36. Katsov AY, Clandinin TR (2008) Motion processing streams in *Drosophila* are behaviorally specialized. Neuron 59(2):322–335
37. Straw AD et al (2011) Multi-camera real-time three-dimensional tracking of multiple flying animals. J R Soc Interface 8(56):395–409
38. Branson K et al (2009) High-throughput ethomics in large groups of *Drosophila*. Nat Methods 6(6):451–457
39. Wolf R, Heisenberg M (1990) Visual control of straight flight in *Drosophila melanogaster*. J Comp Physiol A 167(2):269–283
40. Reiser MB, Dickinson MH (2008) A modular display system for insect behavioral neuroscience. J Neurosci Methods 167(2):127–139
41. Yamaguchi S et al (2008) Motion vision is independent of color in *Drosophila*. Proc Natl Acad Sci U S A 105(12):4910–4915
42. Juusola M, Hardie RC (2001) Light adaptation in *Drosophila* photoreceptors: I. Response dynamics and signaling efficiency at 25 degrees C. J Gen Physiol 117(1):3–25
43. Rust NC, Movshon JA (2005) In praise of artifice. Nat Neurosci 8(12):1647–1650
44. Marmarelis PZ, McCann GD (1973) Development and application of white-noise modeling techniques for studies of insect visual nervous system. Kybernetik 12(2):74–89
45. Chichilnisky EJ (2001) A simple white noise analysis of neuronal light responses. Network 12(2):199–213
46. Sakai HM, Naka K, Korenberg MJ (1988) White-noise analysis in visual neuroscience. Vis Neurosci 1(3):287–296
47. Fairhall AL et al (2001) Efficiency and ambiguity in an adaptive neural code. Nature 412(6849):787–792
48. Brenner N et al (2000) Synergy in a neural code. Neural Comput 12(7):1531–1552
49. Theobald JC, Ringach DL, Frye MA (2010) Dynamics of optomotor responses in *Drosophila* to perturbations in optic flow. J Exp Biol 213(pt 8):1366–1375
50. Lindemann JP et al (2005) On the computations analyzing natural optic flow: quantitative model analysis of the blowfly motion vision pathway. J Neurosci 25(27):6435–6448
51. Kern R, Petereit C, Egelhaaf M (2001) Neural processing of naturalistic optic flow. J Neurosci 21(8):RC139
52. van Hateren JH (1997) Processing of natural time series of intensities by the visual system of the blowfly. Vision Res 37(23):3407–3416
53. Zheng L et al (2009) Network adaptation improves temporal representation of naturalistic stimuli in *Drosophila* eye: I dynamics. PLoS One 4(1):e4307
54. Nikolaev A et al (2009) Network adaptation improves temporal representation of naturalistic stimuli in *Drosophila* eye: II mechanisms. PLoS One 4(1):e4306
55. Chubb C, Sperling G (1989) Two motion perception mechanisms revealed through distance-driven reversal of apparent motion. Proc Natl Acad Sci U S A 86(8):2985–2989
56. Smith AT (1994) The detection of second-order motion. In: Smith AT, Snowden RJ (eds) Visual detection of motion. Academic, London, pp 145–176
57. Smith AT et al (1998) The processing of first- and second-order motion in human visual cortex assessed by functional magnetic resonance imaging (fMRI). J Neurosci 18(10): 3816–3830
58. Theobald JC et al (2008) Flies see second-order motion. Curr Biol 18(11):R464–R465
59. Benzer S (1967) Behavioral mutants of *Drosophila* isolated by countercurrent distribution. Proc Natl Acad Sci U S A 58(3): 1112–1119
60. Pak WL, Grossfield J, White NV (1969) Nonphototactic mutants in a study of vision of *Drosophila*. Nature 222(5191):351–354
61. Gerresheim F (1988) Isolation of *Drosophila melanogaster* mutants with a wavelength-specific alteration in their phototactic response. Behav Genet 18(2):227–246
62. Hotta Y, Benzer S (1970) Genetic dissection of the *Drosophila* nervous system by means of mosaics. Proc Natl Acad Sci U S A 67(3): 1156–1163
63. Pak WL, Grossfield J, Arnold KS (1970) Mutants of the visual pathway of *Drosophila melanogaster*. Nature 227(5257):518–520
64. Zuker CS (1992) Phototransduction in *Drosophila*: a paradigm for the genetic dissection of sensory transduction cascades. Curr Opin Neurobiol 2(5):622–627
65. Clandinin TR et al (2001) *Drosophila* LAR regulates R1-R6 and R7 target specificity in the visual system. Neuron 32(2):237–248
66. Lee CH et al (2001) N-cadherin regulates target specificity in the *Drosophila* visual system. Neuron 30(2):437–450

67. Choe KM, Clandinin TR (2005) Thinking about visual behavior; learning about photoreceptor function. Curr Top Dev Biol 69: 187–213
68. Fischbach KF, Heisenberg M (1984) Neurogenetics and behaviour in insects. J Exp Biol 112(1):65–93
69. Heisenberg M, Buchner E (1977) The role of retinula cell types in visual behavior of *Drosophila melanogaster*. J Comp Physiol A 187:127–162
70. Coombe PE, Heisenberg M (1986) The structural brain mutant vacuolar medulla of *Drosophila melanogaster* with specific behavioral defects and cell degeneration in the adult. J Neurogenet 3(3):135–158
71. Joesch M et al (2010) ON and OFF pathways in *Drosophila* motion vision. Nature 468(7321):300–304
72. Heisenberg M, Wonneberger R, Wolf R (1978) Optomotor-blind[H31]—a *Drosophila* mutant of the lobula plate giant neurons. J Comp Physiol A 124(4):287–296
73. Pflugfelder GO et al (1992) The lethal(1) optomotor-blind gene of *Drosophila melanogaster* is a major organizer of optic lobe development: isolation and characterization of the gene. Proc Natl Acad Sci U S A 89(4): 1199–1203
74. Brunner A et al (1992) Mutations in the proximal region of the optomotor-blind locus of *Drosophila melanogaster* reveal a gradient of neuroanatomical and behavioral phenotypes. J Neurogenet 8(1):43–55
75. Poeck B, Hofbauer A, Pflugfelder GO (1993) Expression of the *Drosophila* optomotor-blind gene transcript in neuronal and glial cells of the developing nervous system. Development 117(3):1017–1029
76. Gengs C et al (2002) The target of *Drosophila* photoreceptor synaptic transmission is a histamine-gated chloride channel encoded by ort (hclA). J Biol Chem 277(44):42113–42120
77. Brand AH, Perrimon N (1993) Targeted gene expression as a means of altering cell fates and generating dominant phenotypes. Development 118(2):401–415
78. Lai SL, Lee T (2006) Genetic mosaic with dual binary transcriptional systems in *Drosophila*. Nat Neurosci 9(5):703–709
79. Potter CJ et al (2010) The Q system: a repressible binary system for transgene expression, lineage tracing, and mosaic analysis. Cell 141(3):536–548
80. Lee T, Luo L (1999) Mosaic analysis with a repressible cell marker for studies of gene function in neuronal morphogenesis. Neuron 22(3):451–461
81. Luan H et al (2006) Refined spatial manipulation of neuronal function by combinatorial restriction of transgene expression. Neuron 52(3):425–436
82. Pfeiffer BD et al (2008) Tools for neuroanatomy and neurogenetics in *Drosophila*. Proc Natl Acad Sci U S A 105(28):9715–9720
83. Suster ML et al (2004) Refining *GAL4*-driven transgene expression in *Drosophila* with a GAL80 enhancer-trap. Genesis 39(4):240–245
84. Bohm RA et al (2010) A genetic mosaic approach for neural circuit mapping in *Drosophila*. Proc Natl Acad Sci U S A 107(37):16378–16383
85. Gordon MD, Scott K (2009) Motor control in a *Drosophila* taste circuit. Neuron 61(3): 373–384
86. Spradling AC et al (1995) Gene disruptions using P transposable elements: an integral component of the *Drosophila* genome project. Proc Natl Acad Sci U S A 92(24):10824–10830
87. Thibault ST et al (2004) A complementary transposon tool kit for *Drosophila melanogaster* using P and piggyBac. Nat Genet 36(3):283–287
88. Horn C et al (2003) piggyBac-based insertional mutagenesis and enhancer detection as a tool for functional insect genomics. Genetics 163(2):647–661
89. Gohl DM, Silies MA, Gao XJ, Bhalerao S, Luongo FJ, Lin C-C, Potter CJ, Clandinin TR (2011) A versatile in vivo system for directed dissection of gene expression patterns. Nat Methods 8:231–237
90. Metaxakis A et al (2005) Minos as a genetic and genomic tool in *Drosophila melanogaster*. Genetics 171(2):571–581
91. Yagi R, Mayer F, Basler K (2010) Refined LexA transactivators and their use in combination with the *Drosophila Gal4* system. Proc Natl Acad Sci U S A 107(37):16166–16171
92. Pfeiffer BD et al (2010) Refinement of tools for targeted gene expression in *Drosophila*. Genetics 186(2):735–755
93. Luo L, Callaway EM, Svoboda K (2008) Genetic dissection of neural circuits. Neuron 57(5):634–660
94. Olsen SR, Wilson RI (2008) Cracking neural circuits in a tiny brain: new approaches for understanding the neural circuitry of *Drosophila*. Trends Neurosci 31(10):512–520
95. Simpson JH (2009) Mapping and manipulating neural circuits in the fly brain. Adv Genet 65:79–143
96. Kitamoto T (2001) Conditional modification of behavior in *Drosophila* by targeted expression of a temperature-sensitive shibire allele in defined neurons. J Neurobiol 47(2):81–92

97. Sweeney ST et al (1995) Targeted expression of tetanus toxin light chain in *Drosophila* specifically eliminates synaptic transmission and causes behavioral defects. Neuron 14(2): 341–351
98. Baines RA et al (2001) Altered electrical properties in *Drosophila* neurons developing without synaptic transmission. J Neurosci 21(5):1523–1531
99. Johns DC et al (1999) Inducible genetic suppression of neuronal excitability. J Neurosci 19(5):1691–1697
100. McGuire SE et al (2003) Spatiotemporal rescue of memory dysfunction in *Drosophila*. Science 302(5651):1765–1768
101. Keller A et al (2002) Targeted expression of tetanus neurotoxin interferes with behavioral responses to sensory input in *Drosophila*. J Neurobiol 50(3):221–233
102. Rister J, Heisenberg M (2006) Distinct functions of neuronal synaptobrevin in developing and mature fly photoreceptors. J Neurobiol 66(12):1271–1284
103. Nitabach MN et al (2006) Electrical hyperexcitation of lateral ventral pacemaker neurons desynchronizes downstream circadian oscillators in the fly circadian circuit and induces multiple behavioral periods. J Neurosci 26(2):479–489
104. Boyden ES et al (2005) Millisecond-timescale, genetically targeted optical control of neural activity. Nat Neurosci 8(9):1263–1268
105. Schroll C et al (2006) Light-induced activation of distinct modulatory neurons triggers appetitive or aversive learning in *Drosophila* larvae. Curr Biol 16(17):1741–1747
106. Pulver SR et al (2009) Temporal dynamics of neuronal activation by Channelrhodopsin-2 and TRPA1 determine behavioral output in *Drosophila* larvae. J Neurophysiol 101(6): 3075–3088
107. Salcedo E et al (2003) Molecular basis for ultraviolet vision in invertebrates. J Neurosci 23(34):10873–10878
108. Lin MZ et al (2009) Autofluorescent proteins with excitation in the optical window for intravital imaging in mammals. Chem Biol 16(11):1169–1179
109. Shcherbo D et al (2010) Near-infrared fluorescent proteins. Nat Methods 7(10):827–829
110. Hamada FN et al (2008) An internal thermal sensor controlling temperature preference in *Drosophila*. Nature 454(7201):217–220
111. Parisky KM et al (2008) PDF cells are a GABA-responsive wake-promoting component of the *Drosophila* sleep circuit. Neuron 60(4):672–682
112. Xiang Y et al (2010) Light-avoidance-mediating photoreceptors tile the *Drosophila* larval body wall. Nature 468(7326):921–926
113. Peabody NC et al (2009) Characterization of the decision network for wing expansion in *Drosophila* using targeted expression of the TRPM8 channel. J Neurosci 29(11):3343–3353
114. Huang H et al (2010) Remote control of ion channels and neurons through magnetic-field heating of nanoparticles. Nat Nanotechnol 5(8):602–606
115. Mank M et al (2008) A genetically encoded calcium indicator for chronic in vivo two-photon imaging. Nat Methods 5(9):805–811
116. Tian L et al (2009) Imaging neural activity in worms, flies and mice with improved GCaMP calcium indicators. Nat Methods 6(12):875–881
117. Joesch M et al (2008) Response properties of motion-sensitive visual interneurons in the lobula plate of *Drosophila melanogaster*. Curr Biol 18(5):368–374
118. Reiff DF et al (2010) Visualizing retinotopic half-wave rectified input to the motion detection circuitry of *Drosophila*. Nat Neurosci 13(8):973–978

Part IV

Futures

Chapter 10

Using Primary Neuron Cultures of *Drosophila* to Analyze Neuronal Circuit Formation and Function

Andreas Prokop, Barbara Küppers-Munther, and Natalia Sánchez-Soriano

Abstract

For many decades, primary neuron cultures of *Drosophila* have been used complementary to work in vivo. Primary cultures were instrumental for the analysis of physiological properties of *Drosophila* neurons and synapses, and they were used for the analysis of developmental processes. Recent developments have established *Drosophila* primary neurons based on Schneider's culture media as a means to investigate the neuronal cytoskeleton, opening up novel opportunities for research into cellular mechanisms of axonal growth, synapse formation, and perhaps even neuronal degeneration. These cell cultures provide readouts for cytoskeletal dynamics that are difficult or impossible to access in vivo, and which turned out to be highly conserved with mammalian or other vertebrate neurons. Therefore, the same genetic manipulations in *Drosophila* can now be studied synergistically in culture and in vivo, to address cell biological principles of neuronal circuit formation and function. Here, we describe in detail how these cell cultures are generated and discuss principal considerations for the experimental design and the solution of common problems. Furthermore, we describe in detail how to generate Schneider's media with adjustable inorganic ion concentrations. These media have been shown to promote the physiological maturation of neurons, thus expanding the use of the primary neuron cultures into the synaptic stage. The culture strategies described here recapitulate in vivo development with impressive accuracy and provide a promising means for *Drosophila* research on neuronal development and function.

Key words: *Drosophila*, Cell culture, Primary neurons, Axonal growth, Neurogenesis, Synapse formation, Cytoskeleton

1. Introduction

The use of neuronal cell cultures provides a powerful means to study neurodevelopmental mechanisms with microscopic detail and amenability to experimental manipulations that cannot be easily achieved in vivo. Accordingly, they have been instrumental

Bassem A. Hassan (ed.), *The Making and Un-Making of Neuronal Circuits in Drosophila*, Neuromethods, vol. 69, DOI 10.1007/978-1-61779-830-6_10, © Springer Science+Business Media, LLC 2012

in expanding our knowledge on mechanisms establishing neuronal polarity (1), mechanisms of synapse formation (2), and the mechanisms of axonal growth and guidance (3), to name but a few outstanding examples. Cells in culture are deprived of their natural signalling environments, and growth under these artificial, low-stimulus culture conditions certainly bears the risk of artefacts. However, they provide unique possibilities to selectively analyse the effects of defined stimuli in isolation, or to test whether cellular processes depend on cell-autonomous mechanisms, on extracellular influences, or a combination of both. Culture work can be carried out using immortalised neuronal cell lines, such as mouse Neuro2A or P19 neurocarcinoma cell lines (4, 5) or *Drosophila* ML-DmBG2 cell lines (6, 7). However, it is debatable whether these cells express true neuronal properties. Alternatively, primary neurons can be used. Primary neurons can be stem cell-derived neurons (8–10), or can be isolated from the organism as post-mitotic neurons which have already undergone a significant period of development and differentiation in vivo (11). Primary neurons tend to better acquire the key features of neurons, such as true axonal or dendritic processes, and the physiological and synaptic properties that usually correspond rather well with the neuronal tissues or brain areas they are derived from (12). This is in contrast to neuronal cell lines. Accordingly, the use of primary neurons has become an important strategy for the study of neurodevelopmental processes in various model organisms, ranging from mammals and other vertebrates to invertebrates, such as snails, crustaceans and various insects including *Drosophila* (12, 13).

Drosophila has been instrumental for the discovery of basic principles of neuronal development and function, and primary cell cultures have played an important part. Thus, neuronal *Drosophila* cultures were used in numerous publications to describe physiological and plastic properties of neurons and underlying regulatory mechanisms (13, 14). Furthermore, work in *Drosophila* primary neurons has provided important information about developmental processes. For example, it has delivered proof that neural cell lineage programmes are intrinsic to neural precursors and their daughter cells (15–18), that cAMP is regulator of growth cone dynamics (19), that 20-hydroxyecdysone enhances axonal growth of pupal Kenyon cells (20), that the timing of the onset of GABA expression during embryonic development is activity dependent (21), that the unipolar organisation of invertebrate neurons is non-cell autonomously determined (22), that axonal membranes are compartmentalised through cell-autonomous mechanisms (23) and that presynaptic differentiation of *Drosophila* interneurons occurs cell-autonomously (24). Importantly, *Drosophila* primary neurons have been used for genetic screens using RNAi (25).

Recently, *Drosophila* primary neuron cultures were established as an excellent means to study cytoskeletal dynamics of neurons at a resolution that cannot be achieved in vivo. Cytoskeletal dynamics are as accessible in these neurons as they are in primary neurons of mammals or other vertebrates, and these dynamics are very well conserved (26–30). Studies in *Drosophila* primary neurons can be carried out in direct comparison to in vivo analyses, using the same genetic manipulations (27, 31). The efficiency and sophistication of fly genetics, the low redundancy of the *Drosophila* genome, the high degree of conservation of its cytoskeletal regulators paired with detailed cellular readouts provide novel opportunities to address the complexity of cytoskeletal regulatory processes at the cellular level (31–33), thus targeting a key problem not only in neurobiology but also in the wider cell biology field (34). Studies of the neuronal cytoskeleton in primary neurons are already being used to investigate molecular mechanisms of axonal growth (26–29). When growing the same type of primary neuron cultures for a longer period, they differentiate synaptic structures, and suitable culture media have been developed that support the maturation of typical physiological properties (22, 24). Therefore, these cultures also provide new opportunities to address the cytoskeletal regulation of synapse formation. Finally, the cytoskeleton lies at the heart of many neurodegenerative processes (35). In the context of current *Drosophila* research into the genetics of neurodegenerative pathways (36, 37), these cultures provide a promising means to address the cell biology downstream of such pathways. Here we describe these techniques in detail to make them accessible for future research.

2. Materials

2.1. Cell Culture Media

The protocols described in this chapter make use of Schneider's culture medium either in its conventional, commercially available composition, or with adjustable inorganic ion concentrations.

2.1.1. Preparing Standard Schneider's Medium

Conventional, commercially available Schneider's medium (38) is ideal for the study of early neural development, since its inorganic ion composition is similar to the haemolymph (39) (see Chap. 4.2). For preparation of standard Schneider's medium, add 10 mL non-heat inactivated foetal calf serum to 40 mL commercial Schneider's medium, filter-sterilise. Wrap the container in aluminium foil to protect against light and incubate for 3 days at 26°C to inactivate the serum's complement system. Thereafter, add 2 μg/mL insulin (e.g. use 10 μL of a 10 mg/mL insulin stock solution for 50 mL medium; note that aliquots of 10 mg/mL H_2O, set to pH 2 with glacial acetic acid, can be stored at −80°C). Check a small aliquot whether it has the correct pH of 6.8–6.9. Otherwise adjust, using sterile 1 N HCl for acid and sterile 1 N NaOH for basic correction. Freeze in 1 mL medium aliquots at −80°C. Frozen aliquots can be used for up to 3–4 months.

2.1.2. Preparing Schneider's Medium with Adjustable Inorganic Ion Concentrations

As explained in Chap. 4.2, the experimental applications of Schneider's medium can be significantly enhanced by adjusting its inorganic ion content. The recipe for ion-adjusted Schneider's media is based on a modular system of organic and inorganic solutions, in which the concentrations of inorganic ions (especially K^+, Na^+, Ca^{2+}, Mg^{2+} and Cl^-) need to be changed only in the inorganic solutions, whereas the labour-intense organic solution is constant and can be stored in frozen aliquots (24, 40). Note that the end concentrations of ions in the final medium include the ions contributed by the serum-supplement. Ion concentrations in the foetal calf serum are considered to be on average 9 mM for K^+, 142 mM for Na^+, 3 mM for Ca^{2+}, 1 mM for Mg^{2+} and 104 mM for Cl^- (40). Naturally, ion concentrations differ between serum batches, and these values need to be measured or obtained from the provider, if exact ion concentration settings are of experimental relevance.

To generate the organic solution, the components listed in Table 1 are weighed and dissolved in the given volumes of commercially available endotoxin-free water, following the detailed instructions in the table legend. Combine all solutions of organic components which add up to an end volume of 789 mL, filter them sterile (using syringe filters, filter bottles or other devices with low protein binding membranes for sterilisation of aqueous solutions; 0.1–0.2 μm pore size) and divide into 32 mL aliquots, which can be frozen and stored at –20°C until use. Three inorganic solutions are prepared and diluted in the end volumes of endotoxin-free water given in Table 2. After sterile filtration, the three inorganic solutions can be stored at 4°C until use.

To prepare 50 mL final medium from these stock solutions, a 32 mL aliquot of organic solution is thawed at 37°C. Then, 3.2 mL endotoxin-free water, 0.8 mL buffer solution, 2 mL chloride solution and 2 mL magnesium solution are added. Like described for commercial Schneider's medium (see Sect. 2.1.1), 40 mL of ion-adjusted Schneider's medium are supplemented with 10 mL non-heat inactivated foetal calf serum (preferably tested on neuronal cells), the medium is sterile filtered, and kept in the dark at 26°C for 2–3 days, before 2 μg/mL insulin are added, and the pH is adjusted to pH 6.8–6.9 (requiring approximately 400 μL sterile 1 N NaOH). The medium can be aliquoted, frozen in liquid nitrogen, and stored at –80°C with a maximum storage time of 3–4 months.

2.2. Dissociation Medium

Two-hundred millilitre dissociation medium contain 167 mL distilled water, 30 mL HBSS (Hanks' Balanced Salt Solutions—without calcium or magnesium, with Phenol Red; Gibco), 3 mL penicillin-streptomycin-solution (10,000 units; Gibco), 0.01 g phenyl-thiourea (Sigma). This buffer can be stored at 4°C for several months. Add 0.5 mg/mL collagenase type 1 (Worthington, Cellsystems) and 2 mg/mL Dispase (Roche) to the dissociation medium. When sterile-filtered, the final solution can be stored at 4°C for only 1 week; therefore, prepare to small amounts (1–2 mL).

Table 1
Preparation of the organic stock solution for ion-adjusted Schneider's medium

Dilution of organic components in endotoxin-free water (789 mL in total)	
Highly soluble amino acids (61 mL)	
L-alanine[a]	0.5 g/11 mL
L-arginine[a]	0.6 g/6 mL
L-cysteine	0.06 g/5 mL
Glycine	0.25 g/5 mL
L-isoleucine[a]	0.15 g/10 mL
L-leucine[a]	0.15 g/12 mL
L-methionine	0.15 g/12 mL
Modestly soluble amino acids (84 mL)	
L-histidine[b]	0.4 g/15 mL
L-lysine	1.65 g/12 mL
L-proline	1.7 g/12 mL
L-serine	0.25 g/15 mL
L-threonine[b]	0.35 g/15 mL
L-tryptophan	0.1 g/15 mL
Little soluble amino acids (569 mL)	
L-aspartic acid[c]	0.4 g/110 mL
L-cystine·2HCl[c]	0.027 g/115 mL
L-glutamic acid[c]	0.8 g/115 mL
L-glutamine[d]	1.8 g/50 mL
L-tyrosine[c]	0.496 g/134 mL (+450 μL 10 N NaOH)
L-valine	0.3 in 45 mL
Diverse components (75 mL)	
Fumaric acid[b]	0.06 g/25 mL
α-Ketoglutaric acid	0.35 g/10 mL
L-malic acid	0.6 g/10 mL
Succinic acid[a]	0.06 g/10 mL
D(+)-trehalose·$2H_2O$ + D-glucose[d]	2.21 g of each/10 mL
Yeast extract[a]	2.0 g/10 mL

[a]Yeast extract, alanine, arginine, isoleucine, leucine and succinic acid require heating to 37°C

[b]Threonine, histidine and fumaric acid can be heated briefly in the microwave (10–15 s at 900 W)

[c]L-aspartic acid, L-cystine·2HCl, L-glutamic acid and L-tyrosine (requires further addition of 10 mM NaOH) require heating in the microwave to just below boiling-point (ca. 1 min at 900 W)

[d]Glutamine and trehalose/glucose require heating to 52°C

2.3. Fixatives

Standard fixations are carried out in 4% paraformaldehyde in 0.1 M phosphate buffer (pH 7.2). Following the method by Sørensen (41), two 0.2 M stock solutions are prepared, which can be stored at 4°C and readily mixed to a set pH when required. Prepare the basic solution A by dissolving 2.84 g Na_2HPO_4 in 100 mL water and the acid solution B by dissolving 2.76 g $NaH_2PO_4 \cdot 1H_2O$ in

Table 2
Preparation of the inorganic solutions for ion-adjusted Schneider's medium

Component	Active Schneider's	Standard Schneider's
Chloride solution (in g) in 50 mL water		
$CaCl_2 \cdot 2H_2O$	0.0690	0.7945
KCl	–	1.6
NaCl	6.9748	2.1
Magnesium solution (in g) in 50 mL water		
$MgSO_4$	0.0843	1.8072
Buffer solution (in g) in 20 mL water		
$NaHCO_3$	0.4	0.4
Na_2HPO_4	0.7	0.7
NaH_2PO_4	0.0696	–
KH_2PO_4	0.3701	0.45

The inorganic solutions determine the essential character of the different Schneider's media and can be changed to produce further variants. Here, two different ionic concentrations are given for active Schneider's medium (first column) and standard Schneider's medium (right column) (21, 24, 40)

100 mL water. For preparing a 4% paraformaldehyde solution in 0.1 M buffer, dissolve 4 g paraformaldehyde in 36 mL of solution A, warm slightly (handwarm) and mix well by shaking. Add 14 mL solution B and fill up to 100 mL with H_2O. Staining for some microtubule-binding proteins requires a different fixative (42) containing 90% methanol, 3% formaldehyde and 5 mM sodium carbonate (pH 9; storable –80°C); keep the fixative on ice before use and incubate for 10 min (can take place at room temperature).

2.4. Needles and Culture Chambers

Cells are harvested from embryos via glass capillaries. We advise to use capillaries with an outer diameter of 1 mm and a wall thickness of 0.1 mm made from soda-lime glass/borosilicate which does not splinter upon bevelling (e.g. from Hilgenberg, Germany). Use a standard electrode puller for electrophysiology to pull a fine needle tip. Using a commercial micro-grinder bevel the needles to a position where the needle has about 30–50 μm diameter and to an angle of about 40°; mark the upper capillary end with a waterproof pen to know the orientation of the bevelled angle when using the capillary at a later stage (Fig. 1b). Check under a compound microscope whether the inner diameter is correct (about 30–40 μm) or compare directly to an embryo. To re-use capillaries, clean with 70% ethanol or undiluted acetone, but dry carefully by pumping air

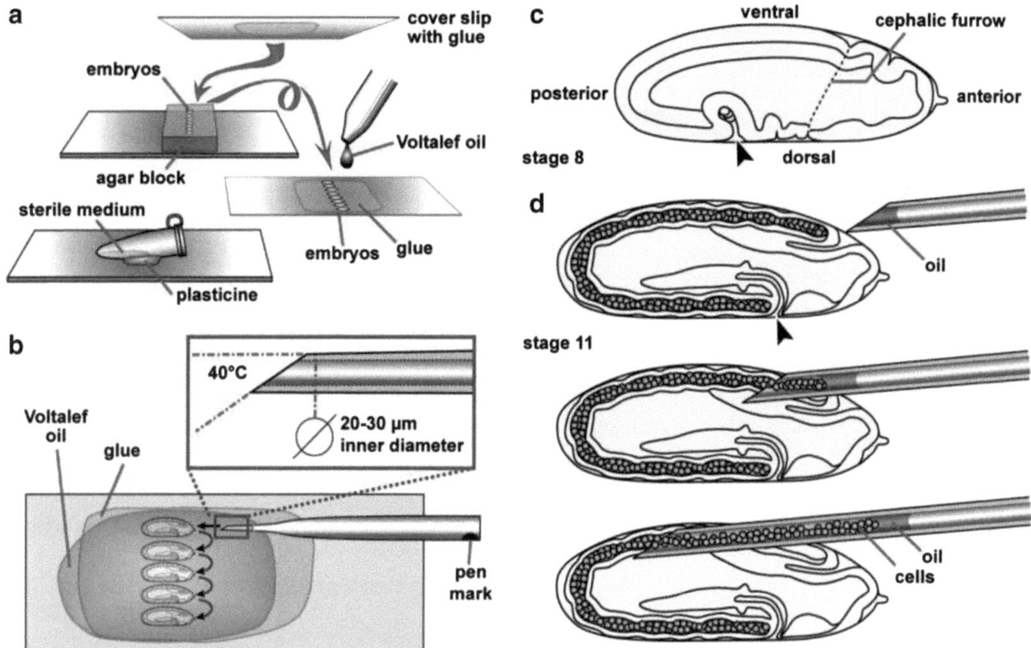

Fig. 1. Procedures for harvesting embryonic cells. (**a**) Properly aligned and oriented embryos on an agar block are transferred to a glue-coated coverslip by gently pressing the coverslip with the glue-bearing side onto the embryos; the embryos are then covered with 10S Voltalef oil. (**b**) The pulled and bevelled capillary is inserted into the needle holder with the pen mark facing towards the experimenter; embryos are accordingly aligned all in the same orientation, anterior left and dorsal facing the experimenter. (**c**) An embryo at stage 8 in the desired orientation; the cephalic furrow and posterior tip (*arrow head*) are indicated. (**d**) Process of cell harvesting in a stage 11 embryo (oriented as embryo in (**c**)): before stabbing into the embryo, the pressure is set so that the capillary has a small amount of oil in its tip; upon entry, suction is applied to drag cells from the ventral germ band into the capillary; whilst sucking, the capillary is gently moved forward to maximise the cell harvest per embryo.

through the capillary before use. Further information on the preparation of capillaries can be found elsewhere (43).

Special culture chambers (Fig. 2) (44, 45) consist of a lead-free glass microscope slide (available for example from Menzel Gläser, Germany) with a 15-mm hole (cut for us in a chemical glass workshop) which is glued with silicone paste (without fungicides) onto an intact lead-free glass slide. Place a weight on the chamber and let the silicone dry for 1 week, scrape off any silicone residues in the well, then clean with warm detergent water, then 100% ethanol and finally 100% acetone.

2.5. Heptane Glue

Fill a 500-mL bottle with as many pieces of brown plastic parcel tape as possible. Fill with about 200 mL of *n*-heptane and shake or rotate for several hours or even days to dissolve the glue from the tape. Pour out the *n*-heptane solution and centrifuge at about 10,000 g. At any stage, the glue supernatant can be concentrated by evaporating or diluted by adding more *n*-heptane (43).

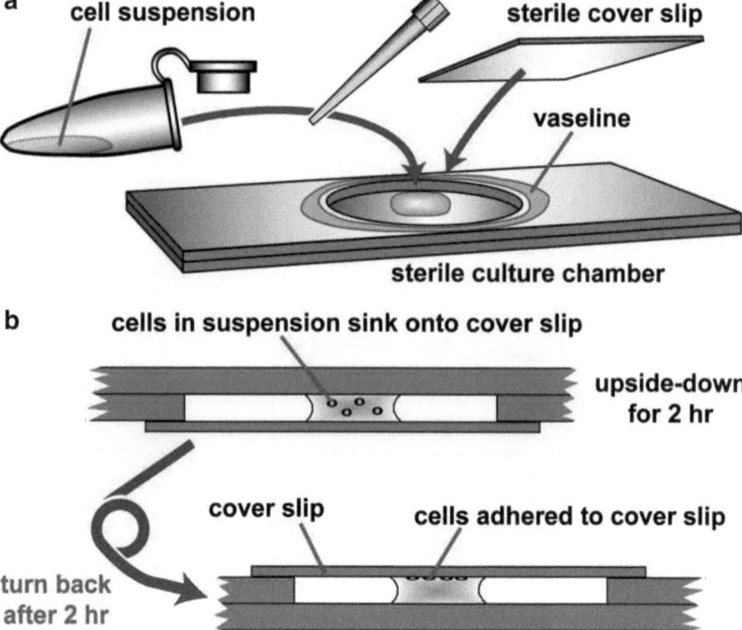

Fig. 2. Culture chambers for hanging drop cultures. (**a**) The culture chamber are made from one slide with a 15-mm hole glued to an intact slide (lead-free glass). The cell suspension is pipetted into the middle of the well, which is then closed with a coverslip using a ring of vaseline as sealant. (**b**) After sealing the culture chamber, turn it upside-down to allow cells to sink onto the coverslip; turn back after 2 h when cells are firmly adhered to the coverslip so that debris can fall off. For live analyses the chamber can be turned either side up so that both upright and inverted microscopes can be used.

2.6. Preparing Coverslip for Growing Primary Neurons

Drosophila primary neurons grow on uncoated 24 × 24 mm lead-free coverslip (VWR international MENZBB024024A123, no. 1). To prepare them, dip coverslip in 100% acetone, let them stand vertically to dry, sterilise in an oven (6 h at 220°C) or in an autoclave. Alternatively, we use other coatings such as concanavalin A or poly-L-lysine. Place 100 µL drops of 0.25 mg/mL concanavalin A solution (in sterile water) in the middle of sterile coverslips. Incubate at 37°C for 90 min, preferentially in a humid chamber. Wash once with sterile water. For poly-L-lysine coating add 0.01% Poly-L-lysine (MW 70,000–150,000, Sigma Chemical Co.) to the sterile coverslips; after 30 min. wash three times with sterile water and let dry. Further synthetic coatings have been used for *Drosophila* neuron cultures, including polyethyleimine in borate buffer (46), or poly-DL-ornithine (23). Importantly, proteins of the extracellular matrix are essential regulators of neuronal differentiation in vertebrates and invertebrates alike (47, 48). However, to our knowledge, fly extracellular matrix components, such as laminin, have rarely been used to culture *Drosophila* neurons (49).

3. Methods

Cells can in principle be grown from all developmental stages of *Drosophila* (see Sect. 4). Here, we describe a detailed protocol for embryonic primary neuronal cultures grown in Schneider's medium.

3.1. Preparing Embryos

Age embryos as required. For the study of axonal growth in developing neurons, cells are extracted from embryos at stage 11 (50), when most neurons are post-mitotic and remain as isolated neurons after dispersion (Fig. 3). For neuroblast cultures, cells are extracted from embryos at stage 7/8, when neuroblasts of the first segregation waves are readily specified (51). Isolated neurons are rare in the latter cultures, since daughter cells of proliferating neuroblasts stay together in clusters (24).

Dechorionate embryos chemically by covering them for 90 s with commercial sodium/calcium hypochlorite (house bleach; diluted 1:1 in H_2O). This can occur straight on the agar plates which have been used for the egg lays. Pour the dechorionated embryos into a sieve and thoroughly wash them with distilled water. Collect embryos with a paint brush from the sieve and spread them carefully onto an agar plate or a block of agar. If required, green-balanced embryos can now be selected under a fluorescent dissection microscope. Using fine forceps, line up selected and staged embryos and orient them on the block of agar (as shown in Fig. 1b, c). Before aligning embryos, use the side of the pipette tip to spread a thin layer of viscous heptane glue (Sect. 2.5) onto a long 24 × 60 mm coverslip and allow to dry. Gently press the coverslip with the glue-bearing side down onto the row of embryos, so that they stick to the dried glue when lifting the coverslip vertically up. Cover embryos with 10S Voltalef oil to prevent their dehydration (Fig. 1a).

About seven embryos will provide sufficient cells for one culture slide, and for a well-trained person, up to 15–20 culture slides can be produced in one experiment. Whenever applying an experimental or genetic manipulation, adequate controls need to be processed in parallel, since the quality of primary cultures can slightly vary from experiment to experiment. An experienced experimenter can process up to four different genotypes in parallel. To perform parallel processing, dechorionate the different embryo batches and carry out the GFP-selection one after the other. Then orient, transfer to coverslips with heptane glue and cover with oil immediately. Try to be as fast as possible to avoid ageing of the embryos.

3.2. Harvesting Cells

From now on, always work with gloves. Prepare the containers for processing the harvested cells. Filter media sterile under a hood, insert 100 μL each into as many 1.5-mL centrifuge tubes as will be

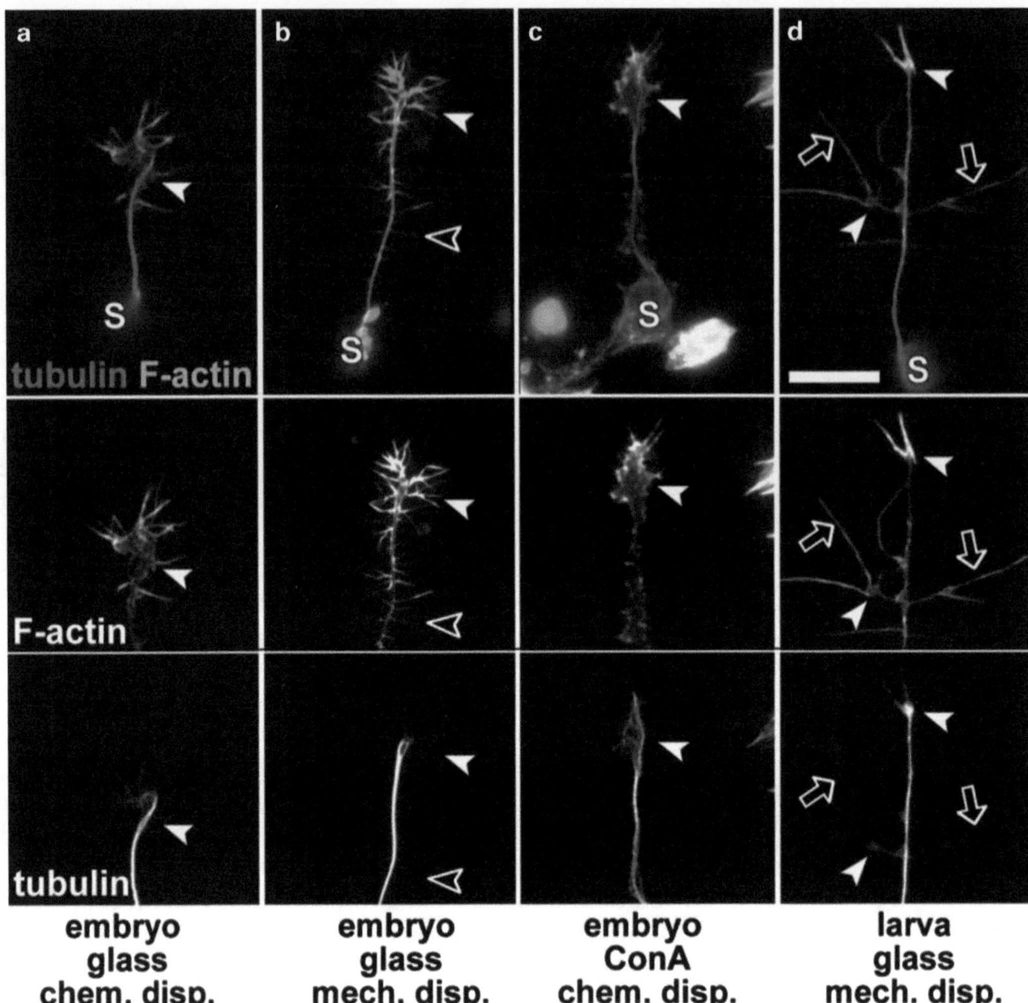

Fig. 3. Examples of embryonic and larval neurons with different treatments. Isolated primary neurons after 6 h in culture, either derived from stage 11 embryos (**a–c**) or from late larval CNS (**d**) stained against tubulin (*bottom row*) and F-actin (*middle row*); "S" indicates somata, *white arrow heads* the growth cones. (**a**) Chemically dispersed neurons display filopodia primary at the growth cone. (**b**) Mechanically dispersed neurons have abundant filopodia also in proximal axon regions (*open arrow head*). (**c**) Neurons grown on concanavalin A usually display fewer filopodia but more prominent lamellipodia, and microtubules at the growth cone have a larger tendency to unbundle. (**d**) Larval neurons tend to branch (*second arrow head*) and display long and straight filopodia (*open arrows*). Scale bar in (**d**) represents 10 μm for all images.

required for your experiment. Fix centrifuge tubes on their side (Fig. 1a). Set the microscope light source to low levels during the whole procedure of cell harvest and processing. Avoid sunny spots in the laboratory for this work. Place the slide with the aligned embryos under the microscope. Since only a ×10 lens is required for the cell harvesting procedure, upright or inverted microscopes can both be used and further specifications, such as Nomarski or phase-contrast optics, are NOT required. Attach the end of the bevelled capillary to a thin plastic hose, which is attached at the other end to a disposable syringe (choose out of a range of 1–10 mL, as works best for you) using a yellow pipette tip or syringe injecting needle as connector. Attach the capillary to the needle holder of a micromanipulator (ideally a micromanipulator that allows fine and coarse movements; we use the micromanipulator from Leica). Orient the needle so that the pen mark (set during grinding; Sect. 2.4) points towards you. This ensures that the needle opening will point in the required direction (Fig. 1b). Focus on the first embryo and, using the micromanipulator, bring the tip of the capillary into focus.

Before harvesting cells, apply pressure to the syringe to force out any oil that has entered the capillary. But leave a bit of oil at the capillary's tip to prevent drying of the cells that are to be harvested (Fig. 1d). Insert the tip of the capillary in a horizontal angle in the area of the cephalic furrow of the embryo (Fig. 1c, d). Apply slight suction force via the syringe to slowly drag cells from the ventral surface of the embryo into the capillary, which is facilitated by steadily moving the capillary tip forward (Fig. 1d). Stop when the darker, granular yolk particles start entering the capillary. Retract the capillary and move it through the Voltalef oil to the next embryo. By setting adequate pressure, take care that no oil is intercalated between the cells harvested from different embryos. Repeat the same procedure for the next embryos. When cells are extracted from the last embryo, seal the capillary tip with a bit of oil. Remove the capillary from the needle holder and carefully insert its tip into one of the prepared centrifuge tubes containing culture medium (Fig. 1a). Avoid touching the tube's wall not to break the capillary tip, and do not force air into the media, since it will kill cells. At this point, cells can be further processed, or stored at room temperature (well protected from light) while cells from other genotype are being harvested. All work from now on should be performed in a sterile hood.

3.3. Processing Harvested Cells for Dispersion Cultures

When cells have been harvested from the embryos of all genotypes, centrifuge the tubes containing the cells for 4 min at 0.1 rcf. Immediately remove supernatant without disturbing the pellet which is visible as a very tiny white dot. Either add dispersion medium to start the culturing procedure, or add fresh Schneider's culture medium to pre-culture cells in the centrifuge tube (in the dark at 26°C for up to 7 days). This pre-culturing procedure is a powerful strategy to deplete cells of maternally contributed gene product (28). To start the culturing procedure (directly or after pre-culturing), add 100 μL of sterile dispersion medium (at room temperature) and re-suspend cells by gently pipetting 15 times up and down. Avoid producing air bubbles, which tend to kill the cells. Do this for each genotype and incubate them together for 4 min at 37°C. Add 200 μL of Schneider's medium to slow down the reaction and centrifuge for 4 min at 0.1 rcf. Note that the pellet is less visible after dispersion and tends to extend along the side of the centrifuge tubes. Remove supernatant and replace with fresh culture medium (5–10 μL of medium per donor embryo or 30–40 μL per slide). Re-suspend cells by gently pipetting up and down.

As an alternative to chemical treatment, cells can be mechanically dispersed by gently pipetting cells up and down with a 100 μL pipette for 15–20 times. However, mechanically dispersed neurons display many more filopodia, most likely due to the fact that non-inactivated growth factors and other extracellular matrix materials remain attached to their surfaces. Note that images obtained from chemically dispersed neurons more closely resemble growth cone morphologies observed in vivo (Fig. 3).

Cells are now dissociated in their appropriate end volume of culture medium. By experience, using 30–40 μL of this cell suspension per culture chamber results in cell densities in the range of 500–2,500 cells/mm^2. Note that the correct cell density is important. If cell densities are too low, this negatively impacts on cell morphology (such as reduced filopodia number and neurite length), whereas too high densities cause increased cell cluster formation, thus reducing the number of isolated neurons. Before transferring the cell suspension, prepare the sterile cell chambers by distributing a thin layer of vaseline around the well, which will serve as an air-tight sealant when closing the chamber. Using a pipette, place the 30–40 μL aliquots centrally at the bottom of their wells (Sect. 2.4) and close the chambers using sterile lead-free coverslips (Sect. 2.4) (Fig. 2a). Flip chamber upside-down, so that the coverslip faces down. Use a fast and steady movement to avoid misplacing the drop of cell suspension to the chamber edge. Instead, it should form a liquid pillar spanning centrally from the bottom of the culture chamber to the attached cover slip (Fig. 2b). Place the chambers in the dark for 1–2 h at the desired temperature (we got satisfactory results for 12, 18 and 26°C) to allow cells to sink down and adhere to the coverslip. After this period, the chambers must be flipped back again (coverslip facing up), so that debris can fall off and does not accumulate between healthy adherent cells (hanging drop culture).

Cell harvest, processing and their initial incubation occurs in standard Schneider's medium, which supports early neuronal development very well. Therefore, studies of developmental processes such as neuronal lineage formation or axonal growth are best carried out in this medium. For studies of mature neurons, cells are usually cultured for 2–3 days (synapse become abundant after ca. 20 h at 26°C) (24) and standard Schneider's medium is replaced after 1–2 days with Schneider's^{active} medium (Sect. 2.1b). In contrast to standard Schneider's medium, Schneider's^{active} promotes the proper physiological maturation of *Drosophila* neurons (24). In any case, if cells are to be kept for extended periods (up to 3 weeks have been reported) (44), media should be replaced about once a week. Changes of the medium are also required, if drugs are to be applied for a restricted period.

To exchange the culture medium, mark the region of the culture droplet on the coverslip using a thin waterproof pen. Rotate coverslip by ~45° and remove from slide. Carefully remove coverslips from the culture chambers. Use a laboratory pipette to remove the old medium from the chamber. The cells adhere to the coverslip and are covered by a thin film of old medium for a short period, enough to place a 40-μL drop of fresh medium in the middle of the chamber. Then, the coverslips are placed back on the culture chambers; take care that the cells (encircled) are in contact with the fresh medium.

3.4. Processing Harvested Cells for Neuroblast Cultures

To process cells obtained from stage 7/8 embryos for neuroblast cultures, simply dissociate cells in the centrifuge tube by pipetting carefully up and down. Place 30–40 μL drops of cell suspension into each culture chamber and continue as explained in Chap. 3.3.

3.5. Fixing and Staining Cultures

To analyse neurons at the growth cone stage, incubate for 6–8 h at 26°C or for 15–20 h at 17°C (27). Standard fixative should be pre-warmed to 37°C, special fixative for microtubule-binding proteins should be kept on ice before use (Sect. 2.3). Mark the region of the culture droplet on the coverslip using a thin waterproof pen. Rotate coverslip by ~45°, remove from culture chamber and place (cells facing up) on a piece of plasticine or a rubber block in a humid chamber containing a wet paper towel. Add 100–200 μL of PF to the cells (always add solution slowly from outside the encircled area containing the cells to not destroy them). Close the lid of the humid chamber and incubate for 30 min with standard fixative and 10 min with special fixative. Remove fixative and wash three times with PBT. For immunocytochemistry, standard procedures can be used (52); usually no blocking agent is required.

3.6. Live Imaging of Primary Neurons

Live imaging experiments can be performed directly on the culture chambers. Since the chambers can be flipped on either side, upright as well as inverted microscopes can be used with highest magnification lenses. However, if using an inverted microscope, change the medium in the culture chamber once (as described in Sect. 3.2.5) or use a fresh culture chamber, to reduce the amount of debris sedimenting on the coverslip and distorting the image. Importantly, the temperature should be constant between experiments (we normally image at 25–26°C). If the imaging temperature differs from the culturing temperature, pre-incubate cells at the imaging temperature for 1 h before starting the experiment.

4. Notes

4.1. Choosing the Developmental Stage of Donor Tissue

As typical of holometabolous insects, *Drosophila* development comprises embryonic, larval and pupal stages, for all of which culture techniques were established (10, 20, 53, 54). It is important to point out that there are significant qualitative differences between neurons in these cultures, rooted in the distinct contexts they are derived from. In the embryo, neural progenitors (neuroblasts) delaminate from the neuroepithelium into the haemolymph where they divide during the first half of embryogenesis (55); their daughter cells mostly differentiate during the second half of embryogenesis, when they wire up into functional circuits. In parallel, neuroblasts and neurons become increasingly shielded from the haemolymph by glia, leading to physiologically favourable ionic changes in their extracellular milieu (56). The later the cells are harvested from embryos, the larger the proportion of post-mitotic neurons over neuroblasts, which has important implications for the dissociation status of cells in the obtained cell cultures (Chapter 3.1). In the larval CNS, many neuroblasts re-assume proliferation (55) and give rise to neurons. In contrast to embryonic stages where single neuroblasts give rise to several tens of cells (24, 57), larval neuroblasts give rise to hundreds of daughter cells each, so that at the end of larval life about ten times more larva-derived than embryonic-born neurons are present (58, 59). Although the larval-born neurons experience the hormonal waves associated with larval moult (60), they do not fully differentiate before metamorphosis (61). Therefore, cells harvested from larval CNSs, are expected to comprise neuroblasts, mature embryo-derived neurons and undifferentiated larval neurons, and the presence of neurons as well as neuroblasts has been reported for these cultures (62). Note that neurons derived from larval CNS are morphologically different from embryo-derived neurons (Fig. 3d). Comparing results for neurons obtained from embryonic vs. larval stages is therefore not straightforward and might require a range of stage specific control experiments to validate conclusions. This is likely to be the case also for cultures obtained from pupal brains. At the pupal stage, neuroblasts continue proliferating for a further period (55). Many neurons degenerate, others undergo massive reorganisation (63), and persisting neurons of embryonic, larval and pupal origins wire up into the circuits of the adult brain. Furthermore, pupal development is accompanied by two strong hormonal peaks (20, 60), which need to be considered for the timing of cell harvest and the use of hormonal supplement in culture media.

Apart from these qualitative differences, different developmental stages provide different access to *Drosophila* genetics. Thus, primary neurons have been used since early days to analyse embryonic lethal mutations (64, 65), whereas genetic analyses in larval and pupal cultures were originally restricted to viable mutant alleles (66). Through the advent of MARCM technology (67), lethal loss-of-function mutations can be analysed also in postembryonic cultures—although usually restricted to simple genetic constellations. In contrast, embryonic cultures are amenable to combinatorial genetics: any genetic combination achievable in *Drosophila* embryos can also be analysed in their primary cultures. On the other hand, mutant analyses in embryonic cultures might be hampered by the presence of maternal gene product, especially when neurons need to be analysed within the first few hours after plating. However, the problem of maternal product can be overcome, either by using donor embryos obtained from germ line clones (the standard technique to suppress maternal product deposition; (68)) or through extended pre-culture strategies (Chap. 3.3).

Finally, the ability to identify specific neuron classes in primary cultures differs between stages. Given the size of *Drosophila* brains, it is impossible to dissect out particular brain areas, as is common procedure for mammalian primary cultures. Furthermore, apart from mushroom bodies and the peripheral optic lobe, no other brain areas can be considered repetitive enough to produce the degree of homogeneity that can be achieved in mammalian cultures. Instead, neuron classes of *Drosophila* can only be identified after cells have been plated, and this is possible through immunohistochemistry and the Gal4 system of targeted gene expression (69). Cell identification in the ready-made cultures is being used routinely for larval and pupal mushroom body neurons (14), and has been used successfully to identify motor neurons, GABAergic interneurons, serotonergic neurons or cell-lineage specific markers in mature embryonic cell cultures (16, 24, 70, 71), to name a few examples. However, these strategies are clearly limited to the stage-specific spatial and temporal patterns of antigens and available Gal4-driver lines. For example, for analyses of developing embryonic neurons at the stage of axonal growth, we have so far not been able to find any markers that would allow us to classify cells by their future transmitter-phenotypes, i.e. to distinguish between motor neurons, sensory neurons and different interneuron classes.

In conclusion, the choice of the developmental stage of the donor animals is not trivial and should be carefully considered in light of the experimental objectives. Procedures for generating cultures from larval and pupal brain have been detailed elsewhere (18, 72). Here, we focus on the newer developments for embryonic primary neuron cultures.

4.2. Choosing the Culture Medium

Different culture media have been used for culturing embryo-derived primary neurons of *Drosophila*, as discussed in greater detail elsewhere (13, 14, 24). In a nutshell, defined, serum-free, bicarbonate-buffered DDM1 medium (73) is ideal for electrophysiological recordings (14), but it does not recapitulate early neural development very well. For example, it does not sufficiently promote neuroblast divisions, and cell bodies of differentiated neurons are larger than in vivo. Furthermore, these cultures need to be kept under CO_2 conditions. In contrast, commercially available serum-supplemented Schneider's medium (38) supports neural development in culture leading to in vivo-like lineage and cell properties (16, 17, 74). Conventional Schneider's medium (SM^{20K}) was designed to resemble aspects of the haemolymph (38), and early stages of neural development take place in the haemolymph. At later stages, when the glial sheath forms in the embryo, the extracellular milieu of neurons changes. Accordingly, conventional Schneider's medium fails to support the late physiological differentiation in culture. This inhibition can be overcome by changing the culture medium after 1 or 2 days to active Schneider's medium in which the inorganic ion concentrations are adjusted accordingly (SM^{active}; Chap. 2.1.2). SM^{active} leads to a dramatic upregulation of neurotransmitter production and supports the generation of action potentials and maturation of fully functional synapses. The combined use of conventional and active Schneider's media therefore provides a strategy to achieve authentic early development and late differentiation in the same culture system (24). The adjustability of inorganic ion concentrations in these Schneider's media can also be used for experimental purposes, as shown for the study of mechanisms regulation the developmental upregulation of GABA in *Drosophila* (21).

4.3. Designing the Genotype of Donor Stocks

For mutant analyses in culture, it is important to be able to identify cells with the right genotype, and fly stocks need to be prepared accordingly. We have successfully used three methods. First, for the analysis of lethal mutations, balancers can be used that carry a combination of *twist-Gal4* with *UAS-GFP* (available from Bloomington for first, second and third chromosome). From about embryonic stage 10 onwards, the green fluorescence of these balancers can be spotted under a fluorescent dissection microscope. This allows pre-selection of mutant embryos as cell donors so that all neurons in the respective cultures will be mutant. However, this method does not work for neuroblast cultures obtained from early embryos, and so far we have not been able to find any fluorescent balancers that can be detected as early as stage 7/8. Alternatively, mutant alleles can be recombined with dominant markers. For example, in one special case we were able to reliably identify neurons carrying a heteroallelic mutant combination on the first chromosome, by having one of the alleles recombined with a ubiquitin-GFP construct carrying a ubiquitin promoter. A more common strategy is to recombine one mutant allele with a suitable pan-neuronal Gal4 construct (*scabrous-Gal4*), and another mutant allele with a *UAS-GFP* construct, so that only mutant neurons show fluorescence (75). MARCM strategies (67) are ideally suited for postembryonic cultures, but the time course of Gal4-deinhibition is too long to be used in embryonic cultures.

Furthermore, targeted gene expression is a powerful strategy to manipulate neurons genetically, carry out structure function analyses, or label defined cellular populations or specific subcellular structures. Targeted gene expression can be used in primary neurons for the same purposes. Essential is the choice of driver lines to achieve reliable expression strength in most or all cultured neurons. Out of the various pan-neuronal driver lines, including *inscuteable/MZ1407-Gal4*, *elav-Gal4* or scabrous-Gal4 (76, 77), *scabrous-Gal4* is the most reliable driver line for analyses at the developmental stage of axonal growth (27), whereas at fully differentiated stages the other two diver lines give excellent results (24). As in vivo, in culture the temperature-sensitivity of the Gal4/UAS system can be used to alter the strength of targeted gene expression (27).

As a final note it should be pointed out that specific labelling of cells allows mixing of different genotypes in the same culture, which is a powerful strategy to carry out mosaic analyses (24). Furthermore, having different genotypes mixed on the same slide provides very precise comparisons between control and genetically modified cells.

4.4. Pharmacological Approaches

Pharmacological treatments are easily executed in the primary neuron cultures described here (21, 27), as is common practice in vertebrate culture systems. Many drugs need to be dissolved in DMSO, and controls should always contain the same concentration of DMSO. This is of particular importance since we have experienced adverse effects of DMSO on *Drosophila* primary neurons. To avoid undesired effects, keep DMSO in the dark under similar storing conditions as the drugs dissolved in DMSO.

4.5. Cell Density

Be aware that cell density has an impact on *Drosophila* primary neurons. Too high or too low density may impact on morphological readouts, such as filopodial number. Therefore, cell densities should be judged during analysis and compared between experimental days as well as between different data sets produced within one experiment.

4.6. Cell Toxicity

Several steps of the culturing protocol can be harmful to the cells. First, reduce the time of chemical dechorionation (Sect. 3.1) to a minimum and wash embryos well afterwards. In case, cultures are of poor quality, test whether the bleach affects embryonic viability. As an alternative, mechanical dechorionation can be used: embryos are collected on a small square of double-sticky tape, briefly sterilised with 70% ethanol, then dechorionated mechanically with a metal needle, as described elsewhere with greater detail (43). Second, heptane glue can affect the viability. When a new batch of heptane glue is to be used, test for any toxicity by assessing embryonic viability. To reduce adverse effects, prepare the coverslip with the heptane glue (Fig. 1a) about 30 min before use.

4.7. Polarity and Lack of Dendrites

In contrast to the mono- or unipolar nature of *Drosophila* neurons in vivo, they display a bipolar organisation in culture, and the reasons for this phenomenon have been discussed elsewhere (22, 23). Another deficit is the lack of properly grown dendrites, likely due to the absence of growth factors that induce dendrites in the dorsal neuropile in vivo (22). This deficit may therefore be overcome, once these growth factors have been identified.

4.8. Further Analysis Strategies

Conventional transmission electron microscopy has been successfully performed on *Drosophila* primary neurons (78–80), including cultures generated with the methods explained here (24). We are not aware of any attempts to carry out more sophisticated ultrastructural studies, such as platinum replica EM, cryo EM, negative staining techniques or immuo-EM, but there is no reason to believe that such techniques should not be applicable to *Drosophila* primary neurons. Electrophysiology is extensively used in *Drosophila* primary neurons, and has also been applied with great success in the type of cultures described here (24). Complementary to electrophysiological approaches, FM-dye uptake studies have been performed to assess neuronal or synaptic activity (24). Calcium imaging has not yet been performed in these cultures, but in other types of *Drosophila* primary neuron cultures (81, 82). In those experiments, addition of cytochalasin B to the culture medium was used, which inhibits cell division and therefore generates larger multinucleate neurons (83). Enlarged growth cones provide a higher subcellular resolution for calcium imaging experiments and have also been used for other experiments, such as measurements of acetyl choline release from *Drosophila* growth cones (81, 84).

4.9. Study of Axonal Growth and the Neuronal Cytoskeleton

Axon growth is executed by highly motile structures at their tips, called growth cones. Growth cones are characterised by an actin-rich peripheral domain (actin networks in veil-like lamellipodia, actin bundles in filopodia and spikes, and condensed networks in actin arcs) and a microtubule-containing central domain from where single splayed microtubules dynamically emanate into the periphery (3, 85). All these features are displayed by growth cones of *Drosophila* primary neurons (Fig. 4a) (27, 31). They provide a number of reliable readouts that can be used for the study of gain- or loss-of-function of genes involved in axonal growth regulation; the readouts are well conserved with those of mammalian or other vertebrate neurons (27). Readouts that have been used and validated in this system are axon length, filopodial parameters (number, length, rate of extension, retraction, bifurcation, backflow), parameters for microtubules (unbundling/disorganisation, frequency of filopodial entry, polymerisation rate), and classifications of growth

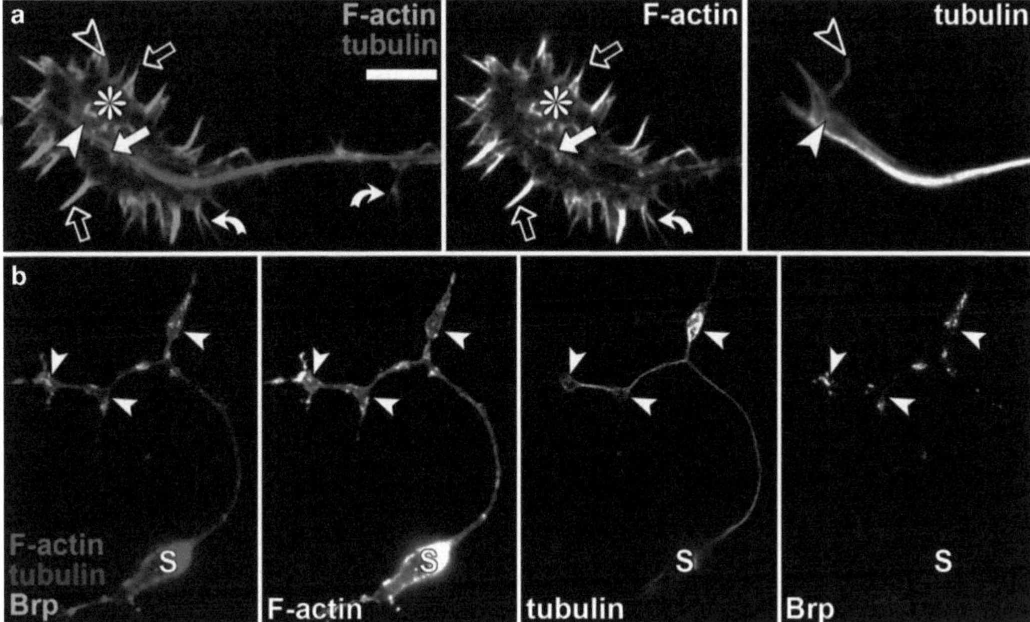

Fig. 4. Subcellular features of cultured neurons. Isolated primary neuron derived from stage 11 embryos and cultured on glass. (**a**) Growth cone after 6 h in culture, stained against F-actin (*middle panel*) and tubulin (*right panel*); various features are indicated: *open arrows*, filopodia; *curved arrows*, bifurcating filopodia; *asterisk*, peripheral domain/lamellipodium; *white arrow head*, microtubules of central domain; *white arrow*, actin arch; *open arrow head*, splayed microtubule in periphery. (**b**) Neuron after 3 days in culture, stained against F-actin (*red*), tubulin (*blue*) and the presynaptic marker Bruchpilot (Brp, *green*); *arrow heads* indicate areas of looped microtubules which seem predestined to harbour presynaptic structures, as indicated by accumulations of Brp and F-actin. Scale bar (in (**a**)) represents 2.5 μm in (**a**) and 10 μm in (**b**).

cone types (26–29). Many of these analyses are accessible through standard fluorescent microscopy in fixed preparations. Live analyses of filopodial and lamellipodial dynamics can be carried out with phase contrast optics, whereas live imaging of fluorescent probes expressed by *Drosophila* primary neurons requires cameras with high sensitivity or spinning disc microscopy to prevent rapid bleaching of the fluorescence. For the analysis of many of the above readouts, ImageJ plug-ins have recently been developed which can be used to speed up data generation (86, 87).

Acknowledgments

We would like to thanks Karin Lüer, Gerd Technau and Robin Beaven for comments on the manuscript. We are grateful to Niki Scaplehorn for sharing her experiences with the generation of Schneider's medium. N.S.S. and A.P. were supported by the Wellcome Trust (092403/Z/10/Z) and the BBSRC (BB/I002448/1).

References

1. Arimura N, Kaibuchi K (2007) Neuronal polarity: from extracellular signals to intracellular mechanisms. Nat Rev Neurosci 8:194–205
2. Craig AM, Graf ER, Linhoff MW (2006) How to build a central synapse: clues from cell culture. Trends Neurosci 29:8–20
3. Lowery LA, van Vactor D (2009) The trip of the tip: understanding the growth cone machinery. Nat Rev Mol Cell Biol 10:332–343
4. Olmsted JB, Carlson K, Klebe R, Ruddle F, Rosenbaum J (1970) Isolation of microtubule protein from cultured mouse neuroblastoma cells. Proc Natl Acad Sci U S A 65:129–136
5. Rasko I, Georgieva M, Farkas G, Santha M, Coates J, Burg K, Mitchell DL, Johnson RT (1993) New patterns of bulk DNA repair in ultraviolet irradiated mouse embryo carcinoma cells following differentiation. Somat Cell Mol Genet 19:245–255
6. Liu T, Sims D, Baum B (2009) Parallel RNAi screens across different cell lines identify generic and cell type-specific regulators of actin organization and cell morphology. Genome Biol 10:R26
7. Ui K, Nishihara S, Sakuma M, Togashi S, Ueda R, Miyata Y, Miyake T (1994) Newly established cell lines from *Drosophila* larval CNS express neural specific characteristics. In Vitro Cell Dev Biol Anim 30A:209–216
8. Song HJ, Stevens CF, Gage FH (2002) Neural stem cells from adult hippocampus develop essential properties of functional CNS neurons. Nat Neurosci 5:438–445
9. Schwartz PH, Bryant PJ, Fuja TJ, Su H, O'Dowd DK, Klassen H (2003) Isolation and characterization of neural progenitor cells from post-mortem human cortex. J Neurosci Res 74:838–851
10. Seecof R, Alleaume N, Teplitz R, Gerson I (1971) Differentiation of neurons and myocytes in cell cultures made from *Drosophila* gastrulae. Exp Cell Res 69:161–173
11. Banker G, Goslin K (1988) Developments in neuronal cell culture. Nature 336:185–186
12. Banker G, Goslin K (1998) Culturing nerve cells, 2nd edn. MIT Press, Cambridge, MA
13. Beadle DJ (2006) Insect neuronal cultures: an experimental vehicle for studies of physiology, pharmacology and cell interactions. Invert Neurosci 6:95–103
14. Rohrbough J, O'Dowd DK, Baines RA, Broadie K (2003) Cellular bases of behavioral plasticity: establishing and modifying synaptic circuits in the *Drosophila* genetic system. J Neurobiol 54:254–271
15. Lüer K, Technau GM (1992) Primary culture of single ectodermal precursors of *Drosophila* reveals a dorsoventral prepattern of intrinsic neurogenic and epidermogenic capabilities at the early gastrula stage. Development 116:377–385
16. Lüer K, Technau GM (2009) Single cell cultures of *Drosophila* neuroectodermal and mesectodermal central nervous system progenitors reveal different degrees of developmental autonomy. Neural Dev 4:30
17. Brody T, Odenwald WF (2000) Programmed transformations in neuroblast gene expression during *Drosophila* CNS lineage development. Dev Biol 226:34–44
18. Ceron J, Tejedor FJ, Moya F (2006) A primary cell culture of *Drosophila* postembryonic larval neuroblasts to study cell cycle and asymmetric division. Eur J Cell Biol 85:567–575
19. Kim YT, Wu CF (1996) Reduced growth cone motility in cultured neurons from *Drosophila* memory mutants with a defective cAMP cascade. J Neurosci 16:5593–5602
20. Kraft R, Levine RB, Restifo LL (1998) The steroid hormone 20-hydroxyecdysone enhances neurite growth of *Drosophila* mushroom body neurons isolated during metamorphosis. J Neurosci 18:8886–8899
21. Küppers B, Sánchez-Soriano N, Letzkus J, Technau GM, Prokop A (2003) In developing *Drosophila* neurones the production of gamma-amino butyric acid is tightly regulated downstream of glutamate decarboxylase translation and can be influenced by calcium. J Neurochem 84:939–951
22. Sánchez-Soriano N, Löhr R, Bottenberg W, Haessler U, Kerassoviti A, Knust E, Fiala A, Prokop A (2005) Are dendrites in *Drosophila* homologous to vertebrate dendrites? Dev Biol 288:126–138
23. Katsuki T, Ailani D, Hiramoto M, Hiromi Y (2009) Intra-axonal patterning: intrinsic compartmentalization of the axonal membrane in *Drosophila* neurons. Neuron 64:188–199
24. Küppers-Munther B, Letzkus J, Lüer K, Technau G, Schmidt H, Prokop A (2004) A new culturing strategy optimises *Drosophila* primary cell cultures for structural and functional analyses. Dev Biol 269:459–478
25. Bai J, Sepp KJ, Perrimon N (2009) Culture of *Drosophila* primary cells dissociated from gastrula embryos and their use in RNAi screening. Nat Protoc 4:1502–1512
26. Sánchez-Soriano N, Travis M, Dajas-Bailador F, Goncalves-Pimentel C, Whitmarsh AJ,

Prokop A (2009) Mouse ACF7 and *Drosophila* short stop modulate filopodia formation and microtubule organisation during neuronal growth. J Cell Sci 122:2534–2542

27. Sánchez-Soriano N, Gonçalves-Pimentel C, Beaven R, Haessler U, Ofner L, Ballestrem C, Prokop A (2010) *Drosophila* growth cones: a genetically tractable platform for the analysis of axonal growth dynamics. Dev Neurobiol 70:58–71
28. Matusek T, Gombos R, Szecsenyi A, Sánchez-Soriano N, Czibula A, Pataki C, Gedai A, Prokop A, Rasko I, Mihaly J (2008) Formin proteins of the DAAM subfamily play a role during axon growth. J Neurosci 28: 13310–13319
29. Gonçalves-Pimentel C, Gombos R, Mihály J, Sánchez-Soriano N, Prokop A (2011) Dissecting regulatory networks of filopodia formation in a *Drosophila* growth cone model. PLoS One 6:e18340
30. Prokop A, Sánchez-Soriano N, Gonçalves-Pimentel C, Molnár I, Kalmár T, Mihály J (2011) DAAM family members leading a novel path into formin research. Commun Integr Biol 4:538–542
31. Sánchez-Soriano N, Tear G, Whitington P, Prokop A (2007) *Drosophila* as a genetic and cellular model for studies on axonal growth. Neural Dev 2:9
32. Conde C, Caceres A (2009) Microtubule assembly, organization and dynamics in axons and dendrites. Nat Rev Neurosci 10:319–332
33. Pak CW, Flynn KC, Bamburg JR (2008) Actin-binding proteins take the reins in growth cones. Nat Rev Neurosci 9:136–147
34. Insall RH, Machesky LM (2009) Actin dynamics at the leading edge: from simple machinery to complex networks. Dev Cell 17:310–322
35. Benitez-King G, Ramirez-Rodriguez G, Ortiz L, Meza I (2004) The neuronal cytoskeleton as a potential therapeutical target in neurodegenerative diseases and schizophrenia. Curr Drug Targets CNS Neurol Disord 3:515–533
36. Hirth F (2010) *Drosophila melanogaster* in the study of human neurodegeneration. CNS Neurol Disord Drug Targets 9:504–523
37. Papanikolopoulou K, Skoulakis EM (2011) The power and richness of modelling tauopathies in *Drosophila*. Mol Neurobiol 44: 122–133
38. Schneider I (1964) Differentiation of larval *Drosophila* eye-antennal discs in vitro. J Exp Zool 156:91–104
39. Stewart BA, Atwood HL, Renger JJ, Wang J, Wu CF (1994) Improved stability of *Drosophila* larval neuromuscular preparations in haemolymph-like physiological solutions. J Comp Physiol A 175:179–191
40. Küppers-Munther B (2004) Optimierung und Verwendung embryonaler Primärkulturen von *Drosophila* zur Untersuchung der Bildung. Struktur und Funktion von Synapsen, Johannes Gutenberg-University, Mainz
41. Glauert AM (1991) Fixation, dehydration and embedding of biological specimens, vol 3, 8th edn. North Holland Publishing Group, Amsterdam
42. Rogers SL, Rogers GC, Sharp DJ, Vale RD (2002) *Drosophila* EB1 is important for proper assembly, dynamics, and positioning of the mitotic spindle. J Cell Biol 158:873–884
43. Prokop A, Technau GM (1993) Cell transplantation. In: Hartley D (ed) Cellular interactions in development: a practical approach. Oxford University Press, London, pp 33–57
44. Shields G, Dübendorfer A, Sang JH (1975) Differentiation in vitro of larval cell types from early embryonic cells of *Drosophila melanogaster*. J Embryol Exp Morphol 33:159–175
45. Dübendorfer A, Eichenberger-Glinz S (1980) Development and metamorphosis of larval and adult tissues of *Drosophila* in vitro. In: Kurstak E, Maramorosch K, Dübendorfer A (eds) Invertebrate systems in vitro. Elsevier, Amsterdam, pp 169–185
46. Matsuura R, Tanaka H, Go MJ (2004) Distinct functions of Rac1 and Cdc42 during axon guidance and growth cone morphogenesis in *Drosophila*. Eur J Neurosci 19:21–31
47. Reichardt L, Prokop A (2011) Introduction: the role of extracellular matrix in nervous system development and maintenance. Dev Neurobiol 71:883–888
48. Broadie K, Baumgartner S, Prokop A (2011) Extracellular matrix and its receptors in *Drosophila* neural development. Dev Neurobiol 71:1102–1130
49. Takagi Y, Nomizu M, Gullberg D, MacKrell AJ, Keene DR, Yamada Y, Fessler JH (1996) Conserved neuron promoting activity in *Drosophila* and vertebrate laminin alpha1. J Biol Chem 271:18074–18081
50. Campos-Ortega JA, Hartenstein V (1997) The embryonic development of *Drosophila melanogaster*. Springer, Berlin
51. Doe CQ (1992) Molecular markers for identified neuroblasts and ganglion mother cells in the *Drosophila* central nervous system. Development 116:855–863
52. Budnik V, Gorczyca M, Prokop A (2006) Selected methods for the anatomical study of *Drosophila* embryonic and larval neuromuscular junctions. In: Budnik V, Ruiz-Cañada C

(eds) The fly neuromuscular junction: structure and function—international review of neurobiology. Elsevier Academic Press, San Diego, pp 323–374

53. Wu C-F, Suzuki N, Poo M (1983) Dissociated neurons from normal and mutant *Drosophila* larval central nervous system in cell culture. J Neurosci 3:1888–1899
54. Su H, O'Dowd DK (2003) Fast synaptic currents in *Drosophila* mushroom body Kenyon cells are mediated by alpha-bungarotoxin-sensitive nicotinic acetylcholine receptors and picrotoxin-sensitive GABA receptors. J Neurosci 23:9246–9253
55. Prokop A, Technau GM (1994) Normal function of the mushroom body defect gene of *Drosophila* is required for the regulation of the number and proliferation of neuroblasts. Dev Biol 161:321–337
56. Prokop A (1999) Integrating bits and pieces—synapse formation in *Drosophila* embryos. Cell Tissue Res 297:169–186
57. Schmidt H, Rickert C, Bossing T, Vef O, Urban J, Technau GM (1997) The embryonic central nervous system lineages of *Drosophila melanogaster*. II. Neuroblast lineages derived from the dorsal part of the neuroectoderm. Dev Biol 198:186–204
58. Truman JW, Bate CM (1988) Spatial and temporal patterns of neurogenesis in the CNS of *Drosophila melanogaster*. Dev Biol 125:145–157
59. Prokop A, Technau GM (1991) The origin of postembryonic neuroblasts in the ventral nerve cord of *Drosophila melanogaster*. Development 111:79–88
60. Truman JW, Talbot WS, Fahrbach SE, Hogness DS (1994) Ecdysone receptor expression in the CNS correlates with stage-specific responses to ecdysteroids during *Drosophila* and *Manduca* development. Development 120:219–234
61. Truman JW, Schuppe H, Shepherd D, Williams DW (2004) Developmental architecture of adult-specific lineages in the ventral CNS of *Drosophila*. Development 131:5167–5184
62. Feiguin F, Llamazares S, Gonzalez C (1998) Methods in *Drosophila* cell cycle biology. Curr Top Dev Biol 36:279–291
63. Technau G, Heisenberg M (1982) Neural reorganization during metamorphosis of the corpora pedunculata in *Drosophila melanogaster*. Nature 295:405–407
64. Donady JJ, Seecof RL (1972) Effect of the gene lethal (1) myospheroid on *Drosophila* embryonic cells in vitro. In Vitro 8:7–12
65. Kuroda Y (1974) In vitro activity of cells from genetically lethal embryos of *Drosophila*. Nature 252:40–41
66. Wu C (1988) Neurogenetic studies of *Drosophila* central nervous system neurons in culture. Academic, New York
67. Lee T, Luo L (1999) Mosaic analysis with a repressible neurotechnique cell marker for studies of gene function in neuronal morphogenesis. Neuron 22:451–461
68. Theodosiou NA and Xu T (1998) Use of FLP/FRT system to study *Drosophila* development. Methods 14, 355–365
69. Duffy JB (2002) GAL4 system in *Drosophila*: a fly geneticist's Swiss army knife. Genesis 34:1–15
70. Küppers B, Sánchez-Soriano N, Prokop A (2001) Regulation of GABA in the developing CNS of *Drosophila* embryos. In: Paper presented at neurobiology of *Drosophila*, Cold Spring Harbor, New York
71. Brody T, Odenwald WF (2005) Regulation of temporal identities during *Drosophila* neuroblast lineage development. Curr Opin Cell Biol 17:672–675
72. Sicaeros B, Campusano JM, O'Dowd DK (2007) Primary neuronal cultures from the brains of late stage *Drosophila* pupae. J Vis Exp 4:200. doi:10.3791/200
73. O'Dowd DK (1995) Voltage-gated currents and firing properties of embryonic *Drosophila* neurons grown in a chemically defined medium. J Neurobiol 27:113–126
74. Lüer K, Schmidt H, Technau GM (1998) Development of neuronal and glial properties within lineages derived from CNS midline precursors in single cell culture. In: Elsner N, Wehner R (eds) New neuroethology on the move; Proceedings of the 26th Göttingen neurobiology conference, Stuttgart, New York, p 696
75. Löhr R, Godenschwege T, Buchner E, Prokop A (2002) Compartmentalisation of central neurons in *Drosophila*: a new strategy of mosaic analysis reveals localisation of pre synaptic sites to specific segments of neurites. J Neurosci 22:10357–10367
76. Luo L, Liao YJ, Jan LY, Jan YN (1994) Distinct morphogenetic functions of similar small GTPases: *Drosophila* Drac1 is involved in axonal outgrowth and myoblast fusion. Genes Dev 8:1787–1802
77. Mlodzik M, Baker NE, Rubin GM (1990) Isolation and expression of *scabrous*, a gene regulating neurogenesis in *Drosophila*. Genes Dev 4:1848–1861
78. Oh HW, Campusano JM, Hilgenberg LG, Sun X, Smith MA, O'Dowd DK (2008) Ultrastructural analysis of chemical synapses and gap junctions between *Drosophila* brain

neurons in culture. Dev Neurobiol 68:281–294

79. Masur SK, Kim YT, Wu CF (1990) Reversible inhibition of endocytosis in cultured neurons from the *Drosophila* temperature-sensitive mutant *shibire*^{ts1}. J Neurogenet 6:191–206
80. Seecof RL, Teplitz RL, Gerson I, Ikeda K, Donady JJ (1972) Differentiation of neuromuscular junctions in cultures of embryonic *Drosophila* cells. Proc Natl Acad Sci U S A 69:566–570
81. Berke BA, Lee J, Peng IF, Wu CF (2006) Sub-cellular Ca^{2+} dynamics affected by voltage- and Ca^{2+}-gated K^{+} channels: regulation of the soma-growth cone disparity and the quiescent state in *Drosophila* neurons. Neuroscience 142:629–644
82. Berke B, Wu CF (2002) Regional calcium regulation within cultured *Drosophila* neurons: effects of altered cAMP metabolism by the learning mutations *dunce* and *rutabaga*. J Neurosci 22:4437–4447
83. Wu CF, Sakai K, Saito M, Hotta Y (1990) Giant *Drosophila* neurons differentiated from cytokinesis-arrested embryonic neuroblasts. J Neurobiol 21:499–507
84. Yao W-D, Rusch J, Poo MM, Wu C-F (2000) Spontaneous acetylcholine secretion from developing growth cones of *Drosophila* central neurons in culture: effects of cAMP-pathway mutations. J Neurosci 20: 2626–2637
85. Dent EW, Gupton SL, Gertler FB (2011) The growth cone cytoskeleton in axon outgrowth and guidance. Cold Spring Harb Perspect Biol 3(3):a001800
86. Fanti Z, Martinez-Perez ME, De-Miguel FF (2010) NeuronGrowth, a software for automatic quantification of neurite and filopodial dynamics from time-lapse sequences of digital images. Dev Neurobiol 71:870–881
87. Stepanova T, Smal I, van Haren J, Akinci U, Liu Z, Miedema M, Limpens R, van Ham M, van der Reijden M, Poot R et al (2010) History-dependent catastrophes regulate axonal microtubule behavior. Curr Biol 20: 1023–1028

Chapter 11

Molecular Profiling of Neural Stem Cells in *Drosophila melanogaster*

Elizabeth E. Caygill, Katrina S. Gold, and Andrea H. Brand

Abstract

The developing *Drosophila melanogaster* central nervous system is populated by asymmetrically dividing neural stem cells called neuroblasts, derived from ectodermal or neuroepithelial precursors. Neuroblasts divide asymmetrically, self-renewing the neuroblast and producing a smaller ganglion mother cell (GMC). Subsequent division of the GMC produces two postmitotic neurons or glial cells. In this chapter, we outline a method for the molecular profiling of neural precursors in the *D. melanogaster* optic lobe, including labeling, extraction, and processing for transcriptome analysis. We have used this strategy to compare the expression profiles of neuroblasts with their neuroepithelial precursors and have identified key genes that regulate the developmental transition from a symmetrically dividing to an asymmetrically dividing stem cell.

Key words: *Drosophila melanogaster*, Molecular profiling, Neural stem cell, Neuroblast, Optic lobe, Transcriptome analysis

1. Introduction

The neural stem cells of the *Drosophila melanogaster* optic lobe give rise to the visual processing centers of the adult brain (1). Division of these neural stem cells, referred to as neuroblasts, occurs along their apicobasal axes, self-renewing the apical neuroblast and producing a smaller, basal, GMC. Further division of each GMC then gives rise to two post-mitotic neurons or glial cells. The neuroblasts of the optic lobe are derived from neuroepithelial precursor cells that proliferate in two distinct regions of the larval brain, the inner proliferation center (IPC) and the outer proliferation center (OPC) (Fig. 1) (2). Divison of these neuroepithelial cells occurs symmetrically, producing two equivalent daughter cells and resulting in an expansion of the cell population. As development progresses, a transient wave of expression of the proneural gene *lethal of scute* (*l'sc*) triggers the transformation from

Bassem A. Hassan (ed.), *The Making and Un-Making of Neuronal Circuits in Drosophila*, Neuromethods, vol. 69,
DOI 10.1007/978-1-61779-830-6_11, © Springer Science+Business Media, LLC 2012

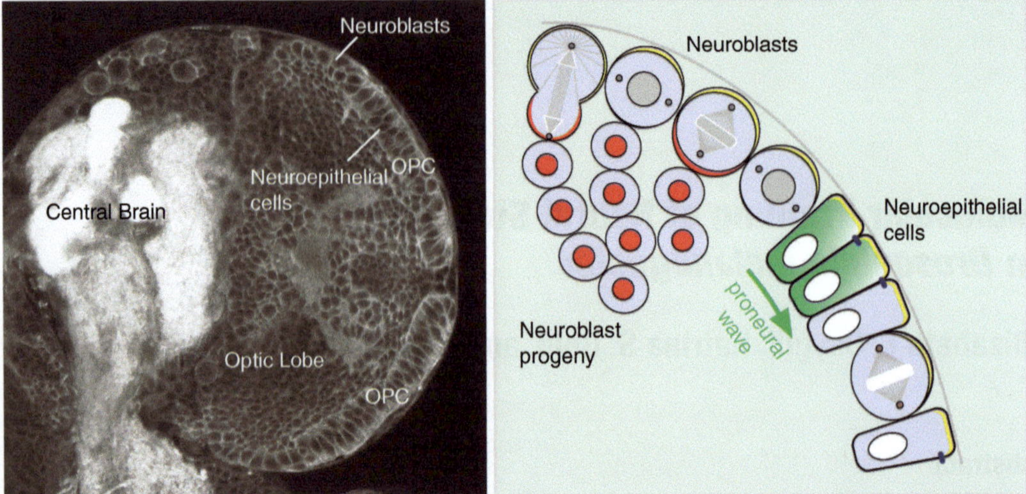

Fig. 1. Neural precursors of the *Drosophila melanogaster* optic lobe. (**a**) A third larval instar brain lobe stained with Dlg (Discs large) to visualize the cells. The lobe consists of the central brain and optic lobe regions. The two cell populations of the outer proliferation centre (OPC) of the optic lobe are indicated: neuroepithelial cells and neuroblasts. (**b**) Schematic of the transition from symmetrically dividing neuroepithelial cells to asymmetrically dividing neuroblasts at the medial edge of the neuroepithelium. Transient expression of the proneural gene *lethal of scute* (*l'sc*) induces the neuroblast formation in medial neuroepithelial cells. Adapted from ref. (2).

neuroepithelial cells into neuroblasts (3, 4). The transforming cells lose their epithelial characteristics and adopt an asymmetric mode of division (2). Cell fate determinants, including the homeodomain transcription factor Prospero (Pros) (5) and the TRIM-NHL domain protein Brain Tumor (Brat) (6), are inherited asymmetrically by the daughter GMC. A similar switch, from a symmetric to an asymmetric mode of division, occurs during neural development in mammals. Neurons of the neocortex are derived from neuroepithelial precursors which initially divide symmetrically then transform into asymmetrically dividing radial glial cells (7, 8).

Research in our lab has recently focused on the mechanisms that control the switch between neuroepithelial cells and neuroblasts (2, 4). We have used transcriptome analysis to compare the expression profiles of isolated neuroepithelial cells and neuroblasts. We found that neuroepithelial cells express genes encoding members of the Notch signaling pathway and that inhibition of Notch activity is required to trigger the switch from symmetrically dividing neuroepithelial cells to asymmetrically dividing neuroblasts (4).

Molecular profiling in the *D. melanogaster* nervous system has been achieved in a variety of different cell types (9–13). The tools available in *D. melanogaster* make it an ideal model in which to examine the profiles of single cells or small groups of cells. The GAL4/UAS expression system (14) can be used to label cells of interest with fluorescent markers for easy identification. The expanding availability of driver lines that drive expression in the nervous system has provided a vast toolset for the study of neural development and function (15, 16). Finding an appropriate GAL4 line with the desired degree of the specificity can be troublesome; however, the binary GAL4/UAS system can be easily modified to increase specificity. Negative regulation by GAL80, complementation with the LexA/LexAop binary transcription factor system (17), or the use of Split GAL4 (18) are all methods that can refine the activity of GAL4, thereby labeling a smaller subset of cells. The generation of either MARCM (19) or flip-out (20) clones can be used to label small groups of clonally related cells with temporal specificity (for details see Chap. 4). Both spatial and temporal specificity of clone induction can be achieved using a GAL4 driver to control spatial expression of flip recombinase while restricting its expression temporally with a temperature sensitive allele of GAL80 (21, 22).

Once appropriately labeled, the cells of interest must be isolated from the surrounding tissue. Enzymatic dissociation with the serine protease trypsin or the cysteine protease papain allows for isolation and collection of a specific cell population based on fluorescence, cell size, and cell shape using fluorescence-activated cell sorting (FACS) (23–25). Similarly, expression of the fusion protein mCD8-GFP under GAL4 control will allow for capture of a cell population using an anti-mCD8-antibody and magnetic-activated cell sorting (MACS) (10, 26). This technique has been used to isolate peripheral dendritic arbrization (da) neurons from the larval body wall (10). Laser capture microdissection (LCM) has been used to isolate sensory organ precursors (SOPs) from wing imaginal discs (12), peripheral da neurons (11), and mushroom body neurons from larval and pupal brians (13). The above techniques have the disadvantage of requiring specialized equipment. Successfull isolation of the cells of interest can also been achieved with a relatively low-tech method for cell extraction in situ using a microcapillary needle (4, 9). In our experience, such manual extraction is faster and less disruptive for the cells (see Note 1).

The molecular profile of the isolated cells can be examined at multiple levels. The choice to look at the level of the genome, the transcriptome, the proteome, protein modifications, or the cellular metabolome will depend both on the question you wish to answer and on the availability of appropriate technology. For example, mass spectrometry has been used to examine the neuropepetide complement of isolated *pdf*- and *hugin*-expressing neurons (9). The majority of molecular profiling is done at the level of the transcriptome. Gene expression profiling has evolved from northern blotting, through Serial Analysis of Gene Expression (SAGE) (29, 30) to microarrays (31, 32) and next generation sequencing technologies (33–35). The lower cost and convenience of microarray analysis means that hybridization technology still predominates.

However, direct sequencing using RNA-seq that, amongst other advantages, offers increased sensitivity requires smaller amounts of starting material (35), continues to grow in popularity.

We outline here a protocol for the isolation of, and the preparation of RNA from, the neural precursors of the optic lobe. We have taken a lead from microinjection and cell transplantation techniques (27, 28) to develop a method for cell extraction in situ using a microcapillary needle (4). Using this technique, we extract approximately 50 GFP-labeled neuroepithelial cells or neuroblasts from individual late-third instar larval brains. We isolate total RNA and use a reverse transcription protocol optimized for single cells to generate cDNA (5, 36–38) before hybridization to whole transcriptome oligonucleotide arrays. In order to analyze the neural precursors of the larval optic lobe we identified GAL4 driver lines that independently label the neuroepithelial cells and neuroblasts. Extraction of these cells is aided by their position, close to the brain surface. This technique can be modified to isolate cells in different locations or at different developmental stages. We have had similar success isolating neuroblasts from the embryonic ventral nerve cord (5) (see Note 2). Any cell or group of cells that can be identified, either by morphology or by reporter gene labeling, can in principle be isolated and profiled.

2. Materials

2.1. Needle Preparation and Setup

1. Borosilicate glass microcapillaries, 1.0 mm outside diameter × 0.78 mm inside diameter (GC100TF-10, Harvard Apparatus, Edenbridge, UK).
2. Micropipette puller (Flaming/Brown P-97 with 2.5 mm × 2.5 mm box filament, Sutter Instrument Company, Novato, USA).
3. Micropipette beveller (Bachhofer, Reutlingen, Germany).
4. Air-filled syringe and polyethylene tubing for controlling pressure in the needle when extracting and expelling cells (27).
5. DEPC-treated ddH_2O.
6. 70% ethanol.

2.2. *Drosophila* Strains (see Note 3)

1. c855a GAL4 (2, 39) driving expression of membrane-tethered mCD8-GFP and histone H2B-mRFP1 to visualize neuroepithelial cells.
2. MZ104 GAL4 (40) (aka *inscuteable* GAL4) driving expression of membrane-tethered mCD8-GFP and histone H2B-mRFP1 to visualize neuroblasts.

2.3. Staged Larval Collections

1. Fly cages.
2. Fresh yeast paste.
3. Apple juice collection plates.
4. Fly food plates or vials.

2.4. Drosophila Larval Brain Dissection

1. Cold phosphate-buffered saline (PBS), pH 7.4.
2. Sharp, fine forceps (Dumont no. 5, Fine Science Tools).
3. Dissection needles (e.g., 0.4 mm × 13 mm syringe needles, BD Microlance, mounted on cotton buds).
4. Lids from plastic Petri dishes (e.g., Nunc).
5. 22 × 50 mm Poly-L-lysine-coated coverslips (no. 1.5, VWR): prepare 20% (v/v) poly-L-lysine solution in ddH_2O from 0.1% (w/v) stock (Sigma Aldrich). Pipette a 5 μL drop onto the centre of each coverslip and leave to dry on a hot plate. Store in a dust-free container.

2.5. Cell Extraction and Lysis

1. Cell lysis buffer (945 μL): prepare stock solution in advance by mixing 100 μL 10× PCR buffer with $MgCl_2$ (Invitrogen), 10 μL NP-40 (American Bioanalytical), 50 μL 0.1 M DTT (Dithiothreitol, Invitrogen), 785 μL DEPC-treated ddH_2O. Store at −20°C.
2. Cell lysis mix (50 μL): 1 μL RNase inhibitor mix (1:1 mixture of RNasin [Promega] and Stop [Flowgen Bioscience]), 1 μL 10 ng/μL Anchor T amplification oligonucleotide primer (HPLC grade and resuspended in ddH_2O, sequence TAT AGA ATT CGC GGC CGC TCG CGA 24 (T)), 1 μL 2.5 mM dNTPs (Takara), 47 μL cell lysis buffer (see Sect. 2.5, step 1). Keep on ice. This should be freshly prepared on the day of cell extraction (see Sect. 3.3, step 3).
3. Mineral oil.
4. Inverted fluorescence microscope (Olympus IX71, Olympus, Japan) with micromanipulator (MN-151 Joystick Micromanipulator, Narishige, Tokyo, Japan) and UV light source.
5. Glass microscope slides.

2.6. Reverse Transcription PCR

1. Reverse transcription (RT) working mix (2.5 μL): 0.3 μL Superscript II [Invitrogen], 0.1 μL RNase inhibitor mix (see 2.5. step 2), 2.1 μL lysis buffer (see 2.5. step 1). Keep on ice.
2. Poly(A) tailing reaction mix (5 μL): 0.15 μL 100 mM dATP (Promega), 0.5 μL 10× PCR buffer with $MgCl_2$ (Invitrogen), 3.85 μL DEPC-treated ddH_2O, 0.25 μL TdT (terminal deoxynucleotidyl transferase, Roche), and 0.25 μL RNaseH (Roche). Keep on ice.
3. PCR mix (50 μL): 5 μL 10× ExTaq buffer (Takara), 5 μL 2.5 mM dNTPs (Takara), 1 μL 1 μg/μL Anchor T primer

(see Sect. 2.5. step 2), 38.5 μL DEPC-treated ddH_2O, 0.5 μL ExTaq polymerase (Takara). Keep on ice.

4. Commercial kit for PCR purification (e.g., Qiagen, Sigma).
5. Equipment for running standard DNA agarose gels.
6. Spectrophotometer.

3. Methods

3.1. Needle Preparation and Setup

Please note that gloves should be worn at all times when handling the microcapillaries.

1. Prepare a number of needles by pulling borosilicate glass microcapillaries and on a micropipette puller (see Sect. 2.1, step 2) with the following settings Heat = ramp – 35; Pull = 120; Velocity = 60; Time = 160 (see Note 4).
2. Cut back the tip of the needle with fine forceps under a dissecting microscope to create a tip with an internal diameter of 12–15 μm (see Note 5).
3. Mount the needle on a micropipette beveller and set the beveling angle to between 10° and 12° from the vertical. Mark the posterior upper surface of the needle with a pen so that the needle has the correct angle when it is being fixed to the micromanipulator for cell extraction.
4. Wash the needles, first with DEPC-treated ddH_2O and then 70% ethanol, to remove glass shards. Store them on plasticine in a dust-free container.

3.2. Staged Larval Collections

1. Set up a fly cross to obtain progeny with the appropriate genotype and culture at 25°C in food vials.
2. Transfer the cross to a fly cage containing an apple juice plate smeared with fresh yeast paste. Allow the flies to lay eggs on the plate over a 4-h time window.
3. Remove any hatched larvae and yeast from the apple juice plates at around ~22–23 h after egg laying.
4. Collect freshly hatched larvae 4 h later. Transfer from apple juice plates to fly food plates or vials.
5. Culture at 25°C until larvae reach the appropriate stage for dissection.

3.3. Preparation of Cell Lysis Buffer and Microscope Setup

1. Prepare fresh cell lysis mix before beginning dissection (see Sect. 2.5. step 2), and keep on ice.
2. Take a clean, sharpened needle and attach it to an air-filled syringe using polyethylene tubing.

3. Mount the needle on the micromanipulator.
4. Place a large drop of mineral oil onto a glass slide, lower the needle into the oil, and slowly draw up the syringe plunger to take up oil into the needle.

3.4. *Drosophila* Larval Brain Dissection

1. Using forceps, pick appropriately staged larvae from vials or food plates.
2. Transfer the larvae to a Petri dish containing tissue paper soaked in water. Leave the larvae to crawl for a few minutes to clean off fly food and yeast.
3. Place a clean larva in a drop of PBS on a Petri dish lid.
4. Tear the larva in half using forceps and discard the posterior half.
5. Gently invert the anterior half of the larvae, steadying the anterior end of the larva against the closed tip of one pair of forceps and using the other pair of forceps to gently roll the body wall back over the mouth hooks. Identify the brain and remove it from the body wall by cutting the nerves and esophagus. Remove attached imaginal discs using dissecting needles, being especially careful not to damage the brain lobes when removing the eye discs.
6. Move the dissected brain to a clean drop of PBS and continue dissecting (see Note 6).
7. Place a drop of PBS onto a poly-L-lysine-coated coverslip. Using the forceps to hold them by the nerves transfer dissected brains into the drop. Orient the brains so that they are dorsal side up and push them down gently onto the coverslip. Check that they adhere securely to the poly-L-lysine.

3.5. Cell Extraction and Lysis

1. Place the coverslip on the inverted microscope stage.
2. Using Nomarski optics, focus on one of the dissected brains.
3. Bring the needle into the same plane of focus. Take up a small volume of PBS by drawing up the syringe plunger so that you can see a clear oil–PBS interface in the needle.
4. Insert the needle into the brain lobe. The easiest point of insertion is usually at the ventral-most level, since this is where the brain adheres most firmly to the coverslip (see Note 7). Once the outer glial sheath has been penetrated, slowly move the needle tip to the region of interest. In the case of the optic lobe, this is the lateral portion of the brain, lying just underneath the surface.
5. Open the UV filter to visualize the exact location of the cells of interest. Position the needle tip next to the GFP-positive cells and slowly draw up the syringe plunger to extract them. Try to limit exposure to UV to reduce damage to the cells.

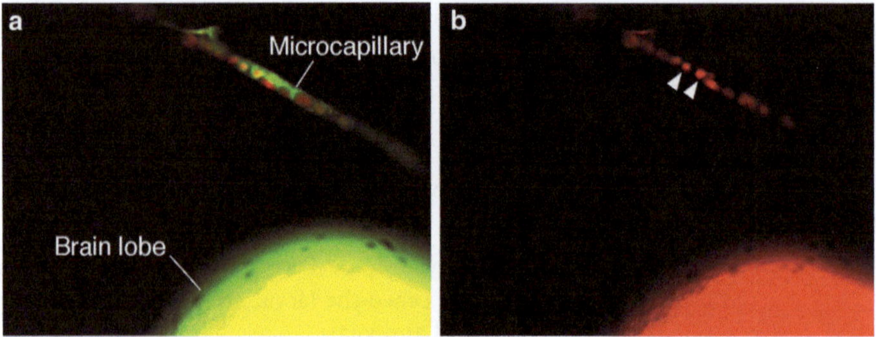

Fig. 2. Neural stem cell extraction in situ. (**a**) Neuroepithelial cells extracted from the optic lobe are visualized inside the glass microcapillary needle. The cells express membrane-tethered GFP and histone-bound RFP, driven by c855*a* GAL4. (**b**) Individual cell nuclei, marked with histone H2B-mRFP1, can be distinguished under UV illumination (*arrowheads*). Picture courtesy of Boris Egger.

6. Remove the needle from the brain and examine the needle using bright-field and UV to confirm the collected cells are a pure population of the desired GFP positive cells (Fig. 2) (see Note 8). Remove the coverslip from the microscope.
7. Pipette 2.5 μL of ice-cold lysis mix from the eppendorf onto a clean glass slide and mount the slide on the microscope.
8. Expel the cells from the needle into the drop of cold lysis mix. Monitor the progress of cell lysis using the UV illumination, to ensure that all the cells lyse.
9. Transfer the lysed material to a PCR tube.
10. Incubate the lysed material at 65°C for 2 min to denature RNA.
11. Following denaturation snap-cool the sample on ice.

3.6. Reverse Transcription

1. Make up the reverse transcription (RT) mix on ice (see Sect. 2.6, step 1).
2. Add 2.5 μL of RT mix to the lysed cells and incubate at 37°C for 90 min.
3. Terminate the RT reaction by heating to 65°C for 10 min, and then cool the samples to 4°C.
4. The first DNA strands synthesized by reverse transcription must be polyadenylated to allow second strand synthesis and PCR amplification. Make up TdT mix for poly(A) tailing on ice (Sect. 2.6, step 2).
5. Add 5 μL of TdT mix to the RT reaction mixture. Place the tube in a PCR machine and incubate the reaction for 20 min at 37°C, followed by 10 min at 65°C.

3.7. PCR Amplification

1. Make up PCR mix on ice (see Sect. 2.6, step 3).
2. Add the 10 μL aliquot of polyA-tailed cDNA to the PCR mix.
3. Amplify the reverse transcribed and tailed samples using the following PCR programme:
   (a) One cycle of 95°C for 1 min, 37°C for 5 min, and 72°C for 20 min.
   (b) 30–34 Cycles of 95°C for 30 s, 67°C for 1 min, and 72°C for 6 min with a 6 s extension per cycle.
   (c) 72°C for 10 min.
   (d) Hold at 4°C, store amplified samples at -20°C.
4. Run out 5 μL of the sample on an agarose gel to check quality. A homogenous smear of DNA should be visible.
5. Purify cDNA with a commercial kit, according to the manufacturer's instructions.
6. Measure cDNA concentration and check sample purity ($A_{260/280}$) with a spectrophotometer. The cDNA is now ready to be used for expression profiling (see Note 9).

4. Notes

1. Our experiments with neuroblast isolation in *Drosophila* embryos using FACS or MACS have shown that while enrichment of neuroblasts can be achieved, these methods can lead to widespread apoptosis (K. Edoff, personal communication).
2. Embryonic neuroblasts can be identified by their size, shape, and position within the ventral nerve cord, in addition to reporter gene expression.
3. A number of different GAL4 lines drive GAL4 expression in the neural stem cells of the embryo and larva. There are also a variety of UAS reporter lines to express fluorescent markers for visualizing cells in living brains. Many of these can be ordered from the major fly stock centers such as the Bloomington Stock Centre (http://flystocks.bio.indiana.edu/) and Kyoto Stock Center (http://www.dgrc.kit.ac.jp/en/index.html). Perdurance of GFP, as well as GAL4 protein, needs to be considered when choosing a driver line to label progenitors. In many cases, perdurance of these proteins can lead to the labeling of more differentiated cells further along the stem cell lineage, resulting in a less pure sample.
4. Generating appropriate needles for cell extraction is one of the most important steps in this protocol. As each needle puller

and filament is different empirical optimization will be required to obtain the perfect needle. The aim is to produce a needle with a long, narrow taper that is still strong enough to pierce the brain. The tip of the needle should be fine enough not to cause too much damage during insertion into the brain but wide enough to take up the desired cells. The heat, pulling strength, velocity, and time are all variables which can be manipulated to produce needles with the right kind of taper. A good guide to micropipette pulling can be found on the Sutter Instruments website (http://www.sutter.com/contact/faqs/pipette_cookbook.pdf). Prokop and Technau's chapter on cell transplantation also contains detailed information on needle preparation (27).

5. The final tip diameter should be slightly larger than that of the cells you are isolating.
6. The number of brains you need will be determined by the abundance of the cells you wish to extract, our experiments have been done using four brains for each experimental condition.
7. The glial sheath surrounding the larval brain can be tough and difficult to penetrate. We recommend experimentation with needle preparation to generate appropriately sharp needles.
8. While our current protocol is to extract around 50 cells per sample, we have also carried out single cell amplification and analysis using this protocol. Advances in single-cell cDNA amplification techniques have made single cell transcriptome analysis feasible (41); however, care must be taken with single cell data as there can be significant transcriptional variability between phenotypically identical cells (42).
9. Diagnostic PCRs can be carried out at this point to assess the likelihood of cDNA library contamination by other cell types. We recommend testing both positive and negative markers for the cell type of interest, using a low number of PCR cycles. For example, neuroblasts express genes such as *asense* and *deadpan*, but not the glial marker *repo*, while neuroepithelial cells should express epithelial markers such as *PatJ*.

Acknowledgments

Thank you to K. Edoff, B. Egger, P. Wu, A. R. Carr, H. Matsunami, and F. J. Livesey for their time and expertise. This work was supported by the Wellcome Trust.

References

1. Truman JW, Bate M (1988) Spatial and temporal patterns of neurogenesis in the central nervous system of *Drosophila melanogaster*. Dev Biol 125:145–157
2. Egger B, Boone JQ, Stevens NR, Brand AH, Doe CQ (2007) Regulation of spindle orientation and neural stem cell fate in the *Drosophila* optic lobe. Neural Dev 2:1
3. Yasugi T, Umetsu D, Murakami S, Sato M, Tabata T (2008) *Drosophila* optic lobe neuroblasts triggered by a wave of proneural gene expression that is negatively regulated by JAK/STAT. Development 135:1471–1480
4. Egger B, Gold KS, Brand AH (2010) Notch regulates the switch from symmetric to asymmetric neural stem cell division in the *Drosophila* optic lobe. Development 137:2981–2987
5. Choksi SP, Southall TD, Bossing T, Edoff K, de Wit E, Fischer BE, van Steensel B, Micklem G, Brand AH (2006) Prospero acts as a binary switch between self-renewal and differentiation in *Drosophila* neural stem cells. Dev Cell 11: 775–789
6. Betschinger J, Mechtler K, Knoblich JA (2006) Asymmetric segregation of the tumor suppressor brat regulates self-renewal in *Drosophila* neural stem cells. Cell 124:1241–1253
7. Noctor SC, Martinez-Cerdeno V, Ivic L, Kriegstein AR (2004) Cortical neurons arise in symmetric and asymmetric division zones and migrate through specific phases. Nat Neurosci 7:136–144
8. Farkas LM, Huttner WB (2008) The cell biology of neural stem and progenitor cells and its significance for their proliferation versus differentiation during mammalian brain development. Curr Opin Cell Biol 20:707–715
9. Neupert S, Johard HA, Nassel DR, Predel R (2007) Single-cell peptidomics of *Drosophila melanogaster* neurons identified by Gal4-driven fluorescence. Anal Chem 79:3690–3694
10. Iyer EP, Iyer SC, Sulkowski MJ, Cox DN (2009) Isolation and purification of Drosophil peripheral neurons by magnetic bead sorting. J Vis Exp Dec 1;(34). pii: 1599. doi: 10.3791/1599
11. Iyer EP, Cox DN (2010) Laser capture microdissection of *Drosophila* peripheral neurons, J Vis Exp May 24;(39). pii: 2016. doi: 10.3791/2016
12. Buffin E, Gho M (2010) Laser microdissection of sensory organ precursor cells of *Drosophila* microchaetes. PLoS One 5:e9285
13. Hoopfer ED, Penton A, Watts RJ, Luo L (2008) Genomic analysis of *Drosophila* neuronal remodeling: a role for the RNA-binding protein Boule as a negative regulator of axon pruning. J Neurosci 28:6092–6103
14. Brand AH, Perrimon N (1993) Targeted gene expression as a means of altering cell fates and generating dominant phenotypes. Development 118:401–415
15. Pfeiffer BD, Jenett A, Hammonds AS, Ngo TT, Misra S, Murphy C, Scully A, Carlson JW, Wan KH, Laverty TR, Mungall C, Svirskas R, Kadonaga JT, Doe CQ, Eisen MB, Celniker SE, Rubin GM (2008) Tools for neuroanatomy and neurogenetics in *Drosophila*. Proc Natl Acad Sci U S A 105:9715–9720
16. Pfeiffer BD, Ngo TT, Hibbard KL, Murphy C, Jenett A, Truman JW, Rubin GM (2010) Refinement of tools for targeted gene expression in *Drosophila*. Genetics 186:735–755
17. Lai SL, Lee T (2006) Genetic mosaic with dual binary transcriptional systems in *Drosophila*. Nat Neurosci 9:703–709
18. Luan H, Peabody NC, Vinson CR, White BH (2006) Refined spatial manipulation of neuronal function by combinatorial restriction of transgene expression. Neuron 52:425–436
19. Lee T, Luo L (1999) Mosaic analysis with a repressible cell marker for studies of gene function in neuronal morphogenesis. Neuron 22: 451–461
20. Basler K, Struhl G (1994) Compartment boundaries and the control of *Drosophila* limb pattern by hedgehog protein. Nature 368: 208–214
21. Jiang H, Grenley MO, Bravo MJ, Blumhagen RZ, Edgar BA (2011) EGFR/Ras/MAPK signaling mediates adult midgut epithelial homeostasis and regeneration in *Drosophila*. Cell Stem Cell 8(1):84–95
22. Jiang H, Patel PH, Kohlmaier A, Grenley MO, McEwen DG, Edgar BA (2009) Cytokine/Jak/Stat signaling mediates regeneration and homeostasis in the *Drosophila* midgut. Cell 137:1343–1355
23. de la Cruz AF, Edgar BA (2008) Flow Cytometric Analysis of *Drosophila* Cells. In: Dahmann C (ed) *Drosophila*: methods and portocols, Methods in molecular biology. Humana Press, Totowa, NJ, pp 373–389
24. Shigenobu S, Arita K, Kitadate Y, Noda C, Kobayashi S (2006) Isolation of germline cells from *Drosophila* embryos by flow cytometry. Dev Growth Differ 48:49–57
25. Tirouvanziam R, Davidson CJ, Lipsick JS, Herzenberg LA (2004) Fluorescence-activated cell sorting (FACS) of *Drosophila* hemocytes

reveals important functional similarities to mammalian leukocytes. Proc Natl Acad Sci U S A 101:2912–2917

26. Wang X, Starz-Gaiano M, Bridges T, Montell D (2008) Purification of specific cell populations from *Drosophila* tissues by magnetic bead sorting, for use in gene expression profiling Protocol Exchange doi:10.1038/nprot.2008.28
27. Prokop A, Technau GM (1993) Cell Transplantation. In: Hartley D (ed) Cellular interactions in development: a practical approach. Oxford University Press, Oxford, UK, pp 33–57
28. Technau GM (1986) Lineage analysis of transplanted individual cells in embryos of *Drosophila melanogaster*. Roux's Arch Dev Biol 195:389–398
29. Velculescu VE, Zhang L, Vogelstein B, Kinzler KW (1995) Serial analysis of gene expression. Science 270:484–487
30. Powell J (2000) SAGE. The serial analysis of gene expression. Methods Mol Biol 99: 297–319
31. Schena M, Shalon D, Davis RW, Brown PO (1995) Quantitative monitoring of gene expression patterns with a complementary DNA microarray. Science 270:467–470
32. Lipshutz RJ, Fodor SP, Gingeras TR, Lockhart DJ (1999) High density synthetic oligonucleotide arrays. Nat Genet 21:20–24
33. Costa V, Angelini C, De Feis I, Ciccodicola A (2010) Uncovering the complexity of transcriptomes with RNA-Seq. J Biomed Biotechnol 2010:853916
34. Metzker ML (2010) Sequencing technologies—the next generation. Nat Rev Genet 11:31–46
35. Wang Z, Gerstein M, Snyder M (2009) RNA-Seq: a revolutionary tool for transcriptomics. Nat Rev Genet 10:57–63
36. Iscove NN, Barbara M, Gu M, Gibson M, Modi C, Winegarden N (2002) Representation is faithfully preserved in global cDNA amplified exponentially from sub-picogram quantities of mRNA. Nat Biotechnol 20:940–943
37. Osawa M, Egawa G, Mak SS, Moriyama M, Freter R, Yonetani S, Beermann F, Nishikawa S (2005) Molecular characterization of melanocyte stem cells in their niche. Development 132:5589–5599
38. Subkhankulova T, Livesey FJ (2006) Comparative evaluation of linear and exponential amplification techniques for expression profiling at the single-cell level. Genome Biol 7:R18
39. Speicher S, Garcia-Alonso L, Carmena A, Martin-Bermudo MD, de la Escalera S, Jimenez F (1998) Neurotactin functions in concert with other identified CAMs in growth cone guidance in *Drosophila*. Neuron 20:221–233
40. Manseau L, Baradaran A, Brower D, Budhu A, Elefant F, Phan H, Philp AV, Yang M, Glover D, Kaiser K, Palter K, Selleck S (1997) GAL4 enhancer traps expressed in the embryo, larval brain, imaginal discs, and ovary of *Drosophila*. Dev Dyn 209:310–322
41. Kurimoto K, Saitou M (2010) Single-cell cDNA microarray profiling of complex biological processes of differentiation. Curr Opin Genet Dev 20:470–477
42. Subkhankulova T, Gilchrist MJ, Livesey FJ (2008) Modelling and measuring single cell RNA expression levels find considerable transcriptional differences among phenotypically identical cells. BMC Genomics 9:268

Chapter 12

Out with the Brain: *Drosophila* Whole-Brain Explant Culture

Marta Koch and Bassem A. Hassan

Abstract

In this chapter, we present a detailed protocol for culturing the adult *Drosophila* brain ex vivo and discuss some of the possibilities this method opens up. Mature *Drosophila* brains can be easily maintained in culture for a long period of time, with very little deterioration.

Explanting and culturing the brains using our technique solves the accessibility, immobilization, and visualization problems inherent in whole animal preparations. To illustrate the utility of this method, we discuss its application as a model to study the mechanisms of axon injury and regeneration. Until recently, axon regeneration has been studied mainly in vertebrate model organisms and neuronal cell cultures. However, the use of invertebrate models, which are much more amenable to genetic manipulation, shows great promise and should greatly accelerate our understanding of the molecular mechanisms of axon regeneration.

Key words: *Drosophila*, CNS, Axon growth, Regeneration, Injury, Brain culture

1. Introduction

Culturing the brain ex vivo is a valuable tool that offers many advantages and a broad range of applications. The idea of culturing the adult fly brain has been based on the work on organotypical mammalian brain slice cultures (1), which represent a well-established system for neuroscience research. Unfortunately, slicing the brain often results in acute damage to the brain tissue, resulting in significant cell death accompanied by reactive gliosis. This cell death presumably explains the poor success rate of culturing slices of mature mammalian brain tissues (1). The adult *Drosophila* brain is approximately 300 μm thick, which is roughly the thickness of a typical mammalian brain slice. Therefore, *Drosophila* whole-brain explants could be used as a viable alternative to mammalian brain slices, with the advantage that the neuronal circuitry remains intact, and therefore preserving the tissue structure and function and consequently improving viability.

Bassem A. Hassan (ed.), *The Making and Un-Making of Neuronal Circuits in Drosophila*, Neuromethods, vol. 69,
DOI 10.1007/978-1-61779-830-6_12, © Springer Science+Business Media, LLC 2012

In vitro culture systems of *Drosophila* whole brains have already been successfully used to study axon remodeling during metamorphosis (2, 3) and to study neuronal activity (4, 5) and more recently to study axonal injury and regeneration (6).

To obtain healthy, immobilized, and easily accessible *Drosophila* brain preparations, we modified the culture media and conditions previously described for use with *Drosophila* tissues (2). Further, to provide mechanical support to the brains we utilize the culture plate inserts used with mammalian brain slices; these inserts consist of a porous cellulose membrane attached to a plastic ring and are placed in a Petri dish filled with medium.

In our experience, fly brains maintained in culture retain much of their morphological and physiological properties for a period of up to three weeks (6).

The brain culture method can be used in a variety of experiments. In our laboratory, we use it as a powerful tool to study axonal injury and regeneration. Mechanisms of axonal regeneration have been traditionally studied in vertebrate models. The small size of the fly brain in combination with the hard cuticle that protects the brain has precluded axon regeneration research and other live reproducible physical manipulations in these organisms. Induction of reliable axonal injury and subsequent visualization is best performed on an immobilized preparation such as ours, which allows precise micromanipulation of the brain. Furthermore, the culture preparation allows the repeated imaging of regenerating axons using simple microscopic techniques over many days.

In combination with the power of genetics inherent in smaller model organisms such as *Caenorhabditis elegans* and *Drosophila*, we hope that such whole brain culture systems will significantly speed up the characterization of the molecular processes that control axonal regeneration in the CNS following traumatic axonal injury.

A significant challenge for the future is the long-term explant culture of the developing CNS for the purpose of the direct and live visualization of developing neuronal circuits. The difficulties inherent in keeping developing brains intact after dissection, asking whether they would support long term culture, finding culture conditions that would support exogenous brain development and combining all that with the proper imaging equipment are considerable. However, the direct visualization of how neurons in fact form neuronal circuits in a developing brain is indispensible for a complete understanding of the logic of neuronal circuit formation. The fact that developing neuronal circuits can be manipulated with high precision in *Drosophila* makes this an even more tempting goal to achieve. Therefore, these challenges need to be met through technical innovations and there is every reason to believe that such achievements may in fact be on the horizon.

2. Materials

2.1. Equipment

Millicell low-height culture plate insert (PICMORG50-Millicell)

Sterile small Petri dishes 35.0/10 mm (e.g., Greiner)

Sharp and fine forceps (e.g., no. 5 Dumont)

Rotating plate

Dissection stereomicroscope

Epi-fluorescence stereomicroscope

Microscope slides and 20/20 mm coverslips

2.2. Solutions and Reagents

2.2.1. For the Brain Culture

Laminin (e.g., BD Biosciences) (make stock 0.33 mg/mL, ready to be diluted 100×; aliquot and store at −20°C)

Poly-D-lysine (e.g., BD Biosciences) (make stock 3.33 mg/mL, ready to be diluted 100×; aliquot and store at −20°C)

Sterile phosphate-buffered saline (PBS) (e.g., Dulbecco)

Schneider's *Drosophila* medium (e.g., Sigma)

Fetal Bovine Serum (e.g., Invitrogen)

Insulin 10 mg/mL (e.g., Sigma 100 mg)

Antibiotics Penicillin and Streptomycin (e.g., Sigma, make stocks of 10,000 U/mL and 10 mg/mL, respectively)

2.2.2. For the Immunolabeling

Formaldehyde fixative: 4% Formaldehyde (made from commercial 37% stock solution, e.g., Sigma) in PBS

Washing solution: PBS with 0.3% (v/v) Triton X-100 (PBT)

Blocking solution: PAX-DG+N3

Antibodies: in blocking solution. *Primary antibodies:* We routinely use the anti-GFP antibody (Invitrogen) at a final dilution of 1:500. For other antigens, we refer the reader to the Developmental Studies Hybridoma Bank (DSHB) (http://dshb.biology.uiowa.edu/) which has a vast collection of primary antibodies directed against a wide variety of *Drosophila* antigens.

Secondary antibodies: We use fluorophore conjugated secondary antibodies obtained from Molecular Probes at a final dilution of 1:500.

Mounting medium: Vectashield (Vector)

2.2.3. For the Injury

Eppendorf Microdissector equipped with foot pedal and cutting tool (Cat No:92000401-6)

Eppendorf MicroChisel (Cat No:930002500)

Micromanipulator

3. Methods

3.1. Preparation of the Culture Medium and Coating of the Filter Membrane Inserts

1. Prepare the culture medium (modified from Gibbs and Truman) (2) by adding 10,000 U/mL penicillin, 10 mg/mL streptomycin, 10% fetal bovine serum, and 10 μg/mL insulin to Schneider's *Drosophila* medium. For 10 mL of medium add 1 mL of FBS, 100 μL of penicillin and 100 μL streptomycin (from a stock solution of 10,000 U/mL and 10 mg/mL, respectively), 10 μL of insulin (from a stock solution of 10 mg/mL), and complete to the volume of 10 mL with Schneider's medium. Filter before use.
2. For the coating, prepare fresh solution of laminin (3.3 μg/mL) and polylysine (33.3 μg/mL) in PBS. For example for ten filters add 100 μL of laminin (stock solution of 0.33 mg/mL) and 100 μL of poly-D-Lysine (stock solution of 3.33 mg/mL) to 9.8 mL of PBS. Mix the solution by vortexing for 10–20 s.
3. Place each filter to be coated on one small sterile Petri dish.
4. Add 1 mL of PBS to the Petri dish to wet the membrane from below.
5. Add 1 mL of polylysine/laminin solution on top of the filter.
6. Place dishes in a plastic box and incubate overnight at 37°C.
7. The next day remove all solution and wash three times for 10 min by adding 1 mL of PBS on top of the filter.
8. Store at 4°C with 200 μL of PBS to keep the membrane humid. The membrane filters can be used for at least one month after preparation.

3.2. Brain Dissection and Culture

1. Dissection and culturing should be performed in a sterile environment. When possible, work should be carried out under a sterile flow hood. When a flow hood is not available, sterilize the bench, forceps, and microscope with 70% (v/v) ethanol before commencing work.
2. Before dissection, submerge the fly in 70% ethanol for a few seconds.
3. Carefully dissect the brains under a stereomicroscope, on a dish filled with ice-cold Schneider's *Drosophila* medium. Several excellent videos and detailed protocols explaining brain removal from the *Drosophila* head are freely available on the Web; therefore, it is not discussed in detail here. In brief, the brain is explanted out of the hard cuticle and all tracheal branches are removed with the help of fine tipped forceps. This procedure demands experience, as it should be performed as fast as possible but with care to avoid mechanical damage. Give yourself plenty of time (and flies) to practice! Leave dissected brains in Schneider's medium until they are transferred to the membrane filter.

4. Place a previously coated membrane filter in a new Petri dish. Add 550 μL of culture medium on the bottom of the dish and a small drop on the top of the membrane filter.
5. Transfer all brains, one by one, with forceps onto the drop of medium placed on top of the filter. You can place up to eight nude brains on a single membrane filter.
6. The brains can be oriented either with the anterior side or with the posterior side up. Rely on morphological differences of the central brain and the optic lobe to orient the brains according to the requirements of your experiment. If using an endogenous fluorescence marker, check under fluorescence for correct orientation of the brains (e.g., for visualization of the axonal pattern of the sLNv dorsal projections, we have to assure that the anterior side of brain is the one in contact with the membrane, thus allowing the posterior side to be imaged.
7. The brains will attach to the membrane and should remain covered by a thin layer of medium. At this air-liquid interface, exchange of nutrients and air takes place, ensuring optimal conditions for the health of the explanted tissue. Remove excess medium on top of the membrane so that no brains are floating anymore. Add a further 550 μL of culture medium to the bottom of the dish.
8. Place the dishes in a humidified fly incubator set at 25°C. It is advisable to exchange half of the total volume of the medium for fresh every 2 days.

3.3. Live Imaging

Depending on the resolution and magnification required, imaging can be performed with a confocal microscope using water immersion objectives, or with a stereomicroscope (with a minimum of ×150 magnification).

(a) If the live imaging is performed with water immersion objectives, the microscope and objective need to be thoroughly disinfected before starting. Culture dish should be filled with medium to completely submerge the brains; medium should be added by letting it drip onto the Petri dish to prevent detachment of the brains from the filter support. After imaging, all medium should be removed and replaced by adding fresh medium to the bottom of the dish.

(b) If a stereomicroscope is used then imaging is performed with a dry objective. The lid of the Petri dish needs to be removed and the brains imaged as fast as possible, to avoid evaporation of the medium.

When we perform injury experiments the imaging is typically done 2–5 h following injury and can be repeated at several further timepoints if desired. Usually the last timepoint is done with fixed and immunostained brains.

3.4. Immunostaining and Mounting of the Brains

Note that all steps should be performed at room temperature, except for the incubation with the primary antibody, which is usually performed at 4°C. All steps are best done on a rotating plate. The protocol we describe here is typical for GFP immunostaining and may have to be adapted depending on the characteristics of the other primary antibodies used.

1. Fix the cultured brains by replacing the culture medium in the Petri dish with fixation solution for 30 min. This step is to assure that the brains remain well attached to the membrane throughout the staining procedure. Then, add 1 mL of fixation solution on top of the filter for 1–2 h.
2. Replace all fixative with PBT and cut the filter out of the plastic insert with a sharp scalpel. Transfer the filter with the brains attached to a new Petri dish.
3. Wash two times for 20 min with PBT.
4. Incubate the preparation in blocking solution for 45 min at room temperature.
5. Incubate with the primary antibody solution overnight at 4°C or for 4–6 h at room temperature.
6. Wash three times for 20 min in PBT.
7. Incubate with the secondary antibody solution for 2 h.
8. Wash again three times for 20 min in PBT.
9. Transfer the brains still attached to the membrane to a clean microscopy slide and remove the excess of solution with a Kim Wipe. Add a drop of anti-fade mounting medium on top of the cut membrane/brains. Seal the edges of the coverslip with nail polish.

3.5. Injury and Regeneration

To perform axonal injury we use an ultrasonic microchisel controlled by a powered device (Eppendorf microdissector). When activated, the fine metal tip of the microchisel, which is guided by a micromanipulator, oscillates and thereby severs the axon(s) where it has been placed (Fig. 1). Our lab has optimized this protocol to sever the sLNv dorsal projections; however, the same principles are valid to other neuronal populations as long as they are located to the surface of the brain. Injuring axon projections that are located deeper in the brain would result in an extensive injury that would be harmful to the brain in culture and have an impact on several other physiological mechanisms. In this case, making use of a laser injury, instead of a mechanical cutting injury, would be more adequate.

1. Disinfect the bench and microscope with 70% (v/v) ethanol to minimize the risk of bacterial contamination. Remove the lid of the Petri dish where the culture membrane has been previously placed. Under visual control of a fluorescence stereomicroscope (minimum ×100 magnification), adjust the

Fig. 1. Schematic showing the setup for the injury of axonal tracts in explanted brains. Brains are placed on a membrane of a low-height organotypic culture plate insert that is placed in a Petri dish containing culture medium. The axonal GFP pattern is visualized using a fluorescence stereomicroscope equipped with a green spectrum specific filter. The microdissector cutting tool is positioned on an axonal tract using a micromanipulator. Schematic not drawn to scale.

micromanipulator to place the tip of the microchisel on the GFP positive axonal tract(s) to be cut.

2. Activate the powered device by using the foot pedal (frequency between 23 and 30 Hz, and amplitude at 100%) to cut the axons. Verify the injury at higher magnification (e.g., ×200 magnification).
3. Injure the axons on only one side of the brain and keep the other side intact as an internal control to access the general health of the brain preparation.
4. Place the Petri dish back in a humidified incubator at 25°C until further analyses (this can either be multi-day imaging or immunostaining).

4. Notes

1. Flipping the brains on the membrane filter can easily damage them. If the brain is oriented incorrectly and needs to be "flipped," add a drop of culture medium to the top of the filter and using forceps slowly bring the brain to the edge of the liquid and then adjust its position.
2. Although there is an overall flattening of the brain after a few days, a phenomenon well known to occur on culture plate

inserts (1), this is not a problem and in fact can even be advantageous for visualization. This process will also accentuate the slight asymmetries in the contact face that occur between the brain and the membrane in culture.

3. If the goal of the experiment is to perform multi-day imaging, bear in mind that GFP intensity tends to decrease with time in culture. Therefore, longer exposure times or higher laser powers may be needed to achieve a signal equivalent to that obtained during the first analysis.

References

1. Stoppini L, Buchs PA, Muller D (1991) A simple method for organotypic cultures of nervous tissue. J Neurosci Methods 37(2):173–182
2. Gibbs SM, Truman JW (1998) Nitric oxide and cyclic GMP regulate retinal patterning in the optic lobe of Drosophila. Neuron 20:83
3. Brown HL, Cherbas L, Cherbas P, Truman JW (2006) Use of time-lapse imaging and dominant negative receptors to dissect the steroid receptor control of neuronal remodeling in *Drosophila*. Development 133:275
4. Wang JW, Wong AM, Flores J, Vosshall LB, Axel R (2003) Two-photon calcium imaging reveals an odor-evoked map of activity in the fly brain. Cell 112:271
5. Gu H, O'Dowd DK (2006) Cholinergic synaptic transmission in adult *Drosophila* Kenyon cells in situ. J Neurosci 26:265
6. Ayaz D et al (2008) Axonal injury and regeneration in the adult brain of *Drosophila*. J Neurosci 28:6010

Index

A

B

Bassem A. Hassan (ed.), *The Making and Un-Making of Neuronal Circuits in Drosophila*, Neuromethods, vol. 69,
DOI 10.1007/978-1-61779-830-6, © Springer Science+Business Media, LLC 2012

D

E

G

H

I

J

K

L

M

N

O

P

MIX
Papier aus verantwortungsvollen Quellen
Paper from responsible sources
FSC® C105338

If you have any concerns about our products, you can contact us on
ProductSafety@springernature.com

In case Publisher is established outside the EU, the EU authorized representative is:
Springer Nature Customer Service Center GmbH
Europaplatz 3, 69115 Heidelberg, Germany

Printed by Libri Plureos GmbH
in Hamburg, Germany